Genomics of Disease

J.P. Gustafson, J. Taylor, and G. Stacey
Editors

Genomics of Disease

 Springer

Editors

J.P. Gustafson
USDA-ARS
University of Missouri
Department of Agronomy
205 Curtis Hall
Columbia MO 65211
USA
pgus@missouri.edu

J. Taylor
University of Missouri
Division Animal Sciences
Columbia, MO 65211
USA
taylorjerr@missouri.edu

G. Stacey
University of Missouri
Plant Sciences Unit
Columbia, MO 65211
USA
staceyg@missouri.edu

ISBN: 978-0-387-76722-2   e-ISBN: 978-0-387-76723-9

Library of Congress Control Number: 2007943040

Printed on acid-free paper

9 8 7 6 5 4 3 2 1

springer.com

To Dr. Gordon Kimber in recognition of his effort, support, and commitment to continuation of the Stadler Genetics Symposium over many years. Gordon has been the strongest supporter of the Stadler Genetics Symposium since he co-organized the very first one with Dr. George Rédei in May 1969. He continued organizing and editing the Symposium series for several years and has always been tireless in his attendance and willingness to help in identifying excellent topics and speakers.

Acknowledgments

The editors gratefully acknowledge the generous support of the USDA-ARS and the following units from the University of Missouri: the College of Agriculture, Food and Natural Resources; the Molecular Biology Program; the Office of Research; the Interdisciplinary Plant Group; the College of Arts and Sciences; the School of Veterinary Medicine; the Plant Sciences Unit; and the School of Medicine. National Research Initiative Grant No. 2006-35205-17455 from the USDA Cooperative State Research, Education, and Extension Service also supported the Symposium. The contributors' continued support made the 24th Stadler Genetics Symposium a success.

The speakers, who spent a tremendous amount of time preparing their presentations and manuscripts, are gratefully acknowledged. Without their expertise and dedication, the Symposium could not have taken place.

We wish to thank all of the local chairpersons for ensuring that all of the speakers were well taken care of during the Symposium.

Sandi Strother from Conferences and Specialized Courses, University of Missouri, who tirelessly handled all of our requirements and made sure that everything was well organized, excellently handled the behind-the-scenes and on-site preparations. Without her, none of this would have been possible

A very special thanks goes to Arturo Garcia, USDA-ARS, Columbia, Missouri, for all of his excellent suggestions in editing the figures and the text to fit a book format.

Contents

Contributors

Frederick M. Ausubel
Department of Molecular Biology, Massachusetts General Hospital and Department of Genetics, Harvard Medical School, Boston, MA, USA

Rajendra Bari
The Sainsbury Laboratory, John Innes Centre, Colney Lane, Norwich NR4 7UH, UK

Laura E. Bartley
Department of Plant Pathology, One Shields Avenue, University of California, Davis, CA 95616, USA

Thomas Brefort
Max Planck Institute for Terrestrial Microbiology, Department Organismic Interactions, Karl-von-Frisch-Strasse, D-35043 Marburg, Germany

Ulla Bonas
Department of Genetics, Martin-Luther-University Halle-Wittenberg, Halle, Germany
ulla.bonas@genetik.uni-halle.de

Jorunn Bos
Department of Plant Pathology, Ohio State University Ohio Agricultural Research and Development Center, Wooster, OH, USA

Catherine Bruce
Department of Plant Pathology, Ohio State University Ohio Agricultural Research and Development Center, Wooster, OH, USA

Daniela Büttner
Department of Genetics, Martin-Luther-University Halle-Wittenberg, Halle, Germany

Edward J. Cargill
Canine Genetics Laboratory, Department of Pathobiology, College of Veterinary Medicine, Texas A&M University, College Station, TX 77843-4467, USA

Hans H. Cheng
USDA, ARS, Avian Disease and Oncology Laboratory, 3606 E. Mount Hope Rd.,
East Lansing, MI 48823, USA

N. Clairoux
Centre for Host-Parasite Interactions, Institute of Parasitology, McGill University,
Ste Anne de Bellevue, Quebec, Canada, H9X 3V9

Leigh Anne Clark
Canine Genetics Laboratory, Department of Pathobiology, College of Veterinary
Medicine, Texas A&M University, College Station, TX 77843-4467, USA

Anne E. Dorrance
Department of Plant Pathology, The Ohio State University, OARDC, 1680 Madison
Ave., Wooster, OH 44691, USA

Gunther Döhlemann
Max Planck Institute for Terrestrial Microbiology, Department Organismic
Interactions, Karl-von-Frisch-Strasse, D-35043 Marburg, Germany

Daolong Dou
Virginia Bioinformatics Institute, Virginia Polytechnic Institute and State
University, Blacksburg, VA 24061, USA

Monique Egler
Department of Genetics, Martin-Luther-University Halle-Wittenberg, Halle,
Germany

E. Estuningsih
Indonesian Veterinary Science Research Institute, Bogor, West Java, Indonesia

Rhonda L. Feinbaum
Department of Molecular Biology, Massachusetts General Hospital and Department
of Genetics, Harvard Medical School, Boston, MA, USA

K.J. Fullard
Reprogen-Centre for Advanced Technologies in Animal Genetics and Repro-
duction, Faculty of Veterinary Science, University of Sydney, Camden NSW,
Australia

Doreen Gürlebeck
Department of Genetics, Martin-Luther-University Halle-Wittenberg, Halle,
Germany

Sang-Wook Han
Department of Plant Pathology, One Shields Avenue, University of California,
Davis, CA 95616, USA

Simone Hahn
Department of Genetics, Martin-Luther-University Halle-Wittenberg, Halle,
Germany

Regina Hanlon

Virginia Bioinformatics Institute, Virginia Polytechnic Institute and State University, Blacksburg, VA 24061, USA

Ina Hoeschele

Virginia Bioinformatics Institute, Virginia Polytechnic Institute and State University, Blacksburg, VA 24061, USA

Rays H.Y. Jiang

Virginia Bioinformatics Institute, Virginia Polytechnic Institute and State University, Blacksburg, VA 24061, USA

Present address: Laboratory of Phytopathology, Wageningen University, Binnenhaven 5, NL6709 PD Wageningen, The Netherlands

Jonathan D.G. Jones

The Sainsbury Laboratory, John Innes Centre, Colney Lane, Norwich NR4 7UH, UK

Sophien Kamoun

Department of Plant Pathology, Ohio State University Ohio Agricultural Research and Development Center, Wooster, OH, USA
kamoun.1@osu.edu

Regine Kahmann

Max Planck Institute for Terrestrial Microbiology, Department Organismic Interactions, Karl-von-Frisch-Strasse, D-35043 Marburg, Germany, kahmann@mpi-marburg.mpg.de

Sabine Kay

Department of Genetics, Martin-Luther-University Halle-Wittenberg, Halle, Germany

N.M. Kingsford

Reprogen-Centre for Advanced Technologies in Animal Genetics and Reproduction, Faculty of Veterinary Science, University of Sydney, Camden NSW, Australia

Konstantinos Krampis

Department of Crop Soil and Environmental Science, Virginia Bioinformatics Institute, Virginia Polytechnic Institute and State University, Blacksburg, VA 24061

Antje Krüger

Department of Genetics, Martin-Luther-University Halle-Wittenberg, Halle, Germany

S.J. Lamont
Department of Animal Science, Iowa State University, 2255 Kildee Hall, Ames, IA 50011, USA
sjlamont@iastate.edu

Daniel G. Lee
Department of Molecular Biology, Massachusetts General Hospital and Department of Genetics, Harvard Medical School, Boston, MA, USA

Minkyoung Lee
Department of Plant Pathology, Ohio State University Ohio Agricultural Research and Development Center, Wooster, OH, USA

Hua Li
Virginia Bioinformatics Institute, Virginia Polytechnic Institute and State University, Blacksburg, VA 24061, USA

Nicole T. Liberati
Department of Molecular Biology, Massachusetts General Hospital and Department of Genetics, Harvard Medical School, Boston, MA, USA

Bing Liu
Virginia Bioinformatics Institute, Virginia Polytechnic Institute and State University, Blacksburg, VA 24061, USA
Present address: The Monsanto Company, 3302 SE Convenience Blvd, Ankeny, IA 50021, USA

Hsin-Yen Liu
Department of Plant Pathology, Ohio State University Ohio Agricultural Research and Development Center, Wooster, OH, USA

Sally A. Leong
USDA, ARS CCRU, Department of Plant Pathology, University of Wisconsin, 1630 Linden Dr., Madison, WI 53706, 608-262-6309, USA
saleong@wisc.edu

Sang-Won Lee
Department of Plant Pathology, One Shields Avenue, University of California, Davis, CA 95616, USA

Christian Lorenz
Department of Genetics, Martin-Luther-University Halle-Wittenberg, Halle, Germany

Leslie A. Lyons
Department of Population Health & Reproduction, School of Veterinary Medicine, University of California, Davis, CA 95616, USA

M.J. Mackinnon
Department of Pathology, University of Cambridge, Tennis Court Road, Cambridge
CB2 1QP and KEMRI-Wellcome Trust Research Programme, Centre for
Geographic Medicine Research – Coast, Kilifi 80108, Kenya
mjm88@cam.ac.uk

E.T. Margawati
LIPI, Cibinong, Indonesia

Yongcai Mao
Virginia Bioinformatics Institute, Virginia Polytechnic Institute and State
University, Blacksburg, VA 24061, USA

Forest Research Institute, Harborside Financial Center, Plaza V, Jersey City, NJ
07311, USA

Steven St. Martin
Department of Horticulture and Crop Science, The Ohio State University,
Columbus, OH 43210-1086, USA

William Morgan
Department of Plant Pathology, Ohio State University Ohio Agricultural Research
and Development Center, Wooster, OH, USA

Santiago X. Mideros
Department of Plant Pathology, The Ohio State University, OARDC, 1680 Madison
Ave., Wooster, OH 44691, USA

Present address: Department of Plant Pathology, 334 Plant Science, Cornell
University, Ithaca, NY, 14853, USA

Keith E. Murphy
Canine Genetics Laboratory, Department of Pathobiology, College of Veterinary
Medicine, Texas A&M University, College Station, TX 77843-4467, USA

Lionel Navarro
The Sainsbury Laboratory, John Innes Centre, Colney Lane, Norwich NR4
7UH, UK; Institut de Biology Moleculaire des Plantes du Centre National de la
Recherche Scientifique, 67084 Strasbourg Cedex, France
lionel.navarro@ibmp-ulp.u-strasbg.fr

Adnane Nemri
The Sainsbury Laboratory, John Innes Centre, Colney Lane, Norwich NR4 7UH,
UK

Sang-Keun Oh
Department of Plant Pathology, Ohio State University Ohio Agricultural Research
and Development Center, Wooster, OH, USA

Elaine A. Ostrander
National Human Genome Research Institute, National Institutes of Health, 50
South Drive, Building 50, Room 5351, Bethesda, MD 20892, USA

Heidi G. Parker
National Human Genome Research Institute, National Institutes of Health, 50
South Drive, Building 50, Room 5351, Bethesda, MD 20892, USA

D. Piedrafita
Centre for Animal Biotechnology, School of Veterinary Science, The University
of Melbourne, Victoria, 3010, Australia

H.W. Raadsma
Reprogen-Centre for Advanced Technologies in Animal Genetics and Repro-
duction, Faculty of Veterinary Science, University of Sydney, Camden NSW,
Australia

Christine A. Rees
Canine Genetics Laboratory, Department of Pathobiology, College of Veterinary
Medicine, Texas A&M University, College Station, TX 77843-4467, USA

Pamela C. Ronald
Department of Plant Pathology, One Shields Avenue, University of California,
Davis, CA 95616, USA

M.A. Saghai Maroof
Department of Crop Soil and Environmental Science, Virginia Polytechnic Institute
and State University, Blacksburg, VA 24061, USA

Kerstin Schipper
Max Planck Institute for Terrestrial Microbiology, Department Organismic
Interactions, Karl-von-Frisch-Strasse, D-35043 Marburg, Germany

Alexandre Seilaniantz
The Sainsbury Laboratory, John Innes Centre, Colney Lane, Norwich NR4 7UH,
UK

Brian M. Smith
Virginia Bioinformatics Institute, Virginia Polytechnic Institute and State
University, Blacksburg, VA 24061, USA

Jing Song
Department of Plant Pathology, Ohio State University Ohio Agricultural Research
and Development Center, Wooster, OH, USA

T.W. Spithill
Institute of Parasitology, Centre for Host-Parasite Interactions, McGill University,
Ste Anne de Bellevue, Quebec, Canada, H9X 3V9

George M. Strain
Canine Genetics Laboratory, Department of Pathobiology, College of Veterinary Medicine, Texas A&M University, College Station, TX 77843-4467, USA

Subandriyo
Indonesian Animal Production Research Institute, Bogor, West Java, Indonesia

Nathan B. Sutter
National Human Genome Research Institute, National Institutes of Health, 50 South Drive, Building 50, Room 5351, Bethesda, MD 20892, USA
Robert Szczesny
Department of Genetics, Martin-Luther-University Halle-Wittenberg, Halle, Germany

Frank Thieme
Department of Genetics, Martin-Luther-University Halle-Wittenberg, Halle, Germany

Sucheta Tripathy
Virginia Bioinformatics Institute, Virginia Polytechnic Institute and State University, Blacksburg, VA 24061, USA

Trudy Torto-Alalibo
Virginia Bioinformatics Institute, Virginia Polytechnic Institute and State University, Blacksburg, VA 24061, USA

Brett M. Tyler
Virginia Bioinformatics Institute, Virginia Polytechnic Institute and State University, Blacksburg, VA 24061, USA

Jonathan M. Urbach
Department of Molecular Biology, Massachusetts General Hospital and Department of Genetics, Harvard Medical School, Boston, MA, USA

Sharon L. Vanderlip
Canine Genetics Laboratory, Department of Pathobiology, College of Veterinary Medicine, Texas A&M University, College Station, TX 77843-4467, USA

Miguel Vega-Sanchez
Department of Plant Pathology, The Ohio State University, OARDC, 1680 Madison Ave., Wooster, OH 44691, USA

Present address: Department of Plant Pathology, The Ohio State University, 2021 Coffey Rd.Columbus, OH 43210, USA

Jacquelyn M. Wahl
Canine Genetics Laboratory, Department of Pathobiology, College of Veterinary Medicine, Texas A&M University, College Station, TX 77843-4467, USA

S. Widjayanti
Research Institute for Veterinary Science, Bogor, West Java, Indonesia

Joe Win
Department of Plant Pathology, Ohio State University Ohio Agricultural Research and Development Center, Wooster, OH, USA

Gang Wu
Department of Molecular Biology, Massachusetts General Hospital and Department of Genetics, Harvard Medical School, Boston, MA, USA
Keying Ye
Department of Statistics, Virginia Polytechnic Institute and State University, Blacksburg, VA24061, USA

Present address: Department of Management Science and Statistics, University of Texasat San Antonio, 6900 North Loop 1604 West, San Antonio, TX 78249-0632, USA

Carolyn Young
Department of Plant Pathology, Ohio State University Ohio Agricultural Research and Development Center, Wooster, OH, USA

Lecong Zhou
Department of Crop Soil and Environmental Science, Virginia Bioinformatics Institute, Virginia Polytechnic Institute and State University, Blacksburg, VA 24061, USA

Roles of Plant Hormones in Plant Resistance and Susceptibility to Pathogens

Lionel Navarro, Rajendra Bari, Alexandre Seilaniantz, Adnane Nemri, and Jonathan D.G. Jones

Abstract Plants and animals trigger an innate immune response upon perception of pathogen-associated molecular patterns (PAMPs) such as flagellin. In *Arabidopsis*, flagellin perception elevates resistance to *Pseudomonas syringae* pv. *tomato* DC3000 (*Pst* DC3000), although the molecular mechanisms involved remain elusive. A flagellin-derived peptide transiently enhances the accumulation of a plant microRNA that directs degradation of mRNA for TIR1, an F-box auxin receptor. The resulting repression of auxin signaling effectively restricts *Pst* DC3000 growth, implicating this previously unsuspected miRNA-mediated switch in bacterial disease resistance. These data suggest that elevation of auxin levels constitute a bacterial pathogenicity strategy that is suppressed during the innate immune response to PAMPs. In a separate work, we showed that DELLA proteins, which are normally associated with gibberellin responses, play a role in the balance between salicylic acid and jasmonic acid–mediated defense signaling pathways. DELLA loss-of-function mutants show reduced growth inhibition in response to flg22, enhanced susceptibility to necrotrophic pathogens, and enhanced resistance to *Pst* DC3000.

1 Introduction

Plants perceive a 22 amino acid–conserved peptide flg22, located in the N-terminal part of eubacterial flagellin (Felix, 1998). In *Arabidopsis*, recognition of flg22 is associated with early changes in host transcript levels, including a rapid down-regulation of a subset of genes (Navarro et al., 2004). This down-regulation is potentially post-transcriptional because promoter analysis does not identify over-representation of common *cis*-regulatory elements within the promoters of flg22-repressed genes (Navarro et al., 2004). Plants use RNA silencing for post-transcriptional gene regulation. This sequence-specific mRNA degradation

L. Navarro
The Sainsbury Laboratory, John Innes Centre, Colney Lane, Norwich NR4 7UH, UK; Institut de Biology Moleculaire des Plantes du Centre National de la Recherche Scientifique, 67084 Strasbourg Cedex, France
e-mail: lionel.navarro@ibmp-ulp.u-strasbg.fr

J.P. Gustafson et al. (eds.), *Genomics of Disease*,
© Springer Science+Business Media, LLC, 2008

mechanism is mediated by small (21-24nt) RNAs known as short interfering (si)RNAs and micro (mi)RNAs. Both siRNAs and miRNAs are derived from double-stranded (ds)RNA by the action of homologues of the RNaseIII-type enzymes called Dicer (Dicer-like, or DCLs). In *Arabidopsis*, miRNAs are excised from intergenic stem-loop transcripts by DCL-1, and direct cleavage of cellular mRNAs carrying miRNA-complementary sequences (Bartel, 2004). Because the steady-state level of several plant miRNAs—-and possibly several plant siRNAs— varies in response to exogenous stresses (Jones-Rhoades and Bartel, 2004; Sunkar and Zhu, 2004), we tested if those molecules could account for some of the rapid post-transcriptional changes elicited by flg22 treatment. We used transgenic *Arabidopsis* expressing plant virus–encoded proteins that suppress miRNA- and siRNA-guided functions, anticipating that transcripts repressed by flg22-stimulated small RNAs would likely be more elevated in those transgenic lines. Comparative transcript profiling identified a subset of mRNAs fulfilling this criterion, among which was an mRNA for the plant auxin receptor *TIR1* (*T*ransport *I*nhibitor *R*esponse *1*).

2 Flg22 Triggers Auxin-Signaling Repression by Inducing a Specific miRNA

TIR1 mRNA was previously identified as a target of miR393, a canonical miRNA conserved across plant species (Bonnet et al., 2004; Jones-Rhoades and Bartel, 2004; Wang et al., 2004). To identify whether miR393 plays a role in flg22-reponse, we examined its levels over a time course of flg22 treatment. Northern analysis revealed a transient and biphasic 2-fold increase in miR393 accumulation in flg22-treated *Arabidopsis* seedlings, whereas the levels of the unrelated miR171 remained unaffected (Navarro et al., 2006). In addition, the miR393 levels were unchanged in seedlings treated with flg22$^{A.tum}$, an inactive peptide derived from *Agrobacterium tumefaciens* flagellin (Felix, 1998). Quantitative RT-PCR (RT-qPCR) analyses (employing primers flanking the miR393 cleavage site) revealed a progressive decrease of *TIR1* transcript accumulation upon flg22 but not flg22$^{A.tum}$ treatments, leading to an overall ∼3-fold reduction in *TIR1* mRNA 60 minutes after elicitation. This progressive flg22-dependent reduction of the *TIR1* mRNA levels was also reflected at the protein level, as assessed in *Arabidopsis* transformants expressing a myc epitope-tagged form of TIR1 under the dexamethasone (Dex)-inducible promoter (Dex::TIR1-Myc) (Navarro et al., 2006). Collectively, these results indicate that flg22 triggers the rapid and specific repression of TIR1 accumulation through transient up-regulation of miR393 levels.

In addition to its role in auxin perception, TIR1 is part of the ubiquitin-ligase complex SCF^{TIR1} that interacts with Aux/IAA transcriptional repressor proteins to promote their ubiquitylation and subsequent degradation by the

26S-proteasome (Gray et al., 2001). We used transgenic lines expressing a heat shock–inducible AXR3/IAA17 protein fused with the β-glucuronidase (GUS) reporter (HS::AXR3NT-GUS) (Gray et al., 2001). Seedlings were exposed to high temperature and treated with either flg22 or flg22$^{A.tum}$ for 2 hours; GUS staining was subsequently performed. Flg22, but not flg22$^{A.tum}$, triggered a strong stabilization of AXR3NT-GUS in roots and leaves. In contrast, no AXR3NT-GUS was observed when seedlings were treated with flg22 at room temperature, indicating that flg22 does not activate the heat shock promoter. Time-course analysis revealed that stabilization of AXR3NT-GUS starts 1.5 hours after flg22 elicitation, consistent with the kinetics of TIR1-Myc protein repression (Navarro et al., 2006).

Aux/IAA proteins act as repressors of auxin signaling through heterodimerization with ARF (Auxin Response Factor) transcription factors (Liscum and Reed, 2002). ARFs bind directly to auxin responsive elements (AuxRE) found in the promoters of primary auxin-response genes, leading to their transcriptional activation (or repression in some cases) (Hagen and Guilfoyle, 2002). The flg22-induced stabilization of AXR3/IAA17, and presumably other Aux/IAA proteins, prompted us to investigate if flg22 inhibits ARF protein function resulting in transcriptional inactivation of primary auxin-response genes. To address this point, *Arabidopsis* seedlings were challenged for 1.5 hours with either flg22 or flg22$^{A.tum}$ and the transcript levels of the primary auxin-response genes *GH3-like*, *BDL/IAA12*, and *AXR3/IAA17* were monitored by RT-qPCR. This time point was chosen based on the flg22-induced stabilization profile of the AXR3/IAA17 protein. We found that all the three auxin-response genes were indeed repressed at this time point (Navarro et al., 2006). Collectively, these results indicate that flg22 triggers, through the action of miR393, a series of molecular events that ultimately lead to the rapid down-regulation of primary auxin-response genes.

To assess if auxin is involved in disease resistance and susceptibility, we used *Arabidopsis* transgenic lines that overexpress myc epitope-tagged versions of the two TIR1 functional paralogs AFB1 and AFB3 (for auxin signaling F-Box proteins 1 and 3). Both act in a redundant manner with *TIR1* to mediate auxin perception and signaling. However, the *AFB1* transcript, unlike the *AFB3* transcript, is partially resistant to miR393-guided cleavage, presumably because of a single nucleotide polymorphism introducing a synonymous mutation in the miRNA complementary site. Therefore, overexpression of AFB1 should have dominant-negative effects upon a putative miR393-mediated defense response. When inoculated with virulent *P. syringae* pv. *tomato* (*Pst*) DC3000, AFB1- but not AFB3-overexpressing plants displayed a ∼100-fold higher bacterial titers compared to non-transformed plants, as assessed at 2 and 4 days post-inoculation (Fig. 1A). In contrast, no difference was observed with avirulent *Pst* DC3000 carrying *AvrRpt2*, which encodes the elicitor of race-specific resistance controlled by the *Arabidopsis* gene *RPS2* (Fig. 1B; Dong et al., 1991; Whalen et al., 1991; Kunkel et al., 1993). These results suggest that miR393 specifically promotes basal resistance to virulent *Pst* DC3000 but is not implicated in race-specific resistance mediated by the RPS2 resistance protein.

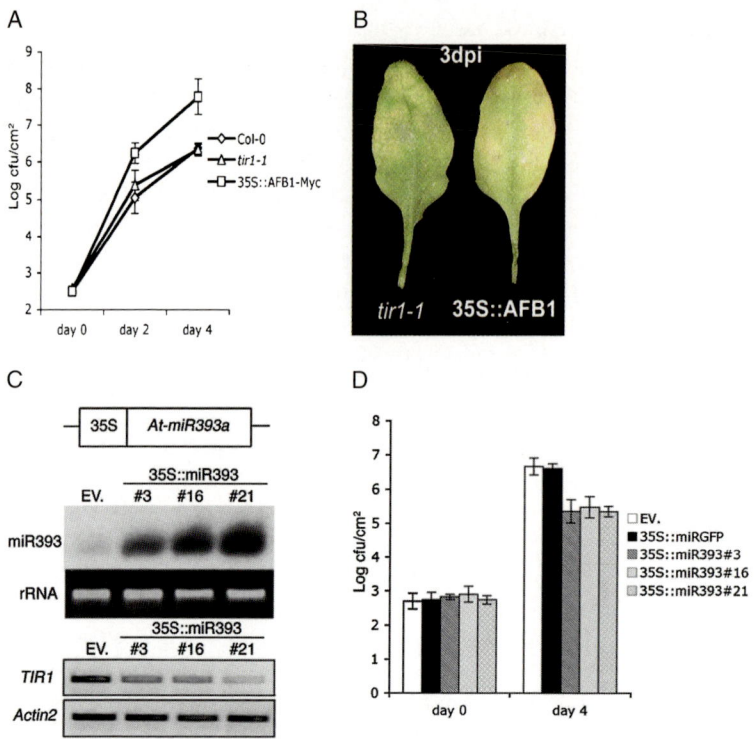

Fig. 1 MiR393-mediated down-regulation of auxin signaling is required for basal resistance to *Pst* DC3000. (**A**) Growth of *Pst* DC3000 on AFB1-overexpressing lines, AFB3-overexpressing lines, or Col-0 plants assessed 2 and 4 days post-inoculation (dpi) of 10^5 colony forming units (cfu/mL) bacterial concentration. Inoculation was performed by syringe infiltration on 5-week old plants. Error bars represent the standard error of log-transformed data from five independent samples. (**B**) Growth of *Pst* DC3000 carrying AvrRpt2 on AFB1-overexpressing lines, AFB3-overexpressing lines, and Col-0 plants. Inoculation was performed as in (**A**) and results are presented as in (**A**). (**C**) Molecular characterization of miR393-overexpressing lines. *Upper panel*: schematic representation of the miR393-overexpressing construct; 35S: strong promoter from Cauliflower Mosaic Virus (CaMV); *At-miR393a*: miR393 precursor derived from *Arabidopsis* chromosome 2. *Middle panel*: Northern analysis of miR393 overexpression in independent T2 transgenic lines; EV: Empty vector; rRNA: ethidium bromide staining of ribosomal RNA. *Bottom panel*: semi-quantitative RT-PCR of the *TIR1* transcript. RT-PCR of the Actin2 transcript was also performed to confirm equal amount of cDNA in each reaction. (**D**) Growth of *Pst* DC3000 in miR393 overexpressing lines. Inoculation was performed as in (**A**) on three independent transgenic lines that overexpress miR393 and display lower *TIR1* mRNA level as depicted in (**C**). Growth was monitored at 4 dpi as described in (**A**)

The *Arabidopsis* genome contains two miR393 precursor loci located on chromosome 2 and 3, termed *At-miR393a* and *At-miR393b*, respectively (Gustafson et al., 2005). Both precursors give rise to an identical mature miRNA. To further assess the role of miR393 and auxin signaling in bacterial disease resistance, we transformed *Arabidopsis* with a construct, in which transcription of the *At-miR393a* precursor is driven by the strong 35S promoter (Fig. 1C). Three independent T2 transgenic lines were selected based on high miR393 accumulation. Accordingly, these transgenic lines had low levels of *TIR1* mRNAs, as compared to lines transformed with an empty vector (Fig. 1C). Upon infection with virulent *Pst* DC3000, all the three miR393-overexpressing lines, but not the empty vector transformants, displayed ∼100-fold lower bacterial titers at 4 dpi, confirming that miR393 restricts *Pst* DC3000 growth (Fig. 1D). Furthermore, no difference in bacterial growth was observed in transgenic lines overexpressing an artificial miRNA directed against the green fluorescence protein (gfp) mRNA (Parizotto et al., 2004), indicating a miR393-specific effect (data not shown).

3 Does Auxin Play a Role in Bacterial Pathogenenity?

Besides its role in plant growth and development, auxin was also reported to affect plant–pathogen interactions (Yamada, 1993). For example, the tumorigenic *P. syringae* pv. *savastanoi* produces high level of IAA that is implicated in the development of oleander knots (Yamada et al., 1991). The role of auxin in disease susceptibility is not restricted to tumorogenic bacteria because exogenously applied auxin was also reported to induce susceptibility of maize to *Helminthosporium* leaf spot and of tobacco to *Tobacco Mosaic Virus* (TMV) (Simons et al., 1972; Hoffman, 1973). Moreover, auxin can suppress the hypersensitive response (HR), a plant-triggered cell death process often induced to restrict pathogen growth (Novacky, 1972; Matthysse, 1987; Robinette and Matthysse, 1990). Interestingly, we note that most *P. syringae* strains produce IAA and *Pst* DC3000 infection triggers higher accumulation of free IAA in *Arabidopsis* (Glickmann et al., 1998; O'Donnell et al., 2003). In addition, the virulent AvrRpt2 type-III secreted protein appears to promote the auxin-signaling pathway in an *Arabidopsis rps2* mutant background, resulting in enhanced bacterial disease symptoms and growth (B. Kunkel, personal communication). Consistent with the role of auxin in bacterial pathogenicity, we found that delivery of the auxin analog 2,4-dichlorophenoxyacetic acid (2,4-D) at 20 µM concentration together with *Pst* DC3000 (10^5 cfu/mL) significantly promotes bacterial disease symptoms as early as 3 days post-inoculation (data not shown). However, only a mild effect on bacterial growth was observed (∼1.5-fold higher bacterial titer compared to plantstreated with *Pst* DC3000 alone). Lower concentrations of 2,4-D still promote *Pst* DC3000 disease symptoms without having a significant effect on bacterial growth (data not shown). These results suggest that exogenous auxin predominantly promotes *Pst* DC3000 disease symptom development.

We show here that a bacterial PAMP down-regulates auxin signaling by enhancing the endogenous levels of miR393. Overexpressing auxin signaling through a TIR1 paralog that is partially refractory to miR393 guided–cleavage enhances susceptibility to virulent *Pst* DC3000 and, conversely, repressing this hormonal pathway through miR393 overexpression induces resistance to the bacterium. These results indicate that down-regulation of auxin signaling is part of a plant-induced immune response. They also suggest that auxin might promote disease susceptibility to bacteria. Consistent with this hypothesis, we found that the auxin-analog 2,4-D promotes significantly *Pst* DC3000 disease symptom development (data not shown). These data suggest that elevation of auxin levels constitute a bacterial pathogenicity strategy that is suppressed during the PAMP-triggered response.

4 Flg22 Triggers Growth Inhibition of *Arabidopsis* Seedlings

Flg22 not only triggers many canonical defense responses, but also causes growth inhibition. DELLA proteins are negative regulators of gibberellin (GA) signaling and are known to restrain plant growth (Harberd, 2003). We wondered whether flg22-triggered growth inhibition involved stabilization of DELLA proteins, whose GA-provoked degradation is important for growth (Harberd, 2003). To test the above hypothesis, *Arabidopsis* DELLA *tetra* mutants (mutants in which 4 out of 5 *Arabidopsis* DELLA genes were mutated: Achard et al., 2006) and wild-type seedlings were examined for flg22-induced growth inhibition. We found that DELLA *tetra* mutants displayed reduced flg22-triggered growth inhibition compared to La-er seedlings (Fig. 2A, B). The *Arabidopsis fls2-17* seedlings, which are defective in flagellin sensing (FLS) receptor, showed no growth inhibition by flg22. This

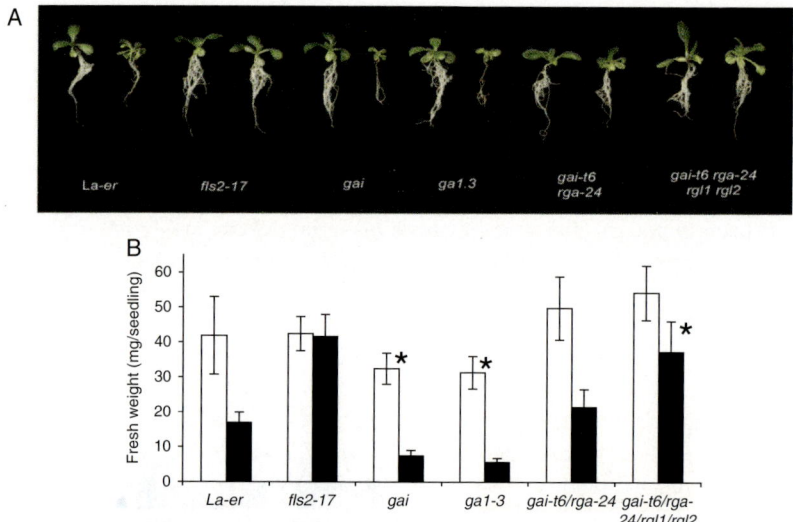

Fig. 2 DELLA proteins act as positive regulators of flg22-triggered growth inhibition. The *Arabidopsis* wild-type La-*er* displays enhanced flg22-triggered growth inhibition, whereas the DELLA *tetra* mutant is partially compromised in this flg22-induced developmental phenotype. *Arabidopsis* wild-type La-*er*, *fls2-17*, and DELLA mutants were grown on solid medium for 7 days and then transferred for another 7 days in liquid medium containing either flg22[A.tum] or flg22 at a final concentration of 10 nM. Growth inhibition was assessed visually (**A**) and quantified by measuring seedling fresh weight ($35 < n < 42$). (**B**) Percentage of growth inhibition of La-er, *fls2-17*, and DELLA *tetra* mutants

suggested that flg22 treatment might stabilize DELLA proteins. In another studies, we found that flg22 delays the GA-induced disappearance of GFP-RGA (data not shown), which is consistent with the induction of DELLA protein stabilization by flg22.

5 Role of DELLA Proteins in Plant Disease Resistance and Susceptibility

The involvement of DELLA proteins in plant defense was examined by challenging DELLA *tetra* mutants with bacterial and fungal pathogens. When DELLA mutant plants were infiltrated with a biotrophic pathogen, *Pst* DC3000, they showed

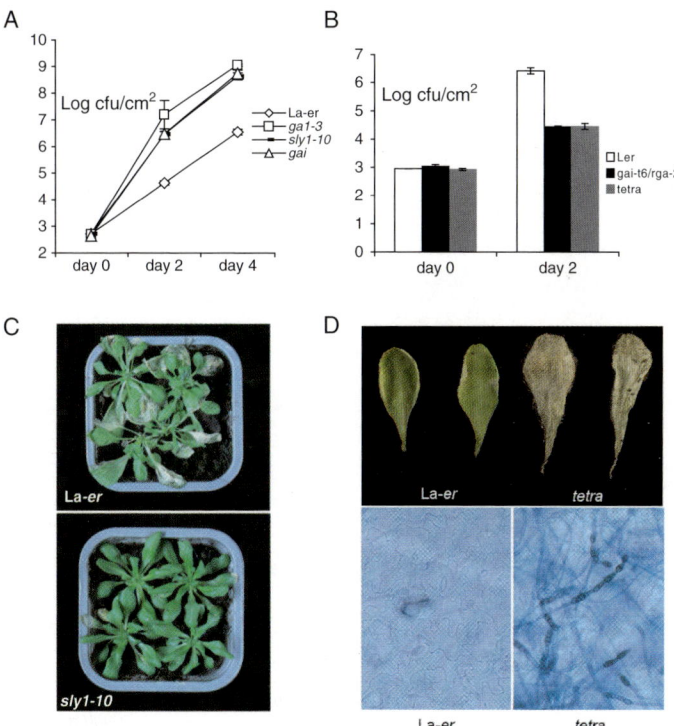

Fig. 3 DELLA proteins act as negative regulators of plant defense against *Pst* DC3000 and as positive regulators of defense against *A. brassicicola*. (**A**) DELLA *tetra* mutants are more resistant to *Pst* DC3000. Growth of *Pst* DC3000 on DELLA *tetra* (*gai-t6/rga-t2/rgl1-1/rgl2-1*) and La-*er* plants was assessed 2 days post-inoculation (dpi) of 10^5 colony-forming units (cfu/ml). Inoculation was performed by syringe infiltration on 4-week old plants grown in long-day conditions. Error bars represent the standard error of log-transformed data from five independent samples. (**B**) DELLA *tetra* mutants are more susceptible to *A. brassicicola*. Four-week old plants (grown in long-day conditions) were sprayed with *A. brassicicola* spores at a concentration of 5×10^5 spores/ml and pictures were taken at 7 dpi

enhanced resistance compared to wild-type plants (Fig. 3A). Salicylic acid (SA) is the major signaling molecule involved in the resistance of plants against biotrophic pathogens (Durner et al., 1997). To check whether DELLA proteins interfere with the SA-dependent plant defense pathway, the expression of SA-dependent marker genes *PR1* and *PR2* were monitored in DELLA *tetra* and wild-type plants challenged with *Pst* DC3000. We found an earlier and stronger induction of both *PR1* and *PR2* genes in *Pst* DC3000-inoculated *tetra* plants compared to wild-type plants (data not shown).

Since the SA-dependent plant defense pathway is mutually antagonistic to the jasmonate/ethylene (JA/ET)-dependent pathway, we next investigated whether the stronger induction of SA-dependent genes, in the *tetra* infected mutant, leads to the suppression of JA/ET-dependent gene expression. The expression of JA/ET-dependent marker genes *PDF1.2* and *VSP1* were reduced in *Pst* DC3000-challenged *tetra* plants compared to wild-type plants (data not shown). This indicates that DELLA proteins are involved in the repression of SA-dependent gene expression and activation of JA/ET-dependent gene expression in *Arabidopsis*.

The involvement of DELLA proteins in the positive regulation of the JA/ET-pathway led us to examine whether DELLA *tetra* mutants were altered in their resistance to the necrotrophic pathogen *Alternaria brassicicola* Interestingly, 7 days after inoculation with *A. brassicicola*, DELLA *tetra* mutant leaves were diseased and heavily colonized with fungal hyphae, indicating enhanced susceptibility to *A. brassicicola* compared to wild-type plants (Fig. 3B). This indicates that DELLA proteins act as positive regulators of plant defense responses to *A. brassicicola*.

6 Are DELLA Proteins Integrators of Plant Defense Pathways?

These data show that DELLA proteins are involved in the repression of SA-dependent plant defense and activation of the JA/ET-dependent plant defense pathway. This suggest that DELLA proteins play an important role in establishing the balance of defense responses mounted by a plant and suggest new mechanisms for crosstalk between different plant-signaling pathways involved in growth and defense (Fig. 4). It is conceivable that abscisic acid (ABA), by antagonizing GA action, could have the opposite effect. ABA and ethylene pathways are involved in plant responses to diverse abiotic and biotic stresses, but particularly in the drought response and wound response respectively. Recently, it has been shown that two independent salt stress-activated phytohormonal signaling pathways (ABA and ethylene) regulate plant development through integration at the level of DELLA function (Achard et al., 2006). These data, together with the data on miR393 and auxin signaling during defense, suggest that crosstalk between hormone pathways has insufficiently been acknowledged as an important determinant in the outcome of plant–pathogen interactions. It seems likely that DELLA proteins will turn out to play an important role in this crosstalk.

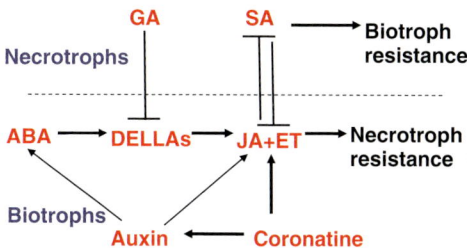

Fig. 4 DELLA proteins are involved in balancing plant defense responses against different biotrophic and necrotrophic pathogens. The phytohormone GA promotes the degradation of DELLA proteins to relieve DELLA restraint. The accumulation of GA is supposed to result in nectrotroph resistance. DELLA proteins activate JA/ET pathway involved in the resistance against necrotrophs. Due to the antagonism between JA/ET and SA pathways, DELLA proteins are involved in the suppression of SA-pathway required for the resistance against biotrophs. Other hormones such as Auxin and Coronatines are also believed to stimulate JA/ET pathway. Auxin stimulates ABA pathway which also acts positively on DELLA proteins. This indicates that DELLA proteins are involved in the mechanism for integration of plant growth and defense responses against various stresses

Acknowledgments We thank Nick Harberd for providing DELLA *tetra* mutant seeds, Patrick Achard for helpful discussion of DELLA protein biology, Olivier Voinnet for advice on small RNAs, and Mark Estelle for advice on auxin experiments. This work was supported by the Gatsby Charitable Foundation and the Biotechnology and Biological Sciences Research Council (BBSRC) grant.

References

Achard P., Cheng H., De Grauwe L., Decat J., Schoutteten H., Moritz T., Van Der Straeten D., Peng J., Harberd N.P., 2006, Integration of plant responses to environmentally activated phytohormonal signals, *Science* **311**:91–94.

Bartel, D.P., 2004, MicroRNAs: genomics, biogenesis, mechanism, and function, *Cell* **116**: 281–297.

Bonnet, E., Wuyts, J., et al., 2004, Detection of 91 potential conserved plant microRNAs in *Arabidopsis thaliana* and *Oryza sativa* identifies important target genes, *Proc. Natl. Acad. Sci. USA* **101**: 11511–11516.

Dong, X., et al., 1991, Induction of *Arabidopsis* defense genes by virulent and avirulent *Pseudomonas syringae* strains and by a cloned avirulence gene, *Plant Cell* **3**:61–72.

Durner J., Shah, J., and Klessig, D.F., 1997, Salicylic acid and disease resistance in plants, *Trends Plant Sci.* **2**:266–274.

Felix, G., et al., 1998, Plants have a sensitive perception system for the most conserved domain of bacterial flagellin, *Plant J.* **18**:265–276.

Glickmann, E., et al., 1998, Auxin production is a common feature of most pathovars of *Pseudomonas syringae*, *Mol. Plant Microbe Interact.* **11**:156–162.

Gray, W.M., et al., 2001, Auxin regulates SCF(TIR1)-dependent degradation of AUX/IAA proteins, *Nature* **414**:271–276.

Gustafson, A.M., et al., 2005, ASRP: the *Arabidopsis* Small RNA Project Database, *Nucleic Acids Res.* **33**:D637–D640.

Hagen, G., and Guilfoyle, T., 2002, Auxin-responsive gene expression: genes, promoters and regulatory factors, *Plant Mol. Biol.* **49**:373–385.

Hoffman, S.E., 1973, Leaf bioassay for *Helminthosporium carbonum* toxin-search for phytoalexin, *Phytopath.* **63**:729–734.

Jones-Rhoades, M.W., and Bartel, D.P., 2004, Computational identification of plant microRNAs and their targets, including a stress-induced miRNA, *Mol. Cell* **14**:787–799.

Kunkel, B.N., et al., 1993, RPS2: an *Arabidopsis* disease resistance locus specifying recognition of *Pseudomonas syringae* strains expressing the avirulence gene avrRpt2, *Plant Cell* **5**:865–875.

Liscum, E., and Reed, J.W., 2002, Genetics of Aux/IAA and ARF action in plant growth and development, *Plant Mol. Biol.* **49**:387–400.

Matthysse, A.G., 1987, A method for the bacterial elicitation of a hypersensitive-like response in plant cell cultures, *J. Microbiol. Methods* **7**:183–191.

Navarro, L., et al., 2004, The transcriptional innate immune response to flg22. Interplay and overlap with Avr gene-dependent defense responses and bacterial pathogenesis, *Plant Physiol.* **135**:1113–1128.

Navarro, L, Dunoyer, P., Jay, F., Arnold, B., Dharmasiri, N., Estelle, M., Voinnet, O., and Jones, J.D., 2006, A plant miRNA contributes to antibacterial resistance by repressing auxin signaling, *Science* **21**:436–439.

Novacky, A., 1972, Suppression of the bacterially induced hypersensitive reaction by cytokinins, *Physiol. Plant Pathol.* **2**:101–104.

O'Donnell, P.J. et al., 2003, Susceptible to intolerance—-a range of hormonal actions in a susceptible *Arabidopsis* pathogen response, *Plant J.* **33**:245–257.

Parizotto, E.A. et al., 2004, In vivo investigation of the transcription, processing, endonucleolytic activity, and functional relevance of the spatial distribution of a plant miRNA, *Genes Dev.* **18**:2237–2342.

Harberd, N.P., 2003, Relieving DELLA restraint, *Science* **21**:1853–1854.

Robinette, D., and Matthysse, A.G., 1990, Inhibition by *Agrobacterium tumefaciens* and *Pseudomonas savastanoi* of development of the hypersensitive response elicited by *Pseudomonas syringae* pv. *phaseolicola, J. Bacteriol.* **172**:5742–5749.

Simons, T.J., et al., 1972, Effect of 2,4-dichlorophenoxyacetic acid on tobacco mosaic virus lesions in tobacco and on the structure of adjacent cell, *Virology* **48**:502–515.

Sunkar, R., and Zhu, J.K., 2004, Novel and stress-regulated microRNAs and other small RNAs from *Arabidopsis, Plant Cell* **16**:2001–2019.

Wang, X.J., et al., 2004, Prediction and identification of *Arabidopsis thaliana* microRNAs and their mRNA targets, *Genome Biol.* **5**: R65.

Whalen, M.C., et al., 1991, Identification of *Pseudomonas syringae* pathogens of *Arabidopsis* and a bacterial locus determining avirulence on both *Arabidopsis* and soybean, *Plant Cell* **3**:49–59.

Yamada, T., 1993, The role of auxin in plant-disease development, *Annu. Rev. Phytopathol.* **31**:253–273.

Yamada, T., et al., 1991, The role of indoleacetic acid biosynthetic genes in tumorigenicity, In: S.S. Patil, S. Ouchi, D. Mills, C. Vance (eds.), *Mol. Strat. Pathogens Host Plants*, New York: Springer-Verlag, pp. 83–94.

Canine Genetics Facilitates Understanding of Human Biology

Elaine A. Ostrander, Heidi G. Parker, and Nathan B. Sutter

Abstract In the past 15 years the field of canine genetics has advanced dramatically. Dense comparative maps, production of ×1.5 and ×7.5 genome sequences, SNP chips, and a growing sophistication regarding how to tackle problems in complex genetics have all propelled the canine system from a backwater to the forefront of the genomics landscape. In this chapter, we explore some of the critical advances in the field that have occurred in the past 5 years. We discuss the implications of each on disease gene mapping. Complex trait genetics and advances related to finding genes associated with morphology are also discussed. Finally, we speculate on what advances will likely define the field in the coming 5 years.

1 Introduction to Dogs and Breeds

The domestic dog is believed to be the most recently evolved species from the family Canidae. Within the Canidae there are three distinct phylogenetic groups (Wayne et al., 1997; 1987a, b). The domestic dog shares a clade with the wolf-like canids such as the gray wolf, coyote, and jackals. Dogs are thought to have arisen in quite recent time, perhaps as little as 40,000 years ago, with the initial domestication events occurring in eastern Asia (Savolainen et al., 2002; Vila et al., 1997).

Most dog breeds arose in the last 200–300 years and many of the most common modern breeds were developed in Europe in the 1800s. Currently, there are over 400 recognized and distinct dog breeds of which 155 are registered by the American Kennel Club (AKC) in the United States (American Kennel Club, 1998). While a breed of dog can be recognized by its physical attributes such as size, shape, coat color, head shape, leg length, etc., the concept of a breed has been formally defined by both dog fanciers and geneticists.

According to registering bodies like the AKC, becoming a registered member of a breed simply requires that both of a dog's parents are documented members of the same breed, and that a small fee be paid. As a result, dog breeds are essentially

E.A. Ostrander
National Human Genome Research Institute, National Institutes of Health, Bethesda MD 20892, USA

J.P. Gustafson et al. (eds.), *Genomics of Disease,*
© Springer Science+Business Media, LLC, 2008

closed-breeding populations with little opportunity for introduction of new alleles. Dog breeds are characterized by a lower level of genetic heterogeneity than that seen in mixed breed dogs as a result of small numbers of founders, population bottlenecks, and the over representation of some males (popular sires) who perform well in dog shows (Parker et al., 2004; Parker and Ostrander, 2005). As a result, the current population of ~10 million purebred dogs in the United States represents an ideal group in which to study the genetics of both simple and complex traits.

Recently, attempts have been made to define the concept of a breed at the genetic level (Koskinen, 2003; Koskinen and Bredbacka, 2000; Parker et al., 2004). For example, Parker et al. (2004) utilized data from 96 (CA)n repeat-based microsatellite markers spanning all dog autosomes on 414 dogs to determine the degree to which dogs could be assigned to their appropriate breed using a clustering algorithm. Only a small set of closely related breed pairs (i.e., Whippet and Greyhound; Alaskan Malamute and Siberian Husky) could not be reproducibly distinguished when compared to other breeds. Similarly, using the Doh assignment test, 99% of the dogs tested were correctly assigned to their distinct breed group using only the microsatellite data.

The above results are interesting in light of studies on genetic diversity in human populations. In the Parker et al. (2004) study, we showed that humans and dogs have similar levels of overall nucleotide diversity, 8×10^{-4}. Genetic variation between dog breeds, however, is much greater than the observed variation between human populations (27.5% versus 5.4% by AMOVA). The degree of genetic homogeneity, not unexpectedly, is much greater within the membership of any given individual dog breed than it is within distinct human populations. So the concept of a dog "breed" is much more definitive, at the genetic level, than is the concept of a human "population" or a human "race."

2 Mapping Disease Genes in Dogs

Because dog breeds represent closed-breeding populations, they offer unique opportunities for disease gene mapping (Ostrander and Kruglyak, 2000). Diseases that are problems for both human and companion animal health are excellent candidates for study, particularly those associated with complicated phenotypes. The mapping of complex traits in humans, such as cancer, diabetes, epilepsy, and heart disease, has been stymied by the lack of large pedigrees, limited statistical methods, and both locus and phenotypic heterogeneity. As a result, the ability to unambiguously identify critical susceptibility loci for diseases like cancer has been problematic (Ostrander et al., 2004). By working with canine families, researchers are able to overcome many of these disadvantages. Dog pedigrees are large, and often permit collection of several generations. For example, the pedigrees used to find the genes for a variety of forms of progressive retinal atrophy (PRA) (Acland et al., 1994, 1998, 1999; Kukekova et al., 2006; Moody et al., 2005; Sidjanin et al., 2002), copper toxicosis (Yuzbasiyan-Gurkan et al., 1997), renal cancer (Jonasdottir et al., 2000), narcolepsy (Lin et al., 1999; Mignot et al., 1991), hyperuricosuria (Safra et al., 2006), pancreatic acinar atrophy (Clark et al., 2005), and epilepsy (Lohi et al., 2005) all involved large, multigenerational families of the sort unheard of in human genetics. In addition, the fact that all the dogs share a common, often inbred genetic background means that phenotypic expression among individuals with the disease is usually very similar. This latter point should prove particularly useful as the community moves from the mapping of single gene Mendelian disorders to identifying loci associated with complex traits such as behavior and morphology.

We have appreciated the importance of genetic predisposition in the occurrence of canine diseases for years (Patterson, 2000; Patterson et al., 1982). Indeed, the dog is second only to human in the attention to which clinicians offer their clients and the number of dollars spent on health care (American Veterinary Medical Association, 2002; Patterson, 2000). As a result, several hundred genetic diseases have been identified in the dog (Sargan, 2004), many of which share strong phenotypic similarities with human diseases. Many of these are collated in an online database called IDID (Inherited Disease in Dogs), which is similar to the Online Mendelian Inheritance of Man (OMIM) database (Sargan, 2004).

To date, dozens of loci have been identified for canine-inherited diseases and in many cases the causative genes have been identified (reviewed in Parker and Ostrander, 2005; Sutter et al., 2004; Switonski et al., 2004). Specific examples include metabolic disorders (van De Sluis et al., 2002; Yuzbasiyan-Gurkan et al., 1997), blindness (Acland et al., 1998, 1999; Aguirre et al., 1978; Aguirre and Acland, 1988, 1998; Kukekova et al., 2006; Moody et al., 2005), cancer (Jonasdottir et al., 2000; Lingaas et al., 2003), neurologic disorders (Lin et al., 1999; Lingaas et al., 1998), hip dysplasia (Chase et al., 2004), osteoarthritis (Chase et al., 2005b), hyperuricosuria (Safra et al., 2006), pancreatic acinar atrophy (Clark et al., 2005), Addison's disease (Chase et al., 2006), and epilepsy (Lohi et al., 2005).

The lessons learned have been plentiful. We have gleaned insight into new genetic mechanisms responsible for disease as well as learned something about the genes and pathways associated with many diseases. In some cases knowledge gained about the underlying disease genes have enlightened us about human conditions for which we had little prior knowledge, such as the inherited sleep disorder narcolepsy. In this case, the underlying mutation found in the genetically susceptible Doberman Pinscher was a splicing defect in the gene for the hypocretin 2 receptor (Lin et al., 1999).

In other cases we have learned about new types of genetic aberrations that can cause disease. Recalling again the example of narcolepsy, the disease has been shown to be caused at the molecular level by insertion of a canine-specific, short interspersed nuclear element (SINE; Bentolila et al., 1999; Minnick et al., 1992; Vassetzky and Kramerov, 2002). These retrotransposons are derived from a tRNA-Lys and occur frequently throughout the canine genome (Bentolila et al., 1999; Coltman and Wright, 1994; Kirkness et al., 2003). In addition to narcolepsy, aberrant insertion of SINEC_Cf elements are associated with centronuclear myopathy in the Labrador Retriever (Pele et al., 2005) as well as the gray merle coat coloring that appears in many breeds (Clark et al., 2006).

Another interesting example is found in a form of epilepsy similar to human Lafora disease. The canine disease affects several breeds including the miniature wirehaired dachshund. Lohi and collaborators have shown that the disease is caused by expansion of an unstable dodecamer repeat in the Epm2b (Nhlrc1) gene (Lohi et al., 2005). While trinucleotide repeat expansion has been reported in association with several human neurologic disorders, this is the first report of a dodecamer repeat expansion causing a disease in any species.

By far, the most interesting advances have been those that highlighted not only new mechanisms of disease, but new genes as well. For instance, extensive progress has been made in understanding the genetic basis of PRA in the dog (Acland et al., 1994, 1998, 1999; Aguirre et al., 1978, 1998; Aguirre and Acland, 1988; Kukekova et al., 2006; Lowe et al., 2003; Moody et al., 2005; Sidjanin et al., 2002). PRA refers to a collection of ocular disorders reminiscent of the constellation of human diseases known as retinitis pigmentosa (reviewed by Petersen-Jones (2005)). Recently, a gene for progressive rod cone degeneration (prcd) was identified in the Poodle, Labrador, and several other breeds (Goldstein et al., 2006; Zangerl et al., 2006). This disease had previously been mapped to a gene-rich region of canine chromosome 9 (CFA9) (Acland et al., 1998). As the disease is present in several related breeds, the authors used linkage disequilibrium (LD) data from a combination of 14 breeds to reduce the disease-associated interval from several megabases (Mb) to just 106 Kb (Goldstein et al., 2006) (Fig. 1). They then identified a single missense mutation that accounted for both the canine disease and the autosomal recessive retinitis pigmentosa in a patient from Bangladesh (Zangerl et al., 2006).

In many cases, disease genes have been found in dogs after identification in humans, or simultaneous with the disease gene in humans. For example, the gene for canine renal cancer in the German Shepherd Dog, although linkage mapped first

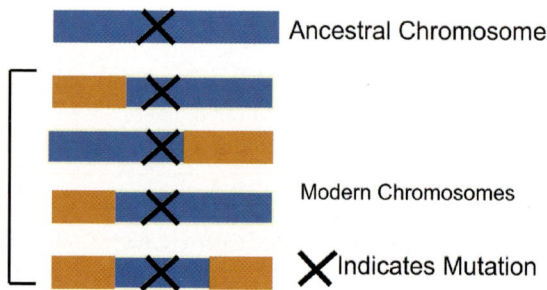

Fig. 1 Linkage disequilibrium. The *top* chromosome represents an ancient chromosome with an initial ancestral mutation as marked by the X. Meiotic recombination whittles away the shared haplotype around the chromosome. Modern day chromosomes will share only a small region of commonality around the mutation. Identification of this shared haplotype by SNP genotyping facilitates fine mapping studies

in the dog (Jonasdottir et al., 2000), was actually found (Lingaas et al., 2003) after the orthologous human gene, which causes a similar disease called Birt-Hogg-Dube Syndrome, was identified (Nickerson et al., 2002). The example of identifying the gene for prcd remains one of the few, together with the identification of the gene for copper toxicosis in the Bedlington Terrier, where the canine community has led the human genetics community in the hunt for truly novel susceptibility genes (van De Sluis et al., 2002).

3 Canine Breed Relationships

The above study by Goldstein et al. provides a nice example of how data can be combined across breeds to identify disease loci of interest (Goldstein et al., 2006). To generalize this concept, Parker et al. have studied over 85 breeds using a clustering algorithm to understand the relatedness of one breed to another (Parker et al., 2004). In their initial analysis, 85 breeds were ordered into four clusters, generating what is now considered to be a new canine classification system for dog breeds (Ostrander and Wayne, 2005) based on similar patterns of alleles, presumably from a shared ancestral pool (Fig. 2). Cluster 1 comprised dogs of Asian and African origin as

Population Structure of 85 Dog Breeds

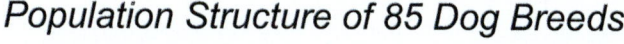
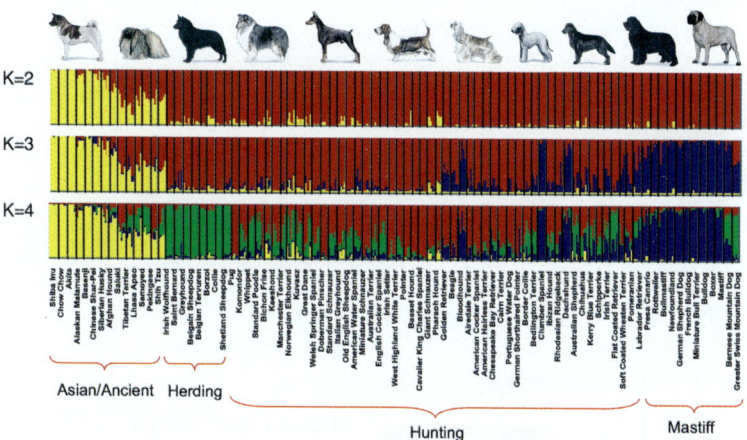

Fig. 2 Population structure of the domestic dog. Figure is derived from the work of Parker et al. (2004). Five dogs from each of 85 breeds were genotyped using 85 (CA)n repeat-based microsatellites. Markers spanned all autosomes at 30 Mb density. Analysis was performed using the computer program *structure*. Analysis at K = 2, 3, and 4 divided the population of 85 breeds into the most likely groups based on allele sharing. Group 1 is comprised largely of Asian breeds such as the Lhasa Apso, Shar Pei, and Akita. Group 2 is the mastiff group and includes, for example, the Boxer, Bull Dog, and Presa Canario. Group 4 includes a mixture of dogs including working breeds. Group 4 is enriched for sight and scent hounds and includes breeds such as the spaniels and retrievers

well as gray wolves. Cluster 2 is typified by mastiff-type dogs with big, boxy heads and strong, sturdy bodies such as the Boxer, Mastiff, and Bulldog. The third and fourth clusters split a group of herding dogs and sight hounds away from the general population of modern hunting dogs comprised of terriers, hounds, and gun dog breeds. Ongoing studies are underway to expand this work to include more breeds. It is expected that this should allow even higher resolution of the breed relationships picture, and a clearer understanding of how to best combine data across breeds for fine resolution mapping studies.

4 Advances in Canine Genomics

While canine genetics has demonstrated significant progress in the past few years, the rate at which we can expect new discoveries will accelerate dramatically in the coming months. This is due almost exclusively to two major advances. First, the publication of a gene dense canine radiation hybrid (RH) map allowed us, for the first time, to understand the evolutionary relationship between the canine and the human genomes (Hitte et al., 2005). In this study, a well-spaced set of 9850 sequence tagged sites (STS) corresponding to a set of evenly spaced human genes selected from the then available ×1.5 poodle sequence (Kirkness et al., 2003) were localized on an RH map using a 9000 rad panel. Mutual-Blast alignments identified the best target (human) gene sequence using the dog sequence as a probe to ensure that we were, in fact, mapping the canine ortholog. A total of 9850 gene fragments were eventually mapped, which corresponds to approximately half of the genes in the dog genome, identifying some 264 conserved segments (CS) between dog and human.

Interestingly, most of these fragments (243) were later identified by the whole genome assembly (CanFam1.0) of the dog (Lindblad-Toh et al., 2005), generated from the ×7.5 sequencing effort. This suggests that a dense RH map provides as much information for comparative genome mapping studies as a ×7–10 whole genome shotgun sequence. In addition, detailed comparison of the canine ×7.5 whole genome assembly (CanFam 1.0) to the 9000 rad RH map showed that 99.3% of the chromosomal assignments predicted by the RH map were in complete agreement with the sequence assembly. Those that were not were quickly resolved and found to represent issues such as the orientation of internal chromosomal fragments. This advance was critical in allowing scientists to move between the canine and the human maps, in assembling the canine genome sequence, and in finding the precise breakpoints between the canine and the human genomes (Murphy et al., 2005).

In addition to the above, the availability of both a ×1.5 poodle survey sequence and a whole genome assembly of a ×7.5 boxer sequence is sure to impact canine genetics research at every level (Kirkness et al., 2003; Lindblad-Toh et al., 2005). We now know that the dog euchromatic genome is approximately 2.4 billion bases and is comprised of about 243 conserved segments when compared to the human genome. The assembled sequence is estimated to cover 98–99% of the genome, with the majority of the sequence contained within two supercontigs per chromosome. That is, on average, two segments of continuous sequence cover each of the dogs' 38 autosomes. The gene count, at ∼19, 000, is less than what has been predicted for the human genome, perhaps due to complexities associated with splicing and gene families. There is a 1-1-1 correspondence between orthologs of human, mouse, and dog for 75% of the genes. The full genome sequence can be accessed through http://www.genome.ucsc.edu; http://www.ncbi.nih.gov, and http://www.ensembl.org. A discussion of mining the canine genome sequence is reviewed in O'Rourke (2005).

In addition to the Boxer sequence, a ×1.5 partial sequence of the Standard Poodle is available (Kirkness et al., 2003). While in itself less complete than the Boxer sequence, together these two resources have enabled the identification of more than 2 million single nucleotide polymorphisms (SNPs). We now know that a SNP occurs about once in every 1000 bases in dogs (Lindblad-Toh et al., 2005) and a first generation canine SNP chip is now available from Affymetrix. The chip contains some 24,000 working SNPs that will change the landscape of whole genome association studies in the dog. While microsatellites have proven sufficient for mapping single gene traits, it has generally not been possible to analyze enough markers to fully interrogate the genome in a complex trait association study. With thousands of SNPs available on a single chip, we believe that it is now possible to identify subtle variants responsible for a host of phenotypic observations.

Key to the development of the canine SNP chip were studies by both Lindblad-Toh et al. (Lindblad-Toh et al., 2005) and Sutter et al. (Sutter et al., 2004) who addressed the issue of how many SNPs are "enough" for doing whole genome association studies in the dog. Sutter and colleagues examined the extent of LD in five breeds with distinct breed histories and reported that the average length of LD in these five breeds is approximately 2 Mb (Fig. 3). This is 40–100 times further

Divergent Population Histories

Fig. 3 Divergent population histories of dog breeds. Breeds were selected by Sutter et al. (2004) in their study of linkage disequilibrium in dogs. Breeds were chosen to represent a variety of morphologic types, levels of present-day and historical popularity, population structure, and history (Wilcox and Walkowicz, 1995)

Bernese Mountain Dog – Small population today, severe population bottlenecks during world wars

Akita – Small population today, mixture of isolated U.S. and Japanese populations. U.S. population bottleneck during world wars

Pekingese – Popular today, but few founders to U.S.

Labrador Retriever – Popular the past century

Golden Retriever – Popular the past century. Created 1800s from Tweed Water Spaniel, Yellow Retriever, Irish Setter, Bloodhound, circus dogs

than the LD that typically extends in the human genome (Fig. 4). Thus, while a typical whole genome association study in humans requires about 500,000 SNPs (Kruglyak, 1999), in dogs the same study would require only about 10,000–30,000 markers. For diseases of interest to both human and canine health such as cancer, heart disease, cataracts, etc., these LD findings argue that it will be far easier to do the initial mapping study in dogs than in humans. These investigators also found that the extent of LD varied over a near 10-fold range between breeds of dog (0.4–3.2 Mb) (Sutter et al., 2004), arguing that breed selection would be important for the initial mapping of any trait of interest.

As part of the canine genome sequencing effort, Lindblad-Toh and colleagues undertook more extensive studies on canine LD, looking at more loci and more SNPs. They concluded that perhaps as few as 10,000 SNPs would be needed to fully cover the genome. They also found that the level of LD between breeds was different, but argued that the levels across the genome will likely vary more than the levels associated, on average, with any one dog breed versus another. Finally, both studies looked at the issue of haplotype sharing and demonstrated that there was low haplotype diversity and high haplotype sharing. Importantly, this means that a single set of SNPs, or a single SNP chip, is likely sufficient for mapping studies in any dog breed.

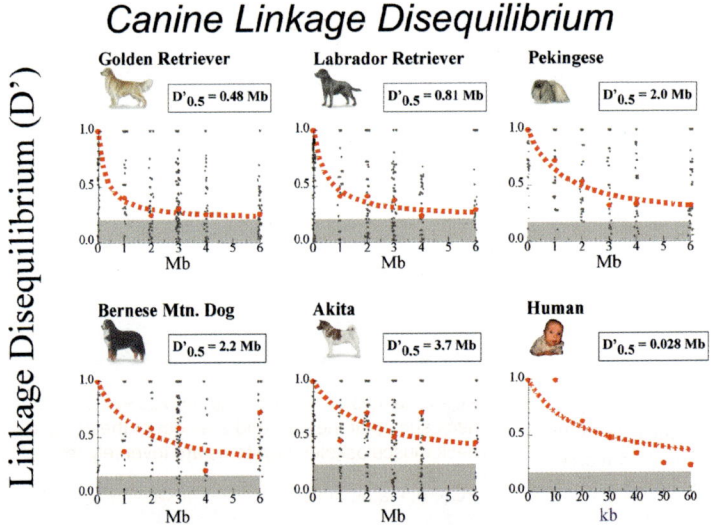

Fig. 4 Canine linkage disequilibrium. Summarized previously in Sutter et al. (2004) and Sutter and Ostrander (2004). Five breeds were analyzed and five loci each and the results averaged for each breed. D' statistic is shown for each breed and for human. Gray shading indicates background level. LD decay at the 50% level is indicated in upper right of each panel. LD extends the farthest for the Akita, at 3.7 Mb, and the shortest for the Golden Retriever, at 0.48 Mb. For human the comparable number is about 0.028 Mb. LD in dogs is about 50 times greater than that observed in most human populations

5 Mapping Genes for Morphology in the Dog

Breeds of dog differ by over 40-fold in size and display an amazing level of morphologic variation. Indeed, Wayne et al. (1986a, b) have argued that the diversity in skeletal size and proportion of dogs is greater than that observed in any other terrestrial mammal. Studies to map quantitative trait loci (QTLs) in the dog associated with body conformation have been led by Gordon Lark, Kevin Chase and collaborators and are based upon their work with the Portuguese Water Dog (PWD) (Chase et al., 1999, 2002). They chose the PWD because the breed offers several advantages for complex trait mapping. There are only about 10,000 living AKC registered PWD, and they derive historically largely from just two kennels (Chase et al., 1999). The breed standard allows for significant variation, offering a greater opportunity for mapping traits associated with morphology than would studies of other breeds.

To initiate their studies these researchers collected DNA samples, health information, pedigree data, and five standard X-rays of over 500 dogs (Chase et al., 1999). They undertook a genome wide scan using over 500 microsatellite markers. Analysis of the data suggested four sets of correlated traits or principal components (PC) (Fig. 5). Each PC described a set of correlated phenotypic features for which QTLs could be identified. PC1 regulates overall body size; PC2 describes the relationship between pelvis, head, and neck; PC3 is shown in the inverse relationship between the cranial volume and the length of skull and limbs; PC4 is the length versus width of the skull and axial skeletons, representing a tradeoff between speed and strength as illustrated by the hound-type dogs on the left and the mastiff-type dogs on the right of Fig. 5.

Of particular interest to Chase and colleagues has been an understanding of the loci which control body size in males versus females (Chase et al., 2005a). In an analysis of 42 metrics derived from the five X-rays, they show that there are five QTLs controlling overall body size in the PWD. Differences in skeletal size between

Fig. 5 Four principal components for morphology. Figure summarizes work of Chase and colleagues in their analysis of the Portuguese Water Dog (Chase et al., 2002). Data are based on analysis of 90 metrics derived for five X-rays taken from each of several hundred dogs. Analysis of each principal component allows identification of QTLs controlling each set of correlated traits

Four Principal Components

PC1 regulates the overall size of the skeleton.

PC2 regulates the relationship between pelvis, head and neck such that the size and strength of the pelvis and head-neck musculoskeletal systems are inversely related.

PC3 regulates the inverse relationship between the metrics of cranial volume and length of skull and limbs.

PC4 controls the skull and axial skeletons, representing a tradeoff between speed and strength.

females and males are due to an interaction between a QTL on CFA15, adjacent to the insulin-like growth factor-1 (IGF-1) gene, and a locus on the X-chromosome defined by the CHM marker. The locus on CFA15 is defined by marker FH2017. Analysis of FH2017 genotypes suggests that in females the CFA15 allele controlling small size is dominant. However, in males the reverse is true and the genotype associated with large size appears dominant. The situation is partly explained by consideration of the QTL on the X-chromosome. Females that are homozygous at the CHM marker and homozygous for the large size CFA15 genotype are, on average, as large as the very largest males in the breed. However, any female that is heterozygous at the CHM locus will be small, regardless of her FH2017 genotype. Overall, this interaction explains about 50% of sexual dimorphism in the breed.

Why are these observations so important? First, they demonstrate that the canine system is amenable to mapping of complex traits that are of interest to all mammalian biologists. Second, these studies highlight the value of studying complex traits first in a single breed, especially one with small numbers of founders, but a large amount of phenotypic variation. Finally, the results demonstrate that the number of genes controlling complex traits is not so large as to be intractable. That is, body size is likely controlled by a small number of QTLs that are identifiable in the canine system. This has important implications for the mapping of complex diseases as well as truly complex phenotypes such as those associated with behavior.

6 Summary and Future Aims

Until recently, advancement in the study of companion animal health has relied on data from human and mouse studies. With the development of a ×7.5 whole genome sequence assembly of the dog, a SNP chip, studies of canine breed relationships, and a growing understanding of the architecture of the canine genome we are, for the first time, mapping genetic traits of interest first in the dog. Our understanding of disease genes important for both simple and complex traits is advancing at a rapid rate, informing us about the underlying biology of diseases critical to both humans and companion animals.

In the coming decade, the dog is likely not only to lead man in the discovery of disease genes but to provide novel insights into our understanding of truly complex phenotypes. How many of those will be relevant to the human condition? Only time will tell, but we can rest assured that the dog, ever man's faithful companion, will be by our side as we unravel the mysteries of disease susceptibility, behavior, and morphology.

Acknowledgments We thank our colleagues who read the chapter during preparation and made excellent suggestions and the many dog owners, breeders, and supporters who provided us with samples and information about their beloved pets. This work is supported by the Intramural Program of the National Human Genome Research Institute, a Burroughs Wellcome Innovation Award, and The American Kennel Club Canine Health Foundation.

During preparation of this chapter, the 12-year-old Border Collie of one of the authors died after a short and unexpected illness. We dedicate this chapter to the many pet owners who, in similar situations, have shown us how to deal with our loss with grace and dignity.

References

Acland, G.M., Blanton, S.H., Hershfield, B., and Aguiree, G.D., 1994, XLPRA: a canine retinal degeneration inherited as an X-linked trait, *Am. J. Med. Genet.* **52**:27–33.

Acland, G.M., Ray, K., Mellersh, C.S., Gu, W., Langston, A.A., Rine, J., Ostrander, E.A., and Aguirre, G.D., 1998, Linkage analysis and comparative mapping of canine progressive rod-cone degeneration (prcd) establishes potential locus homology with retinitis pigmentosa (RP17) in humans, *Proc. Natl. Acad. Sci. USA* **95**:3048–3053.

Acland, G.M., Ray, K., Mellersh, C.S., Gu, W., Langston, A.A., Rine, J., Ostrander, E.A., and Aguirre, G.D., 1999, A novel retinal degeneration locus identified by linkage and comparative mapping of canine early retinal degeneration, *Genomics* **59**:134–142.

Aguirre, G., Lolley, R., Farber, D., Fletcher, T., and Chader, G., 1978, Rod-cone dysplasia in Irish Setter dogs: a defect in cyclic GMP metabolism in visual cells, *Science* **201**:1133.

Aguirre, G.D., and Acland, G.M., 1988, Variation in retinal degeneration phenotype inherited at the prcd locus, *Exp. Eye Res.* **46**:663–687.

Aguirre, G.D., Baldwin, V., Pearce-Kelling, S., Narfstrom ,K., Ray, K., and Acland, G.M., 1998, Congenital stationary night blindness in the dog: common mutation in the RPE65 gene indicates founder effect. *Mol. Vis.* **4**:23.

American Kennel Club, 1998, The Complete Dog Book, Howell Book House, New York, NY.

American Veterinary Medical Association, 2002, U.S. Pet Onwership and Demographics Source-book, pp. 1–136, American Veterinary Medical Association, Schauburg.

Bentolila, S., Bach, J.M., Kessler, J.L., Bordelais, I., Cruaud, C., Weissenbach, J., and Panthier, J.J., 1999, Analysis of major repetitive DNA sequences in the dog (*Canis familiaris*) genome, *Mamm. Genome* **10**:699–705.

Chase, K., Adler, F.R., Miller-Stebbings, K., and Lark, K.G., 1999, Teaching a new dog old tricks: identifying quantitative trait loci using lessons from plants, *J. Hered.* **90**:43–51.

Chase, K., Carrier, D.F., Adler, F.R., Ostrander, E.A., and Lark, K.G., 2005a, Size sexual dimor-phism in Portugese Water Dogs: interaction between an autosome and the X chromosome, *Genome Res.* **15**:1820–1824.

Chase, K., Carrier, D.R., Adler, F.R., Jarvik, T., Ostrander, E.A., Lorentzen, T. D., and Lark, K.G., 2002, Genetic basis for systems of skeletal quantitative traits: principal component analysis of the canid skeleton, *Proc. Natl. Acad. Sci. USA* **99**:9930–9935.

Chase, K., Lawler, D.F., Adler, F.R., Ostrander, E.A., and Lark, K.G., 2004, Bilaterally asym-metric effects of quantitative trait loci (QTLs): QTLs that affect laxity in the right versus left coxofemoral (hip) joints of the dog (*Canis familiaris*), *Am. J. Med. Genet.* **124**:239–247.

Chase, K., Lawler, D.F., Carrier, D.R., and Lark, K.G., 2005b, Genetic regulation of osteoarthritis: a QTL regulating cranial and caudal acetabular osteophyte formation in the hip joint of the dog (*Canis familiaris*), *Am. J. Hum. Genet.* **135**:334–335.

Chase, K., Sargan, D., Miller, K., Ostrander, E.A., and Lark, K.G., 2006, Understanding the genet-ics of autoimmune disease: two loci that regulate late onset Addison's disease in Portuguese Water Dogs, *Int. J. Immunogenet.* **33**:179–84.

Clark, L.A., Wahl, J.M., Rees, C.A., and Murphy, K.E., 2006, Retrotransposon insertion in SILV is responsible for merle patterning of the domestic dog, *Proc. Natl. Acad. Sci. USA* **103**: 1376–1381.

Clark, L.A., Wahl, J.M., Steiner, J.M., Zhou, W., Ji, W., Famula, T.R., Williams, D.A., and Murphy, K.E., 2005, Linkage analysis and gene expression profile of pancreatic acinar atrophy in the German Shepherd Dog, *Mamm. Genome* **16**:955–962.

Coltman, D.W., and Wright, J.M., 1994, Can SINEs: a family of tRNA-derived retroposons specific to the superfamily Canoidea, *Nucleic Acids Res.* **22**:2726–2730.

Goldstein, O., Zangerl, B., Pearce-Kelling, S., Sidjanin, D.J., Kijas, J.W., Felix, J., Acland, G.M., and Aguirre, G.D., 2006, Linkage disequilibrium mapping in domestic dog breeds narrows the progressive rod-cone degeneration interval and identifies ancestral disease-transmitting chromosome, *Genomics* **88**:551–563.

Hitte, C., Madeoy, J., Kirkness, E.F., Priat, C., Lorentzen, T.D., et al., 2005, Facilitating genome navigation: survey sequencing and dense radiation-hybrid gene mapping, *Nat. Rev. Genet.* **6**:643–8.

Jonasdottir, T.J., Mellersh, C.S., Moe, L., Heggebo, R., Gamlem, H., Ostrander, E.A., and Lingaas, F., 2000, Genetic mapping of a naturally occurring hereditary renal cancer syndrome in dogs, *Proc. Natl. Acad. Sci. USA* **97**:4132–4137.

Kirkness, E.F., Bafna, V., Halpern, A.L., Levy, S., Remington, K., Rusch, D.B., Delcher, A.L., Pop, M., Wang, W., Fraser, C.M., and Venter, J.C., 2003, The dog genome: survey sequencing and comparative analysis, *Science* **301**:1898–1903.

Koskinen, M.T., 2003, Individual assignment using microsatellite DNA reveals unambiguous breed identification in the domestic dog, *Anim. Genet.* **34**:297–301.

Koskinen, M.T., and Bredbacka, P., 2000, Assessment of the population structure of five Finnish dog breeds with microsatellites, *Anim. Genet.* **31**:310–317.

Kruglyak, L., 1999, Prospects for whole-genome linkage disequilibrium mapping of common disease genes, *Nat. Genet.* **22**:139–144.

Kukekova, A.V., Nelson, J., Kuchtey, R.W., Lowe, J.K., Johnson, J.L., Ostrander, E.A., Aguirre, G.D., and Acland, G.M., 2006, Linkage mapping of canine rod cone dysplasia type 2 (rcd2) to CFA7, the canine orthologue of human 1q32, *Invest. Ophthalmol. Vis. Sci.* **47**:1210–1215.

Lin, L., Faraco, J., Li R., Kadotani, H., Rogers, W., Lin, X., Qiu, X., Jong, P.J., Nishino, S., and Mignot, E., 1999, The sleep disorder canine narcolepsy is caused by a mutation in the hypocretin (orexin) receptor 2 gene, *Cell* **98**:365–376.

Lindblad-Toh, K., Wade, C.M., Mikkelsen, T.S., Karlsson, E.K., Jaffe, D.B., et al., 2005, Genome sequence, comparative analysis and haplotype structure of the domestic dog, *Nature* **438**:803–819.

Lingaas, F., Aarskaug, T., Sletten, M., Bjerkas, I., Grimholt, U., Moe, L., Juneja, R.K., Wilton, A.N., Galibert, F., Holmes, N.G., and Dolf, G., 1998, Genetic markers linked to neuronal ceroid lipofuscinosis in English setter dogs, *Anim. Genet.* **29**:371–376.

Lingaas, F., Comstock, K.E., Kirkness, E.F., Sorensen, A., Aarskaug, T., Hitte, C., Nickerson, M.L., Moe, L., Schmidt, L.S., Thomas, R., Breen, M., Galibert, F., Zbar, B., and Ostrander, E.A., 2003, A mutation in the canine BHD gene is associated with hereditary multifocal renal cystadenocarcinoma and nodular dermatofibrosis in the German Shepherd dog, *Hum. Mol. Genet.* **12**:3043–3053.

Lohi, H., Young, E.J., Fitzmaurice, S.N., Rusbridge, C., Chan, E.M., Vervoort, M., Turnbull, J., Zhao, X.C., Ianzano, L., Paterson, A.D., Sutter, N.B., Ostrander, E.A., Andre, C., Shelton, G.D., Ackerley, C.A., Scherer, S.W., and Minassian, B.A., 2005, Expanded repeat in canine epilepsy, *Science* **307**:81.

Lowe, J.K., Kukekova, A.V., Kirkness, E.F., Langlois, M.C., Aguirre, G.D., Acland, G.M., and Ostrander, E.A., 2003, Linkage mapping of the primary disease locus for collie eye anomaly, *Genomics* **82**:86–95.

Mignot, E., Wang, C., Rattazzi, C., Gaiser, C., Lovett, M., Guilleminault, C., Dement, W.C., and Grumet, F.C., 1991, Genetic linkage of autosomal recessive canine narcolepsy with a mu immunoglobulin heavy-chain switch-like segment, *Proc. Natl. Acad. Sci. USA* **88**:3475–3478.

Minnick, M.F., Stillwell, L.C., Heineman, J.M., and Stiegler, G.L., 1992, A highly repetitive DNA sequence possibly unique to canids, *Gene* **110**:235–238.

Moody, J.A., Famula, T.R., Sampson, R.C., and Murphy, K.E., 2005, Identification of microsatellite markers linked to progressive retinal atrophy in American Eskimo Dogs, *Am. J. Vet. Res.* **66**:1900–1902.

Murphy, W.J., Larkin, D.M., Everts-van der Wind, A., Bourque, G., Tesler, G., et al., 2005, Dynamics of mammalian chromosome evolution inferred from multispecies comparative maps, *Science* **309**:613–617.

Nickerson, M., Warren, M., Toro, J., Matrosova, V., Glenn, G., Turner, M., Duray, P., Merino, M., Choyke, P., Pavlovich, C., Sharma, N., Walther, M., Munroe, D., Hill, R., Maher, E., Greenberg, C., Lerman, M., Linehan, W., Zbar, B., and Schmidt, L., 2002, Mutations in a novel gene lead to kidney tumors, lung wall defects, and benign tumors of the hair follicle in patients with the Birt-Hogg-Dube syndrome, *Cancer Cell* **2**:157.

O'Rourke, K., 2005, Mining the canine genome. Identification of genes helps breeders and researchers, *J. Am. Vet. Med. Assoc.* **226**:863–864.

Ostrander, E.A., and Kruglyak, L., 2000, Unleashing the canine genome, *Genome Res.* **10**: 1271–1274.

Ostrander, E.A., Markianos, K., and Stanford, J.L., 2004, Finding prostate cancer susceptibility genes, *Annu. Rev. Genomics Hum. Genet.* **5**:151–175.

Ostrander, E.A., and Wayne, R.K., 2005, The canine genome, *Genome Res.* **15**:1706–1716.

Parker, H.G., Kim, L.V., Sutter, N.B., Carlson, S., Lorentzen, T.D., Malek, T.B., Johnson, G.S., DeFrance, H.B., Ostrander, E.A., and Kruglyak, L., 2004, Genetic structure of the purebred domestic dog, *Science* **304**:1160–1164.

Parker, H.G., and Ostrander, E.A., 2005, Canine genomics and genetics: Running with the pack, *PLoS Genet.* **1**:e58.

Patterson, D., 2000, Companion animal medicine in the age of medical genetics, *J. Vet. Internal. Med.* 14:1–9.

Patterson, D.F., Haskins, M.E., and Jezyk, P.F., 1982, Models of human genetic disease in domestic animals, *Adv. Hum. Genet.* **12**:263–339.

Pele M., Tiret L., Kessler J. L., Blot S., and Panthier J. J. 2005. SINE exonic insertion in the PTPLA gene leads to multiple splicing defects and segregates with the autosomal recessive centronuclear myopathy in dogs, *Hum. Mol. Genet.* **14**:1417–1427.

Petersen-Jones S. 2005. Advances in the molecular understanding of canine retinal diseases, *J. Small Anim Pract.* **46**:371–380.

Safra N., Schaible R.H., and Bannasch D.L. 2006. Linkage analysis with an interbreed backcross maps Dalmatian hyperuricosuria to CFA03, *Mamm. Genome.* **17**: 340–345.

Sargan D.R. 2004. IDID: inherited diseases in dogs: web-based information for canine inherited disease genetics, *Mamm. Genome* **15**:503–506.

Savolainen, P., Zhang, Y.P., Luo, J., Lundeberg, J., and Leitner, T., 2002, Genetic evidence for an East Asian origin of domestic dogs, *Science* **298**:1610–1613.

Sidjanin, D.J., Lowe, J.K., McElwee, J.L., Milne, B.S., Phippen, T.M., Sargan, D.R., Aguirre, G.D., Acland, G.M., and Ostrander, E.A. 2002, Canine CNGB3 mutations establish cone degeneration as orthologous to the human achromatopsia locus ACHM3, *Hum. Mol. Genet.* **11**: 1823–1833.

Sutter, N.B., Eberle, M.A., Parker, H.G., Pullar, B.J., Kirkness, E.F., Kruglyak, L., and Ostrander, E.A., 2004, Extensive and breed-specific linkage disequilibrium in *Canis familiaris*, *Genome Res.* **14**:2388–2396.

Sutter, N.B., and Ostrander, E.A., 2004, Dog star rising: the canine genetic system, *Nat. Rev. Genet.* **5**:900–910.

Switonski, M., Szczerbal, I., and Nowacka, J., 2004, The dog genome map and its use in mammalian comparative genomics, *J. Appl. Genet.* **45**:195–214.

van De Sluis, B., Rothuizen, J., Pearson, P.L., van Oost, B.A., and Wijmenga, C., 2002, Identification of a new copper metabolism gene by positional cloning in a purebred dog population, *Hum. Mol. Genet.* **11**:165–173.

Vassetzky, N.S., and Kramerov, D.A., 2002, CAN—a pan-carnivore SINE family, *Mamm. Genome* **13**:50–57.

Vila, C., Savolainen, P., Maldonado, J.E., Amorim, I.R., Rice, J.E., Honeycutt, R.L., Crandall, K.A., Lundeberg, J., and Wayne, R K., 1997, Multiple and ancient origins of the domestic dog, *Science* **276**:1687–1689.

Wayne, R.K., 1986a, Cranial morphology of domestic and wild canids the influence of development on morphological change, *Evolution* **40**:243–261.

Wayne, R.K., 1986b, Limb morphology of domestic and wild canids: the influence of development on morphologic change, *J. Morphol.* **187**:301–319.

Wayne, R.K., Geffen, E., Girman, D.J., Koeppfli, K.P., Lau, L.M., and Marshall, C.R., 1997, Molecular systematics of the Canidae, *Syst. Biol.* **46**:622–653.

Wayne, R.K., Nash, W.G., and O'Brien, S.J., 1987a, Chromosomal evolution of the Canidae. I. Species with high diploid numbers, *Cytogenet. Cell Genet.* **44**:123–133.

Wayne, R.K., Nash, W.G., and O'Brien, S.J., 1987b, Chromosomal evolution of the Canidae. II. Divergence from the primitive carnivore karyotype, *Cytogenet. Cell Genet.* **44**:134–141.

Wilcox, B., and Walkowicz, C., 1995, "Atlas of Dog Breeds of the World," T.F.H. Publications, Neptune City, NJ.

Yuzbasiyan-Gurkan, V., Blanton, S.H., Cao, V., Ferguson, P., Li, J., Venta, P.J., and Brewer, G.J., 1997, Linkage of a microsatellite marker to the canine copper toxicosis locus in Bedlington terriers, *Am. J. Vet. Res.* **58**:23–27.

Zangerl ,B., Goldstein, O., Philp, A.R., Lindauer, S.J., Pearce-Kelling, S.E., Mullins, R.F., Graphodatsky, A.S., Ripoll, D., Felix, J.S., Stone, E.M., Acland, G.M., and Aguirre, G.D., 2006, Identical mutation in a novel retinal gene causes progressive rod-cone degeneration in dogs and retinitis pigmentosa in humans, *Genomics* **26**:26.

Xanthomonas oryzae pv. oryzae AvrXA21 Activity Is Dependent on a Type One Secretion System, Is Regulated by a Two-Component Regulatory System that Responds to Cell Population Density, and Is Conserved in Other Xanthomonas spp.

Sang-Won Lee, Sang-Wook Han, Laura E. Bartley, and Pamela C. Ronald

Abstract The rice pathogen recognition receptor, XA21, confers resistance to *Xanthomonas oryzae* pv. *oryzae* (*Xoo*) strains expressing the pathogen-associated molecule, AvrXA21. *XA21* codes for a receptor-like kinase consisting of an extracellular leucine rich repeat (LRR) domain, a transmembrane domain, and a cytoplasmic kinase domain (Ronald et al., 1992; Song et al., 1995). We show that AvrXA21 activity requires the presence of *rax* (*r*equired for AvrXA21) *A, raxB,* and *raxC* genes that encode components of a type one secretion system (TOSS). In contrast, an *hrpC*$^-$ strain deficient in type three secretion maintains AvrXA21 activity. *Xanthomonas campestris* pv. *campestris* (*Xcc*) can express AvrXA21 activity if *raxST*, encoding a putative sulfotransferase, and *raxA* are provided in trans. Expression of *rax* genes is dependent on population density and other functioning *rax* genes, suggesting that AvrXA21 is involved in quorum sensing and that the AvrXA21 pathogen-associated molecule represents an entirely new class of Gram-negative bacterial signaling molecules. We discuss the implications of these results for models of plant innate immunity.

Here, we provide a brief overview of some of the major concepts and molecular features of plant and animal innate immune system perception. We then describe new results from our studies of the XA21–AvrXA21 interaction and discuss how these results call for some modifications in the way we think about plant innate immunity strategies.

S.-W. Lee
Department of Plant Pathology, One Shields Avenue, University of California, Davis
CA 95616, USA

J.P. Gustafson et al. (eds.), *Genomics of Disease*,
© Springer Science+Business Media, LLC, 2008

1 Detection of Pathogens by Plants and Animal Hosts

Animals and plants both have well-developed immune systems for protection against pathogen challenges. Adaptive immunity system, specific to animals, depends on somatic gene rearrangements for generation of antigen receptors with random specificities. In contrast, innate immunity is common to metazoans and plants and involves perception of pathogen associated molecular patterns (PAMPs) by pathogen recognition receptors (PRRs) (Girardin et al., 2002). PAMPs for plants and/or animals have been defined as microbe-associated molecules that are relatively conserved and required for the microbe's lifecycle (Medzhitov et al., 1997; Janeway et al., 2002). Representative PAMPs that have been identified to date are flagellin, a proteinasceous component of bacterial polar flagella (Ramos et al., 2004), the peptidoglycan of Gram-positive bacteria (Leulier et al., 2003), lipopolysaccharide of Gram-negative bacteria (Erbs et al., 2003), single-stranded viral RNA (Jurk et al., 2002), and oomycete transglutaminase (Brunner et al., 2002).

In animals, recognition of PAMPs in extracellular compartments is largely carried out by the Toll-like receptor (TLR) family, which contains extracellular leucine rich repeats (LRR) that act in ligand recognition and an intracellular Toll-Interleukin 1 (TIR) domain (Werling et al., 2003). Although TLRs recognize diverse molecules, they activate a common signaling pathway to induce a core set of defense responses (Barton et al., 2003). Intracellular recognition is largely carried out by the cytoplasmic nucleotide-binding oligomerization domain (NOD)–protein family. The NOD family contains a large number of proteins from animals, plants, fungi, and bacteria (Inohara et al., 2003).

Evidence has accumulated that plants also detect PAMPs, but unlike animal systems, which have PRRs for cytoplasmic and extracellular perception of PAMPs, all biochemically characterized phytopathogen PAMPs are active at the cell surface (Nurnberger et al., 2004). With notable exceptions, another general characteristic of PAMP recognition by plants is that, while promoting expression of pathogenesis-related proteins and other characteristics of pathogen response, it does not lead to clear disease resistance or a related hypersensitive response (HR), which includes localized plant cell death.

Surprisingly, little is known about PAMP receptors in plants. The only well-characterized plant PRR for a PAMP is the *Arabidopsis thaliana* receptor-like kinase (RLK), FLS2 (flagellin sensing 2), which includes an extracellular LRR ligand-binding domain and an intracellular serine/threonine kinase, and directly recognizes a conserved N-terminal fragment of bacterial flagellin (Felix et al., 1999; Chinchilla et al., 2006). Stimulation of FLS2 by flagellin activates pathogenesis-related gene expression and pretreatment with it leads to resistance (Gomez-Gomez et al., 1999; Sun et al., 2006). Recent studies have shown that the presence of FLS2 decreases host susceptibility to a pathovar of *Pseudomonas syringae* when the pathogen is applied to leaves by spraying, but not infiltration (Zipfel et al., 2004; Sun et al., 2006). This demonstrates a direct link between disease resistance and PAMP perception that had previously been lacking in plants.

In addition to recognition of conserved PAMPs, plant PRRs also recognize strain-specific molecules produced by phytopathogens, termed pathogen avirulence (Avr) factors (van't Slot et al., 2002). This specific recognition generally triggers a strong defense response, often including the HR. To explain the observation of interactions between dominant PRR-genes and bacterial *avr* genes, Flor proposed the gene-for-gene hypothesis in which a single plant-gene product recognizes a single bacterial *avr* gene product and blocks disease formation (Flor, 1971).

The majority of plant PRR-genes cloned to date code for cytoplasmically localized NOD family members, with a domain that contains a nucleotide-binding site and a domain with leucine rich repeats (NBS-LRRs). All characterized NBS-LRRs recognize Avr proteins from pathogenic bacteria that are secreted through a large bacterial complex called a type three secretion system (TTSS). The TTSS is thought to function by transporting molecules directly into the cytoplasm of the host cell and it is known to be an essential transport system for disease development and bacterial multiplication (Henderson et al., 2004; Staskawicz et al., 2001). As recently reviewed (Mudgett, M.B., 2005; Chisholm, 2006), examples of the profusion of Avr molecules that rely on the TTSS for secretion, type III effectors, are AvrRpt2, AvrB, AvrRpm1, HopPtoD2, AvrPphB, and AvrPto from *P. syringae*; and XopD, AvrXv4, and AvrBsT from *X. campestris*. However, only a few type III effectors have known biochemical functions. Of those that have been characterized, several are enzymes, such as proteases and phosphatases, that act on host protein substrates to interfere with, suppress, and manipulate host basal defense and signaling by defense hormones such as salicylic acid, jasmonic acid, and ethylene (Thomma et al., 2001). A recent elegant example of this comes from the human pathogen *Yersinia*, in which the type III effector YopJ acetylates a mitogen activate kinase (MAPK) protein kinase, blocking the site of its activation by phosphorylation (Mukherjee, 2005). Others encode proteins with a nuclear localization signal and more directly modulate host transcription. For example, the AvrBs3/PthA family has a nuclear localization signal and an acidic transcription activation domain that is required for AvrBs3-dependant HR (Szurek et al., 2001). Few direct interactions have been reported between NBS-LRR PRRs and their corresponding effectors (Dodds et al., 2006). Instead, the defense response is often triggered by interaction between the NBS-LRR and another plant protein, which is targeted or modified by the type III effector (Dangl, 2001).

In addition to the cytoplasmic NBS-LRRs, two other classes of plant PRR that recognize Avr proteins have been described. These are the RLKs, composed of various putative ligand binding extracellular domains and an intracellular kinase, and receptor-like proteins (RLPs), composed only of a membrane anchored extracellular domain or a presumed secreted extracellular domain (Shiu, 2003; Wang, 1998). Relatively few plant RLKs and RLPs have been cloned and characterized to date, though, like NBS-LRRs, sequence analyses indicate that there are a large number of genes of these classes in plants (Shiu et al., 2003). The best-studied PRR RLK that confers a race-specific response is the rice XA21 protein, which recognizes *Xoo* strains carrying AvrXA21 activity (Song et al., 1995). The three other dominant RLKs cloned and characterized to date are the Xa26 and Pi-d2 proteins in rice and

RPG1 in barley (Sun et al., 2004; Chen et al., 2006; Brueggeman et al., 2002). The potential ligand-binding capabilities of the extracellular domains of the RLK PRRs and the fact that XA21 is present in microsomal fractions (Xu, 2006) and that Pi-d2 localizes to the cell membrane (Chen et al., 2006) suggest a simple model in which XA21 and other RLKs recognizes Avr proteins in the extracellular space, directly or indirectly, like the tomato Cf RLPs (Rooney et al., 2005).

Recent discoveries, including the results described here, call for a blurring of the distinctions between PRR proteins that recognize PAMPs and those that recognize Avr proteins and other conceptual dichotomies that have been established in plant pathology. Rather, a continuum of classes seems to exist, with plants making use of a diversity of strategies for innate immunity (McDowell et al., 2003). Here we show that the PRR XA21 recognizes a race-specific molecule(s), AvrXA21, that nonetheless has PAMP-like qualities. Although the AvrXA21 molecule(s) itself has not yet been identified, we show here that AvrXA21 activity is (1) dependent on a bacterial type one secretion system (TOSS), not a TTSS, (2) regulated by a two-component regulatory system that responds to *Xoo* cell population density, and (3) may be conserved in most *Xanthomonas* spp. These data suggest that AvrXA21 represents a new class of signaling molecules, which is used by *Xoo* for quorum sensing and which does not fall into any of the previously described classes of PAMPs or Avr factors.

2 The PRR XA21 Represents a Large Class of Kinases Predicted to Be Involved in Innate Immunity

In our recent survey of kinases in yeast, fly, worm, human, *Arabidopsis*, and rice show a correlation between a function in innate immunity and the absence of a conserved arginine adjacent to a conserved asparatate in the activation loop in domain VII of IRAK (interleukin-1 receptor-associated kinase)-family kinases, called non-RD kinases (Dardick et al., 2006). Of the 38 characterized IRAK-family receptor kinases in plants, all six RLKs associated with pathogen recognition fall into the non-RD class as do animal kinases associated with innate immunity, IRAK, and RIP (receptor-interacting protein) kinases. While there are only seven non-RD IRAK-family kinases in humans, there are 47 in *Arabidopsis* and 371 in rice (Dardick et al., 2006). Though certainly all of these receptors will not recognize the same class of molecule; nonetheless, characterization of the molecule that XA21 recognizes could have significant impact toward understanding this large but poorly understood class of receptors, and major effort has been ongoing in our lab toward describing the AvrXA21 molecule(s).

3 AVRXA21 Activity Requires a Type One Secretion System

As previously described, we have cloned eight genes, *raxC* and three operons (*raxSTAB*, *raxPQ*, *raxRH*), which are required for AvrXA21 (*rax*) activity of *Xoo* strain, Philippine race 6 (PXO99) (Shen et al., 2002; Goes da Silva et al., 2004;

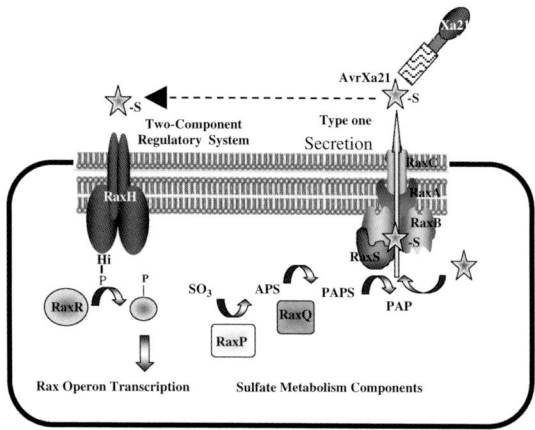

Fig. 1 Working model for the synthesis, regulation, and function of AvrXA21. Functions assigned to each of the *rax* gene products based on sequence homology and/or functional studies are as follows RaxH, histidine kinase; RaxR, response regulator; RaxP, ATP sulfurylase; RaxQ, adenosine-5′-phosphosulphate kinase; RaxST, sulfotransferase; RaxA, membrane fusion protein, spanning the inner membrane and the periplasmic space; RaxB, ABC (ATP binding cassette) transporter; and RaxC, outer-membrane protein. See text for elaboration

Burdman et al., 2004). Mutations in any of the eight *rax* genes allow this normally avirulent strain to form lesions when inoculated onto XA21-containing rice plants. Figure 1 shows our working model for the action of the *rax* gene products in producing AvrXA21 activity. Based on sequence analysis and functional studies, the *rax* gene products can be grouped into three functional classes as follows: RaxP, RaxQ, and RaxST are sulfur metabolism enzymes; RaxA, RaxB, and RaxC form a TOSS; and RaxH and RaxR form a two-component regulatory system (Shen et al., 2002; Goes da Silva et al., 2004; Burdman et al., 2004). Because phylogenic analysis suggests that the RaxB protein belongs to a specific family of ABC transporters that secretes peptides (Goes da Silva et al., 2004), we hypothesize that the AvrXA21 molecule is a type one secreted peptide.

To test the model that AvrXA21 is produced and secreted due to the action of the *rax* genes, we have developed a bioassay that detects AvrXA21 activity, consisting of cutting the tips off of 6-week-old rice plant leaves and pretreating the leaves by dipping them for 5 hours into supernatant prepared from media in which *Xoo* has been grown. Pretreated leaves are then inoculated by cutting immediately below the first cut site with scissors that have been dipped in a suspension of *Xoo* (Kauffman et al., 1973), and disease progression is monitored for 2–3 weeks. We used this bioassay to test the effect of mutations in each of the *rax* genes on AvrXA21 activity. Without pretreatment (first four leaves from left in Fig. 2), *Xoo* wild-type strain carrying AvrXA21 activity (AvrXA21[+]) and *raxST* knockout (*raxST[−]*) strain were inoculated onto leaves of *japonica* rice varieties, Taipei 309 (TP309, susceptible rice line) and a TP309 transgenic line carrying XA21 (TP309-XA21). This shows that TP309 is susceptible to both the strains and the XA21 plants are susceptible to the *raxST[−]* strain. In contrast, XA21 leaves inoculated with the wild-type strain expressing AvrXA21 activity are resistant. The next four leaves show the effects of pretreatment with supernatants from strain supernatant onto various *Xoo* genotypes. Pretreatment of wild-type leaves in XA21 rice plants prevents infection by the *raxST⁻* strain. In contrast, pretreatment with supernatant from the AvrXA21⁻, *rax*-gene knock-out mutants, including *raxST⁻* (Fig. 2), *raxA⁻* (Fig. 2), and *raxB⁻*, *raxC⁻*, *raxP⁻*, and *raxQ⁻* strains (data not shown), does not prevent disease by the *raxST⁻* strain. However, a *hrpC⁻*, TTSS-deficient strain (Zhu et al., 2000) has no effect on AvrXA21 activity as pretreatment with supernatant from this strain blocks lesion development by the *raxST⁻* strain. These results indicate that AvrXA21 activity is dominant, i.e., it can block lesion development by AvrXA21 minus strains, and is secreted by a TOSS but not the TTSS, leading to recognition by XA21 at the cell surface.

This is the first discovery of type I secreted factor that can trigger innate immunity. While there are five systems (type I–V) for protein secretion in Gram-negative bacteria (Henderson et al., 2004), only the TTSS has previously been shown to secrete Avr factors, which are then detected intracellularly. We hypothesize that

Fig. 2 Bioassay showing that AvrXA21 activity is present in medium of PXO99 wild-type and TTSS deficient mutant (*hrpC⁻*) strains, but not in a TOSS-deficient mutant (*raxA⁻*) strain or a sulfuryltransferase-deficient (*raxST⁻*) strain. Rice leaves from cultivar TP309 and TP309 transgenic for XA21 (TP309-XA21) were inoculated with PXO99 wild type (first and third leaves from left) or *raxST⁻* (second and fourth leaves from left) strains using the standard clipping method in which scissors are dipped in a solution of bacteria (OD$_{600}$ = 0.5), the rice leaves are clipped, and then lesions are measured 2 weeks later. To measure AvrXA21 activity, the *raxST⁻* strain was used for inoculations onto TP309-XA21 following pretreatment with cell-free supernatant of PXO99 wild-type, *raxST⁻*, *raxA⁻*, and *hrpC⁻* strains. Shown are representative leaves from one of three independent experiments

other members of the large complement of plant non-RD IRAK-family receptor kinases will be found to detect molecules secreted by other systems (Dardick et al., 2006). Indeed, indirect evidence suggests that type II secreted molecules are involved in eliciting plant defense (Jha et al., 2005).

4 The AVRXA21 Pathogen-Associated Molecule Is Conserved in *Xanthomonas campestris* pv. *campestris*

XA21 confers resistance to 29 out of 32 tested *Xoo* strains, which suggests that all 29 strains carry AvrXA21 activity (Wang et al., 1996). In our previous report, the *Xoo* strain KR1, which lacks AvrXA21 activity, acquired this activity when the *raxSTAB* operon was provided on a plasmid (Goes da Silva et al., 2004). We hypothesized that other *Xanthomonas* species carry the cognate molecules that confers AvrXA21 activity but lack the appropriate sulfation and secretion systems. To test this possibility, we introduced the PXO99 *raxSTAB* region into a bacteria that is non-pathogenic on rice, *Xcc* ATCC 33913. We chose this strain because it carries closely related homolog of all the *rax* genes except for the *raxSTAB* operon.

We carried out the AvrXA21 activity assay with supernatant prepared from *Xcc* transformed with the PXO99 *raxSTAB* genes and compared the lesion lengths (Table 1). Pretreatment with the supernatant of *Xoo* wild-type strain and the *Xcc* strain carrying the *raxSTAB* genes induces resistance against infection by the *raxST*⁻ *Xoo* (lesion lengths of 1.3 ± 0.43 and 1.8 ± 0.8 cm, respectively); whereas, leaves pretreated with the supernatant of *Xoo raxST*⁻ strain and *Xcc* wild-type strain have long lesions (10.1 ± 5.2 and 10.0 ± 4.8 cm). Thus, the supernatant of *Xcc* carrying *raxST*, *raxA*, and *raxB* genes possesses AvrXA21 activity. Furthermore, *Xcc* carrying *Xoo raxST* and *raxA* genes showed full AvrXA21 activity but not with *raxST* alone (data not shown), suggesting that AvrXA21 modification and secretion may be a sequential process requiring both the *raxST* and the *RaxA* gene products or that the activity or stability is altered in the absence of *RaxA*.

Importantly, these results suggest that the core AvrXA21 molecule is conserved among different species, a key component of the definition of a PAMP. Nonetheless, the requirement for *raxST and raxA* suggests that XA21-mediated recognition of the AvrXA21 molecule requires a specific TOSS inner membrane protein (RaxA) and/or a post-translational modification, sulfation, that is likely catalyzed by the putative sulfotransferase, RaxST (Goes da Silva et al., 2004). There are many examples of post-translational modification affecting extracellular recognition. Typically,

Table 1 Comparison of lesion lengths on rice leaves. The data are averages from 10 scored leaves

Pretreatment of supernatant	PXO99	*raxST*⁻	*Xcc*	*Xcc* (*raxSTAB*)
Inoculation		*raxST*⁻		
Lesion length [a]	1.3 ± 0.43	10.1 ± 5.2	10.0 ± 4.8	1.8 ± 0.8

[a] cm, average ± standard deviation

sulfated molecules are directed outside of the cell to serve as modulators of cell–cell interactions (Bowman et al., 1999). A notable example pertinent to agriculture is sulfation of the *Sinorhizobium meliloti* Nod factor that is required for specific recognition by its host alfalfa (Roche et al., 1991). In plants, tyrosine sulfation of phytosulfokine is required for recognition by the phytosulfokine receptor kinase, a plasma membrane RLK for proliferation of plant cells (Matsubayashi et al., 2002).

The other element of the definition of a PAMP is that it is essential for the pathogen. The fact that the core of AvrXA21 appears to be conserved in *Xcc* suggests that the molecule(s) has a function that makes it selectively advantageous. In addition, while we did not observe differences in virulence in the *raxSTAB* and *raxC* knockout strains under controlled conditions, in a field study, 37 *Xoo* Korean strains lacking AvrXA21 activity appeared to have reduced fitness (Choi, 2003). An additional study showed that *Xoo* strains that lost AvrXA21 activity did not persist at the same field site into the next season suggesting that these strains were at a competitive disadvantage (C.M. Vera Cruz, personal communication). These results suggest both that the core AvrXA21 molecule is conserved between *Xoo* and *Xcc* and that loss of AvrXA21 results in a fitness cost to *Xoo*. Therefore, AvrXA21, the detection of which by PRR XA21 confers dominant resistance, shares characteristics of PAMPs and Avr proteins.

5 Cell Density Dependent Expression of *Rax* Genes

What is the function of AvrXA21? Like other living organisms, bacteria have developed multiple systems for responding to environmental variation, such as changes in temperature, osmolarity, pH, nutrient availability, and even population size. Two-component systems, composed of histidine kinases (HK) and response regulators (RR), have an important role in environment sensing. Phosphorylation of the RR by the HK regulates gene expression and governs the response to an environmental stimulus (Charles et al., 1992). Bacteria themselves produce some stimuli. In a process called quorum sensing (QS), small molecules serve as signals to recognize cell population, leading to changes in expression of specific genes when the signal has accumulated to some threshold concentration (Fuqua, et al., 1994). QS molecules regulate their own expression and are also called autoinducers (Waters et al., 2005).

Preliminary studies using the *raxST* promoter in front of GFP suggested that *rax* gene expression may be dependent on cell density (data not shown). To test this more thoroughly, the wild-type strain (PXO99) was cultured for 72 hours and then diluted. The diluted cultures were returned to the incubator and, as they grew, aliquots were removed for RNA isolation. Gene expression was measured using real-time quantitative PCR and primers specific for the following four *rax* genes, one in each operon: *raxST*, in *raxSTAB* (Fig. 3); *raxP*, in *raxPQ*; *raxR*, in *raxRH*; and *raxC* (data not shown). Remarkably, the expression of all four *rax* genes was only observed in the wild-type strain at high cell densities; whereas, levels of a ribosomal RNA were unaffected. We also observed *rax* gene expression in a strain constitutively expressing RaxR. Figure 3 shows that the density dependence of *rax*

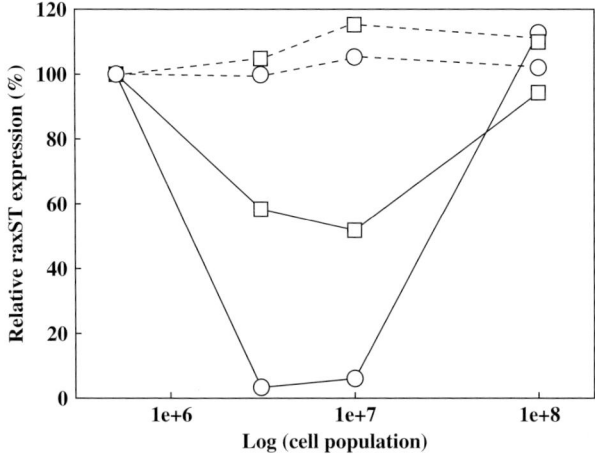

Fig. 3 Cell-density-dependent expression of the *raxST* gene. Cultures of wild-type (*circles*) and *raxR* constitutive expression mutant (PXO99*raxR⁻*/pML122*raxR*: *squares*) strains were grown to saturation in NB media for 72 hours at 28 °C (cell density at OD$_{600}$ ≥ 1×10^8 colony forming units (CFU)/mL and then diluted to 1×10^5 CFU/mL. We measured expression of the *raxST* gene (*closed symbols, solid lines*) by preparing RNA from samples at various cell densities after dilution and performing real-time quantitative PCR with three primers pair corresponding to different region of the gene. Ribosomal RNA (*open symbols, dashed lines*) in wild-type and the mutant strains were used as a control. Percent expression at each cell density was calculated relative to the copy number of the sample at 1×10^5. Data are from one of two independent experiments. Expression is high at 10^5 CFU/ml because the strains were grown to high density and diluted, i.e., the bacteria in this sample behave as if they are still at high density

gene expression in the *raxR* overexpressing strain is partially lost. This result indicates that RaxR is involved in regulation of the density-dependent expression of the *rax* genes, including itself. This auto-induction is consistent with a roll of the product of the *rax* gene pathway, namely AvrXA21, in QS (Waters et al., 2005).

The characteristics of the *Xoo rax*-genes suggest a QS system with features similar to that typically found in Gram-positive bacteria; thus, the AvrXA21 QS system represents a novel class of QS in Gram-negative bacteria. Oligpeptides are the predominant signaling molecules for genetic competence in *Bacillus substilis* (Tortosa et al., 1999), virulence in *Staphylococcus aureus* (Novick et al., 1999), and the production of antimicrobial peptides, inducing bacteriocins and lantibiotics, in lactic acid bacteria (Risoen et al., 2000; Kleerebezem et al., 1997). As QS molecules, these peptides also activate their own expression and have been given the name auto-inducible peptide (AIP). AIPs are known to be a secreted by ABC-transporters that are similar to RaxB and are sensed by two-component systems. Significantly, while QS via two-component systems is not unique to Gram-positive bacteria, secretion of the putative AvrXA21 peptide by a TOSS would be the first described occurrence of an AIP in Gram-negative bacteria.

In summary, our data suggest that the AvrXA21 PAM is a secreted peptide, and hint that the biological function of the AvrXA21 PAM is as a QS signal molecule, the production of which is regulated in a cell-density-dependent manner in the Gram-negative bacteria, *Xoo*, by a two-component system. Thus, the rice XA21 PRR effectively eavesdrops on the AvrXA21-mediated molecular conversation among *Xoo*, using AvrXA21 as an indicator of the presence of a pathogenic bacterial population of significant size and leading to plant responses that effectively halt infection (Fig. 1). Sequence analysis of the genomes of several phytopathogenic bacteria that are similar to *Xoo* reveals the abundance of the *rax* classes of molecular components. *Xoo*, *Xcc*, *Xanthomonas axonopodis* pv. *citri*, and *Xylella fastidiosa* each possess genes for approximately 30 two-component systems together with three to five TOSSs that, like RaxB, are predicted to transport peptides. Further-more, it is apparent that *X. fastidiosa* does not use TTSS for pathogenesis because genome analysis has revealed the absence of these components in this bacterium. Rather, *X. fastidiosa* contains TOSS genes similar to *raxA*, *raxB*, and *raxC* as well as sequences encoding as many as 10–12 peptides that are candidates for secretion (C. Dardick and P.C. Ronald, unpublished results). These observations highlight that the interactions among phytopathogens, their hosts, and the environment are more sophisticated than currently recognized.

Our data suggesting that the function of the AvrXA21 pathogen-associated molecule is a QS signaling molecule may present a critical clue to answering the question, why is AvrXA21 maintained in *Xoo* regardless of its use for detection by a host protein? Based on others' results that suggest decreased fitness of *rax* mutants, AvrXA21 may be essential for *Xoo* as a signal molecule for cell–cell com-munication. We carefully hypothesize that *Xoo* may use QS to coordinate infec-tion and production of virulence factors. QS has global effects on bacteria growth, survival, and interactions with eukaryotes as QS-responsive genes compose up to 10% of the *Pseudomonas aeruginosa* trascriptome (Wagner et al., 2003). In that report, a large portion of QS-promoted genes code for membrane proteins and pro-tein export apparatuses involved in secretion of virulence factors. A relationship between QS and virulence has also been described for the human pathogen *Vibrio cholerae* (Miller et al., 2002). Interestingly, in another case a host can produce QS mimic molecules that interfere the QS regulated behaviors of the infecting bacteria (Teplitski et al., 2004). However, given the abundance of other QS peptides and small molecules such as bacteriocins and lantibiotics in the host vicinity, it will not be surprising if these peptides are shown to play a role in the interaction of other bacteria with their hosts. In principle, recognition of TOSS elicitors could be exploited to generate new types of specific and environmentally benign pesticides from those elicitors.

6 Perspective

The data presented here as well as results of others in recent years establish that there is not a clear dichotomy between PAMP recognition and Avr recognition; rather, plant innate immunity functions as a continuum between the two general types (Table 2) (Sun et al., 2006).

Table 2 General characteristics of PAMPs (pathogen-associated molecular patterns) and the PRR (pathogen recognition receptors) responsible for their recognition as opposed to the features of Avr (avirulence) proteins and their corresponding PRR proteins. Characteristics of AvrXA21 and XA21 are shown in bold if they appear to be unique compared with known systems

PAMP·PRR General characteristics	AvrXA21·XA21 characteristics	Avr·PRR General characteristics
PAMP (e.g., flagellin)	AvrXA21	Avr (e.g., AvrRpt2)
- Conserved among diverse species - Required for lifecycle - Not typically actively secreted	- Strain specific - Activity is regulated in a cell-density dependent by a two-component system - Activity requires modification enzymes - Synthesized in absence of plant host (thus part of lifecycle) - Absence compromises virulence in field - Secreted by TOSS	- Strain specific - Many synthesized only in presence of host - Absence compromises virulence - From bacteria, secreted by TTSS
PRR (e.g., FLS2)	XA21	PRR (e.g., Rps2)
- Extracellular, (intracellualr recognition only shown in animals) - Recognition and response does not lead to resistance in most cases	- Recognition likely extracellular - Confers dominant resistance	- In dicots, recognition is mostly intracellular. - Confers dominant resistance

PAMPs have been defined as being derived from conserved structures required for pathogen function; whereas, Avr factors are thought to be maintained by specific strains as virulence factors. The features of AvrXA21 form a hybrid between these two categories. The AvrXA21 molecule appears to be conserved in *Xcc*, but sulfation seems to provide specificity to the system, just as flagellin recognition by rice is modulated by glycosylation (Taguchi et al., 2003, 2003). Further evidence that plant-recognized PAMPs are not entirely conserved comes from a recent paper showing that sequence variation in *Xcc* flagellin modulates FLS2-dependent pathogen recognition by *Arabidopsis* (Sun et al., 2006). Also, compromised virulence and survival of *avrXA21⁻* strains in the field is consistent with a crucial role for AvrXA21 in Xanthomonad quorum sensing. On the other hand, AvrXA21 and XA21 fulfill the standard gene-for-gene model of the dominant resistance gene interacting with the bacterial virulence factor as well. Thus, AvrXA21 has features that are similar to both PAMPs and other Avr's. For this reason, we propose the terminology pathogen-associated molecule (PAM) to describe both PAMPs and Avr factors.

The fusion of PRR recognition of Avr and PAMP classes proposed here has implications for the mechanism of plant innate immunity. Recognition of PAMPs by the animal innate immune system has been proposed to act as guide for the adaptive immune system as to the nature of the infecting pathogen (Medzhitov et al., 1997). There is not an obvious need for such a function in plants since each receptor can be tied to a specific response through signal transduction. Nonetheless, it is clear that the plant innate immune system recognizes both conserved and less conserved molecules. As articulated by Ausubel in 2005, the fact that plant responses to PAMPs often do not lead to resistance may be due to evolutionary selection for plant innate immune perception to require "priming" or "two-hits" so that resources are not unnecessarily wasted in responding to nonpathogenic or small populations of bacteria. In such a model, PAMP perception constitutes an early warning, and indeed pretreatment with lipopolysaccharide or flagellin-derived peptide leads to resistance to subsequent treatments with normally pathogenic bacteria (Erbs et al., 2003; Sun et al., 2006; Zipfel et al., 2004). However, it may be that the necessity for the factor that primes the plant innate immune system to be a PAMP rather than an Avr molecule is simply a matter of experimental history. Because they are less useful in an agricultural context, we have not looked for PRRs and corresponding Avr molecules that do not lead to significant resistance. Though the XA21D receptor, which confers only partial resistance (Wang et al., 1998), may be such a molecule. Conversely, as mentioned above, FLS2-dependent resistance to bacteria that express flagellin has also been observed (Zipfel et al., 2004). Physiologically relevant experiments in diverse genetic backgrounds may lead to further evidence that responses mediated by PRRs for both more and less conserved PAMs (i.e., PAMPs and Avr's) are also part of a continuum, with some functioning more often in priming and others more often leading to resistances. Already among described Avr PRRs there is a diversity of strengths, with some leading to super-resistance (Greenberg et al., 2004). Thus, in possessing receptors for more conserved molecules, it may be that plants are simply taking advantage of the conserved nature of PAMPs as targets for recognition, rather than making a functional distinction in terms of responses.

Several examples support the idea that plants gain a selective advantage by making use of a variety of strategies for pathogen recognition beyond the classes of molecules that have typically been discussed (Chisholm, 2006; Dangl, 2001), both in terms of the PAM detected and the cognate PRRs. AvrXA21 is currently the only known PAM that is secreted by a TOSS. Until recently LRRs constituted the only signal recognition domain described for a cloned PRR RLK; however, a recently cloned RLK conferring rice blast resistance, PID2, breaks from this by possessing an extracellular lectin-binding domain (Chen et al., 2006). Moreover, sequence analyses suggest that RLKs and RLPs with other, diverse extracellular binding domains are also involved in plant innate immunity (Dardick et al., 2006; Shiu et al., 2003). Similarly, plants appear to use other entirely different classes of molecules as PRRs as well. Indeed, the first cloned PAMP high-affinity binding site was the soybean β-glucan elicitor binding protein (Umemoto, 1997). Localized to the extracellular side of the cell membrane and containing two glucan binding sites, it is predicted to be tethered to the membrane via interactions in a receptor complex (Mithofer, 2000). Yet another perception strategy is represented by Xa27, which codes for a novel defense protein the promoter of which serves as the PRR by directly or indirectly interacting with the cognate nuclear-localized AvrXA27 type III effector (Gu, 2005).

In conclusion, our recent results suggest that the AvrXA21 PAM is a TOSS-dependent secreted peptide that represents a new family of signaling molecules for QS via a two-component system. Although QS molecules are known to be widely conserved and essential for bacterial communication, they have never before been shown to trigger innate immune responses. Our continuing work to identify the AvrXA21 molecule and understand the physical basis of its recognition may allow identification of similar PAMs from plant and animal pathogens and their cognate host receptors.

References

Ausubel, F.M., 2005, Are innate immune signaling pathways in plants and animals conserved? *Nat. Immunol.* **6:**973–979.

Barton, G.M., and Medzhitov, R., 2003, Toll-like receptor signaling pathways, *Science* **300:** 1524–1525.

Bowman, K.G., and Bertozzi, C.R., 1999, Carbohydrate sulfotransferases: mediators of extracellular communication, *Chem. Biol.* **6:**R9.

Brueggeman, R., Rostoks, N., Kudrna, D., Kilian, A., Han, F., Chen, J., Druka, A., Steffenson, B., and Kleinhofs, A., 2002, The barley stem rust-resistance gene *Rpg1* is a novel disease-resistance gene with homology to receptor kinases, *Proc. Nat. Acad. Sci. USA* **99:**9328–9333.

Brunner, F., Rosahl, S., Lee, J., Rudd, J. J., Geiler, C., Kauppinen, S., Rasmussen, G., Scheel, D., and Nuernberger, T., 2002, Pep-13, a plant defence-inducing pathogen-associated pattern from *Phytophthora* transglutaminases, *EMBO* **21:**6681–6688.

Burdman, S., Shen, Y., Lee, S.-W., Xue, Q., and Ronald, P., 2004, RaxH/RaxR: A Two-component regulatory system in *Xanthomonas oryzae* pv. *oryzae* required for AvrXa21 activity, *MPMI* **17:**602–612.

Charles, T.C., Jin, S., and Nester, E.W., 1992, Two-component sensory transduction systems in phytobacteria, *Ann. Rev. Phytopath.* **30:**463–484.

Chen, X., Shang, J., Chen, D., Lei, C., Zou, Y., Zhai, W., Liu, G., Xu, J., Ling, Z., Cao, G., Ma, B., Wang, Y., Zhao, X., Li, S., and Zhu, L., 2006, A B-lectin receptor kinase gene conferring rice blast resistance, *Plant J.* **46:**794–804.

Chinchilla, D., Bauer, Z., Regenass, M., Boller, T., and Felix, G., 2006, The Arabidopsis receptor kinase FLS2 binds flg22 and determines the specificity of flagellin perception, *Plant Cell* **18:**465–476.

Chisholm, S. T., Coaker, G., Day, B., and Staskawicz, B.J., 2006, Host-microbe interactions: shaping the evolution of the plant immune response, *Cell* **124:**803–814.

Dangl, J.L., and Jones, J.D.G., 2001, Plant pathogens and integrated defence responses to infection, *Nature* **411:**826.

Dardick, C., and Ronald, P., 2006, Plant and animal pathogen recognition receptors signal through non-RD kinases, *PLoS Pathog* **2:**e2.

Dodds, P.N., Lawrence, G.J., Catanzariti, A.M., Teh, T., Wang, C.I., Ayliffe, M.A., Kobe, B., and Ellis, J.G., 2006, Direct protein interaction underlies gene-for-gene specificity and coevolution of the flax resistance genes and flax rust avirulence genes, *Proc. Nat. Acad. Sci. USA* **103:** 8888–8893.

Erbs, G., and Newman, M.-A., 2003, The role of lipololysaccharides in induction of plant defence responses, *Mol. Plant Path.* **4**:421–425.

Fuqua, W.C., Winans, S.C., and Greenberg, E.P., 1994, Quorum sensing in bacteria: the LuxR-Lux1 family of cell density-responsive transcriptional regulators, *J. Bacteriol.* **176**:269–275.

Felix, G., Duran, J.D., Volko, S., and Boller, T., 1999, Plants have a sensitive perception system for the most conserved domain of bacterial flagellin, *Plant J.* **18**:265–279.

Flor, H.H., 1971, Current status of the gene for gene concept. *Ann. Rev. Phytopath.* **9**:275–296.

Girardin, S.E., Sansonetti, P.J., and Philpott, D.J., 2002, Intracellular vs. extracellular recognition of pathogens—common concepts in mammals and flies, *Trends Microbiol.* **10**:193–199.

Goes da Silva, F., Shen, Y., Dardick, C., Burdman, S., Yadav, R., Sharma, P., and Ronald, P., 2004, Bacterial genes involved in type I secretion and sulfation are required to elicit the rice *Xa21*-mediated innate immune response, *MPMI* **17**:593–601.

Gomez-Gomez, L., Felix, G., and Boller, T. (1999) A single locus determines sensitivity to bacterial flagellin in *Arabidopsis thaliana, Plant J.* **18**:277–284.

Greenberg, J.T., and Yao, N., 2004, The role and regulation of programmed cell death in plant–pathogen interactions, *Cell. Microbiol.* **6**:201–211.

Henderson, I.R., Navarro-Garcia, F., Desvaux, M., Fernandez, R.C., and Ala'Aldeen, D., 2004, Type V protein secretion pathway: the autotransporter Story. *Microbiol. Mol. Biol. Rev.* **68**:692–744.

Inohara, N., and Nunez, G., 2003, NODs: intracellular proteins involved in inflammation and apoptosis, *Nat. Rev. Immunol.* **3**:371–382.

Janeway, C.A., Jr., and Medzhitov, R., 2002, Innate immune recognition, *Annu. Rev. Immunol.* **20**:197–216.

Jha, G., Rajeshwari, R., and Sonti, R.V., 2005, Bacterial type two secretion system secreted proteins: double edged swords for plant pathogens, *Mol. Plant In.* **18**:891–898.

Jurk, M., Heil, F., Vollmer, J., Schetter, C., Krieg, A. M., Wagner, H., Lipford, G., and Bauer, S., 2002, Human TLR7 and TLR8 independently confer responsiveness to the antiviral compound R-848, *Nat. Immunol.* **3**:499.

Kauffman, H.E., Reddy, A.P.K., Hsieh, S.P.Y., and Merca, S.D., 1973. An improved technique for evaluating resistance of rice varieties to *Xanthomonas oryzae, Plant Dis. Rep.* **57**:537–541.

Kleerebezem, M., Quadri, L.E., Kuipers, O.P., and de Vos, W.M., 1997, Quorum sensing by peptide pheromones and two-component signal-transduction systems in Gram-positive bacteria. *Mol. Microbiol.* **24**:895–904.

Leulier, F., Parquet, C., Pili-Floury, S., Ryu, J.H., Caroff, M., Lee, W. J., Mengin-Lecreulx, D., and Lemaitre, B., 2000, The *Drosophila* immune system detects bacteria through specific peptidoglycan recognition, *Nat. Immunol.* **4**:478–484.

Matsubayashi, Y., Ogawa, M., Morita, A., and Sakagami, Y., 2002, An LRR receptor kinase involved in perception of a peptide plant hormone, phytosulfokine, *Science* **296**:1470–1472.

McDowell, J.M., and Woffenden, B.J., 2003, Plant defense, one post, multiple guards? *Trends Biotechnol.* **21**:178–183.

Medzhitov, R., and Janeway, C.A., Jr., 1997, Innate immunity: the virtues of a nonclonal system of recognition, *Cell* **91**:295–298.

Miller, M.B., Skorupski, K., Lenz, D.H., Taylor, R.K., and Bassler, B.L., 2002, Parallel quorum sensing systems converge to regulate virulence in *Vibrio cholerae, Cell* **110**:303–314.

Mudgett, M.B., 2005, New insights into the function of phytopathogenic bacterial type III effectors in plants, *Ann. Rev. Plant Biol.* **56**:509–531.

Novick, R.P., and Muir, T.W., 1999, Virulence gene regulation by peptides in staphylococci and other Gram-positive bacteria, *Curr. Opin. Microbiol.* **2**:40–45.

Nurnberger, T., Brunner, F., Kemmerling, B., and Piater, L., 2004, Innate immunity in plants and animals: striking similarities and obvious differences, *Immunol. Rev.* **198**:249–266.

Ramos, H.C., Rumbo, M., and Sirard, J.C., 2004 Bacterial flagellins: mediators of pathogenicity and host immune responses in mucosa, *Trends Microbiol.* **12**:509–517.

Risoen, P.A., Brurberg, M.B., Eijsink, V.G., and Nes, I.F., 2000, Functional analysis of promoters involved in quorum sensing-based regulation of bacteriocin production in *Lactobacillus*, *Mol. Microbiol.* **37**:619–628.

Roche, P., Debelle, F., Maillet, F., Lerouge, P., Faucher, C., Truchet, G., Denarie, J., and Prome, J.-C., 1991, Molecular basis of symbiotic host specificity in *Rhizobium meliloti*: *nodH* and *nodPQ* genes encode the sulfation of lip-oligosaccharide signals, *Cell* **67**:1131

Ronald, P.C., Albano, B., Tabien, R., Abenes, L., Wu, K.S., McCouch, S., and Tanksley, S.D., 1992, Genetic and physical analysis of the rice bacterial blight disease resistance locus, Xa-21, *Mol. Gen. Genet.* **236**:113–120.

Rooney, H.C., Van't Klooster, J.W., van der Hoorn, R.A., Joosten, M.H., Jones, J.D., and de Wit, P.J., 2005, Cladosporium Avr2 inhibits tomato Rcr3 protease required for Cf-2-dependent disease resistance, *Science* **308**:1783–1786.

Shen, Y., Sharma, P., da Silva, F.G., and Ronald, P., 2002, The *Xanthomonas oryzae* pv. *oryzae* RaxP and RaxQ genes encode an ATP sulphurylase and adenosine-5'-phosphosulphate kinase that are required for AvrXa21 avirulence activity, *Mol. Microbiol.* **44**:37–48.

Shiu, S.H., and Bleecker, A.B., 2003, Expansion of the receptor-like kinase/Pelle gene family and receptor-like proteins in *Arabidopsis*, *Plant Physiol.* **132**:530–543.

Song, W.Y., Wang, G.L., Chen, L.L., Kim, H.S., Pi, L.Y., et al., 1995, A receptor kinase-like protein encoded by the rice disease resistance gence, *Xa-21*, *Science* **270**:1804–1806.

Staskawicz, B.J., Mudgett, M.B., Dangl, J.L., and Galan, J.E., 2001, Common and contrasting themes of plant and animal diseases, *Science* **292**:2285–2289.

Sun, X., Cao, Y., Yang, Z., Xu, C., Li, X., Wang, S., and Zhang, Q., 2004, *Xa26*, a gene conferring resistance to *Xanthomonas oryzae* pv. *oryzae* in rice, encodes an LRR receptor kinase-like protein, *Plant J.* **37**:517–527.

Sun, W., Dunning, F.M., Pfund, C., Weingarten, R., and Bent, A.F., 2006, Within-species flagellin polymorphism in *Xanthomonas campestris* pv. *campestris* and its impact on elicitation of *ArabidopsisFLAGELLIN SENSING2*-dependent defense, *Plant Cell* **18**:764–779.

Szurek, B., Marois, E., Bonas, U., and Van den Ackerveken, G., 2001, Eukaryotic features of the *Xanthomonas* type III effector AvrBs3: protein domains involved in transcriptional activation and the interaction with nuclear import receptors from pepper, *Plant J.* **26**:523–534.

Taguchi, F., Shimizu, R., Inagaki, Y., Toyoda, K., Shiraishi, T., and Ichinose, Y., 2003, Post-translational modification of flagellin determines the specificity of HR production, *Plant Cell Physiol.* **44**:342–349.

Takeuchi, K., Taguchi, F., Inagaki, Y., Toyoda, K., Shiraishi, T., and Ichinose, Y., 2003, Flagellin glycosylation island in *Pseudomonas syringae* pv. *glycinea* and its role in host specificity, *J. Bacteriol.* **185**:6658–6665.

Thomma, B.P., Penninckx, I.A., Broekaert, W.F., and Cammue, B.P., 2001, The complexity of disease signaling in *Arabidopsis*, *Curr. Opin. Immunol.* **13**:63–68.

Teplitski, M., Chen, H., Rajamani, S., Gao, M., Merighi, M., Sayre, R.T., Robinson, J.B., Rolfe, B.G., and Bauer, W.D., 2004, *Chlamydomonas reinhardtii* secretes compounds that mimic bacterial signals and interfere with quorum sensing regulation in bacteria, *Plant Physiol.* **134**:137–146.

Tortosa, P., and Dubnau, D., 1999, Competence for transformation: a matter of taste, *Curr. Opin. Microbiol.* **2**:588–592.

van't Slot, K., and Knogge, W,K., 2002, A dual role for microbial pathogen-derived effector proteins in plant disease and resistance, *Crit. Rev. Plant Sci.* **21**:229–271.

Wagner, V.E., Bushnell, D., Passador, L., Brooks, A.I., and Iglewski, B.H., 2003, Microarray analysis of *Pseudomonas aeruginosa* quorum-sensing regulons: effects of growth phase and environment, *J. Bacteriol.* **185**:2080–2095.

Wang, G.L., Ruan, D.L., Song, W.Y., Sideris, S., Chen, L., Pi, L.Y., Zhang, S., Zhang, Z., Fauquet, C., Gaut, B.S., Whalen, M.C., and Ronald, P.C., 1998, Xa21D encodes a receptor-like molecule with a leucine-rich repeat domain that determines race-specific recognition and is subject to adaptive evolution, *Plant Cell* **10**:765–779.

Wang, G.L., Song, W.Y., Ruan, D.L., Sideris, S., and Ronald, P.C., 1996, The cloned gene, *Xa21*, confers resistance to multiple *Xanthomonas oryzae* pv. *oryzae* isolates in transgenic plants, *Mol. Plant Microbe Interact.* **9:**850–855.

Waters, C.M., and Bassler, B.L., 2005, Quorum sensing: Cell-to cell communication in bacteria, *Ann. Rev. Cell Develop. Biol.* **21:**319–346.

Werling, D., and Jungi, T.W., 2003, TOLL-like receptors linking innate and adaptive immune response, *Vet. Immunol. Immunop.* **91:**1–12.

Zhu, W., MaGbanua, M.M., and White, F.F., 2000, Identification of two novel hrp-associated genes in the hrp gene cluster of *Xanthomonas oryzae* pv. *Oryzae*, *J. Bacteriol.* **182:**1844–1853.

Zipfel, C., Robatzek, S., Navarro, L., Oakeley, E.J., Jones, J.D., Felix, G., and Boller, T., 2004, Bacterial disease resistance in Arabidopsis through flagellin perception, *Nature* **428:**764–767.

Unraveling the Genetic Mysteries of the Cat: New Discoveries in Feline-Inherited Diseases and Traits

Leslie A. Lyons

Abstract The domestic cat is one of man's most beloved species, living in house-holds as companions, working on farms for vermin control, and co-habitating in urban environments as semi-feral occupants. Advances in veterinary medicine provide health care and diagnostics for the domestic cat on a comparable level to humans. Fancy breeds result in the selection of aesthetically pleasing traits and, sometimes, undesired health conditions, both of which can be useful as models for human development, physiology, and health. Reproductive techniques have allowed research in the cat to expand into cloning studies and genomic tools are assisting the cat with the same research scrutiny as the more prominent research animal models, rodents and humans. This chapter explores the genetic mysteries of the cat and provides a current state of the union for cat genetic research.

Cat \\'kat\\ – a long domesticated carnivorous mammal that is usually regarded as a distinct species though probably ultimately derived by selection, among the hybrid progeny of several small Old World wildcats that occurs in several varieties distinguished chiefly by length of coat body and presence or absence of tail and that makes the pet valuable in controlling rodents and other small vermin but tends to revert to a feral state if not housed and cared for (*Webster 3rd New International Dictionary*).

1 Cat Phenotypes

The domestic cat is one of 36 extant species of felids (Kitchener, 1991; Seidensticker and Lumpkin, 1991; Sunquist and Sunquist, 2002), grouping with four other Old World wildcat species in the domestic cat "*Felis*" lineage, including, *Felis bieti*, the Chinese desert cat, *Felis margarita*, the sand cat, *Felis nigripes*, the black-footed cat, and *Felis chaus*, the jungle cat (Johnson et al., 2006). Two additional wildcats, the European wildcat, *Felis silvestris*, and the African wildcat, *Felis libyca*, are usually considered separate species and distinct from the domestic cat. However, these two

L.A. Lyons
Department of Population Health & Reproduction, School of Veterinary Medicine, University of California, Davis, CA 95616, USA

J.P. Gustafson et al. (eds.), *Genomics of Disease*,
© Springer Science+Business Media, LLC, 2008

wildcats produce fertile hybrids with domestic cats and the speciation of these three cats remains a mystery, as genetic studies have not yet been able to focus on their recent divergence. Many other sub-species of European, Asian, and African wildcats have been described; however, the true origins and the number of domestication events that led to the domestication of the cat has not been resolved and also remain a mystery.

The term "domesticate" has various definitions, including "to adapt to life in intimate association with and to the advantage of man." Signs of domestication include docility, retention of juvenile features, smaller brain size, raised tail carriage, droopy ears, and pelage coloration, fur types, and morphologies that would likely have a selective disadvantage in the wild (Hemmer, 1972, 1976, 1978, 1990; Kukekova et al., 2006; Wayne, 1986). Cats have been recognized to have "seven ancient mutations" that were evident before the advent of breeds (Wastlhuber, 1991). These mutations, melanism, dilution, long fur, orange, white spotting, the classic tabby markings, and dominant white, all affect pelage coloration or length and were likely the focus of early phenotypic selections by humans (Fig. 1). Many studies have examined the distribution of coat colors in various populations throughout the world, which revealed a "cline" of color variation frequencies across the Old World (Clark, 1975; Davis and Davis, 1977; Todd and Lloyd, 1984; Todd et al., 1974, 1975; Wagner and Wolsan, 1987). These same phenotypic mutations segregate in a variety of modern cat breeds (Robinson, 1991; Vella et al., 1999). The genes controlling these cat phenotypes and their associated mutations are presented in Table 1. Two of the mutations, dominant white and long fur, are yet to have direct scientific study in the cat, but candidate genes such as *KIT* (Brooks and Bailey, 2005; Cooper et al., 2006; Johansson Moller et al., 1996; Marklund et al., 1998; Terry et al., 2001) and *fibroblast growth factor 5* (*FGF5*) (Housley and Venta, 2006; Sundberg et al., 1997) have been shown to cause similar phenotypes in other species. The mutations for melanism and dilution have been determined (Eizirik et al., 2003; Ishida et al., 2006); however, only genes or genetic regions have been implicated for orange, white spotting, and tabby (Cooper et al., 2006; Grahn et al., 2005; Lyons et al., 2006).

The advent of the cat fancy in England during the late 1800s fostered the development of cat breeds (CFA, 2004). Although over 50 cat "breeds" can be demarcated by breeders today (Morris, 1999a b), many of the cat breeds are single gene variants of a parent breed and hence these different breeds are considered "variants" or "classes" in one cat fancy association or a distinctive breed by a different cat fancy association (http://www.cfainc.org/ and http://www.tica.org/). For example, Exotic Shorthairs are a short-haired variety of Persians, long hair being the recessive trait. Himalayans are "pointed" Persians, often representing a color class or variety. One breed that had early recognition by the cat fancy is the Siamese. These cats have a distinctive temperature-sensitive pattern to their coloration; melanin is produced where the cats body temperature is lower. Hence, the "points" of a Siamese cat is a result of proper color production at the extremities, including the face, ears, paws, and tail. A close cousin to the Siamese, the Burmese, has a less-severe temperature-sensitive color restriction; hence lower production of eumelanin occurs, resulting in

Fig. 1 The seven ancient mutations of the cat: (**a**) wild type, (**b**) melanism, (**c**) dilution, (**d**) long fur, (**e**) white spotting, (**f**) classic tabby, (**g**) orange, and (**h**) dominant white. The recessive non-agouti mutation changes wild type to melanistic. The recessive dilution mutation affects all pigment, presented is a melanistic cat that is diluted to blue, a.k.a. gray. Long fur is recessive and is independent of coloration. White spotting is dominant and additive. Heterozygotes have ventral white as presented but homozygotes display coloration mainly at the ears and tail. Orange affects all pigments, producing an orangish hue, a.k.a. red or ginger. Dominant white is independent of background coloration and is associated with deafness and blue or odd-eyed color

a sable-colored cat. The mutations for Siamese, Burmese, and complete albinism have been identified (Imes et al., 2006; Lyons et al., 2005b) and form an allelic series, $C > c^b = c^s > c$ at the *Color* locus, which codes for the gene, *tyrosinase* (*TYR*). Brown color variants also segregate or are fixed in many cat breeds. The DNA mutations for black, brown (chocolate or chestnut), and light brown (cinnamon or red) have been identified (Lyons et al., 2005a) in the gene *tyrosinase-related protein* (*TYRP1*), also forming an allelic series, $B > b > b^l$. Many other single gene traits distinguish the extant cat breeds; however, brown variant phenotypes are recognized in the wildcat species, as is melanism, tabby pattern variants, and long fur. Thus these four loci may represent the most "ancient mutations" in the cat. Phenotypes that are becoming more frequent in cat breeds, such as silver, golden, and caramel,

Table 1 The seven ancient and coat color mutations in the domestic cat

Locus	Phenotype	Mutation	Reference
Agouti	A- = banded fur, aa = melanism	del122-123	4
Dilution	D- = wild type, dd = bluish grey	T83del	13
Long fur	L- = short, ll = long fur	Unknown	None
Orange	O = orange, o = wild type	Unknown	(Grahn et al., 2005)
Dominant white	W- = all white, w = wild type	Unknown	None
White spotting	SS = high white, Ss = bicolor,		(Cooper et al., 2006)
	ss = wild type	Unknown	
[1]**Tabby**	T^a- = no pattern, t^b t^b = blotched	Unknown	(Lyons et al., 2006)
Brown	B- = wild type, bb or		15
	bb^l = brown,	b = C8G	
	$b^l b^l$ = light brown	b^l = C298T	
Color	C- = wildtype, $c^b c^b$ or		12, 16
	$c^b c$ = sable,	c^b = G715T	
	$c^b c^s$ = Tonkinese,	c^s = G940A	
	$c^s c^s$ OR $c^s c$ = pointed,		
	cc = albino	c = C975del	

[1] Additional alleles such as mackerel and spotted are recognized at the *tabby* locus; however, multiple genes are considered for the *tabby* phenotypes. Loci in bold are the "seven ancient mutations."

could also represent variation seen in wildcats, but these colorations were not as popularized in the early breeds.

Although cat breeds are beautiful aesthetic pleasures, the usual population dynamics associated with any animal breed development, including selection, population bottlenecks, reduced migration, inbreeding, founder, and popular sire affects, can lead to the increase of frequency of many undesirable traits and health conditions within the small breed populations. Thus, the domestic cat has become an important model for inherited diseases that are also found in humans. Combining the cat's importance in human health, comparative genomics, and evolutionary studies (http://www.genome.gov/Pages/Research/Sequencing/SeqProposals/CatSEQ.pdf), the National Institutes of Health – National Human Genomics Research Institute (NIH-NHGRI) has not only supported the production of a low coverage, 2×, sequence of the cat genome, which is publicly available, but an additional 7× coverage has been scheduled for completion (http://www.genome.gov/19517271), which will provide a deeper coverage draft sequence of the cat within the coming years.

2 Cat Diseases

Over 277 disorders have been documented in the cat that have been shown to have a heritable component in other species and at least 46 have been suggested as single gene traits (http://omia.angis.org.au/). To date, the phenotypic and disease mutations that have been identified in the cat, Table 2, have benefited from the candidate gene approach, as similar traits had been previously identified in other species and the

Table 2 Genetic diseases and coat color traits in the cat with identified mutations

Disease/coat color	Gene	Mutation	Breeds	References
AB Blood Type (type B)	CMAH	18indel-53	All breeds	2
Gangliosidosis 1	GBL1	G1457C	Korat, Siamese	3
Gangliosidosis 2	HEXB	15bp del (intron)	Burmese	Up
Gangliosidosis 2	HEXB	inv1467-1491	DSH	18
Gangliosidosis 2	HEXB	C667T	DSH (Japan)	14
Gangliosidosis 2	HEXB	C39del	Korat	21
Gangliosidosis 2	GM2A	del390-393	DSH	19
Glycogen storage disease IV	GBE1	230bp ins 5′ – 6kb del	Norwegian forest	Up
Hemophilia B	F9	G247A	DSH	9
Hemophilia B	F9	C1014T	DSH	9
Hypertrophic Cardiomyopathy	MYBPC	G93C	Maine Coon	20
Lipoprotein lipase deficiency	LPL	G1234A	DSH	8
Alpha mannosidosis	LAMAN	del1748-1751	Persian	1
Mucolipidosis II	GNPTA	C2655T	DSH	7
Mucopolysaccharidosis I	IDUA	del1047-1049	DSH	10, 11
Mucopolysaccharidosis VI	ARSB	T1427C	Siamese	24
Mucopolysaccharidosis VI	ARSB	G1558A	Siamese	25
Mucopolysaccharidosis VII	GUSB	A1052G	DSH	5
Muscular dystrophy	DMD	900bp del M promoter - exon 1	DSH	23
Niemann-Pick C	NPC	G2864C	Persian	22
Polycystic kidney disease	PKD1	C10063A	Persian	17
Pyruvate kinase deficiency	PKLR	13bp del in exon 6	Abyssinian	Up
Spinal muscular atrophy	LIX1	140kb del, exons 4-6	Norwegian forest	6

'Up' are mutations that are unpublished to date.

genes and mutations were known. Although one of the first traits to be mapped to a particular chromosome in any species was the sex-linked orange coloration of the cat (Ibsen, 1916), the first autosomal trait linkage in the cat was recognized in 1986 (O'Brien et al., 1986) between the polymorphisms for hemoglobin and the Siamese pattern caused by mutations in *tyrosinase*. Two of the earliest DNA mutations identified in the cat were in 1994. A promoter mutation in the gene that causes Duchene muscular dystrophy (DMD), *dystrophin*, was identified in cats with muscular dystrophy (Winand et al., 1994) and a cat mutation was identified in *beta-hexominidase A* (*HEXB*) for a well-known lysosomal storage diseases (LSD) in humans, Sandhoff's disease (Muldoon et al., 1994). Several additional inborn errors of metabolism or LSDs have been identified (Table 2) and are maintained as research colonies at the University of Pennsylvania, leading to the establishment of one of the finest comparative genetics programs in veterinary medicine. Several breeds, such as Korats and Persians, have more than one metabolism defect, and the MPS VI group of cats are compound heterozygotes for mutations in *ARSB* (Crawley et al., 1998). Several feline lysosomal storage disease models have matured and advanced to gene therapy trials for the condition, pioneering efforts to provide corrective measures that will be of benefit to humans with the same conditions (for reviews see Casal and Haskins, 2006; Desnick et al., 1982; Haskins, 1996; Haskins et al., 2002). The mutations for several types of LSDs have been identified and different therapeutic strategies have been tested in afflicted cats, including enzyme replacement therapy (ERT), heterologous bone marrow transplantation (BMT), and gene therapy. Localized ERT in the joint space has successfully N-acetylgalactosamine-4-sulfatase deficiency in the MPS VI feline model (Auclair et al., 2006). Adeno-associated virus vectors containing the normal cDNA of alpha-mannosidase have been used to intracranially administer corrected enzymes in successful therapeutic trials for cats with mannosidosis, an LSD that attacks the central nervous system (Vite et al., 2005). Several feline disorders, including the mannosidosis, mucolipidoses II, and MPS VI cats, have all been models for BMT (Dial et al., 1997; Haskins et al., 1984; Simonaro et al., 1999), which have been some of the earliest attempts to corrective therapies. Thus, as the genetic causes of feline diseases are identified, many are translating into therapeutic models to correct both the feline and the human homologous diseases.

Approximately 16 different genes account for the 22 mutations causing diseases in cats. However, only eight disease-causing mutations (Table 2) are segregating in cat breeds and not maintained only in research colonies. Two of the most prevalent mutations cause cardiac and renal disease. Hypertrophic cardiomyopathy (HCM) is a heterogeneous cardiac disease that has been recognized is several breeds, including Bengals, Ragdolls, and Sphynx; however, HCM is scientifically documented as heritable only in the Maine Coon cat (Kittleson et al., 1999). A mutation in *myosin C binding protein* (*MYCBP*) is highly correlated with clinical presentation of HCM in the Maine Coon cat (Meurs et al., 2005) and is considered the causative mutation. Although the frequency of the disease has not been clearly established for the Maine Coon breed, combined with echocardiogram evaluations, an active genetic typing program is assisting the reduction of this disease within the Maine Coon breed. The same mutation has been evaluated in other breeds with HCM but, as in humans, HCM appears to be heterogenic in cats and the mutation is not correlated with disease in other breeds (K. Meurs, personal communication, L. Lyons, data not shown).

Other diseases found in the cat, such as polycystic kidney disease (PKD) in Persian cats, have benefited from a combined genetic linkage analysis and candidate gene approach (Lyons et al., 2004; Young et al., 2005). Feline PKD was first clinically described as an autosomal dominant inherited trait in 1990 (Biller et al., 1990, 1996). Persian cats are the most popular breed worldwide, and together with other breeds that use Persians as a parent breed (Exotics and Himalayans) or for outcrosses (Selkirk Rex, Scottish Folds and American Shorthairs), Persians represent nearly 60% of the cats in the cat fancy. Approximately 38% of Persians have PKD worldwide (Barrs et al., 2001; Barthez et al., 2003; Cannon et al., 2001), making PKD the most common inherited disease in the domestic cat. PKD affects 600,000 people in the US alone, and 12.5 million worldwide. There are more people afflicted with PKD than cystic fibrosis, muscular dystrophy, hemophilia, and Downs syndrome and sickle cell anemia combined. Over 90% of PKD is heritable. More than 60% of the individuals with PKD develop kidney failure, or end-stage renal disease (ESRD). Patients with ESRD are generally on dialysis for approximately

7 years, significantly lower than their life span (see reviews in Sessa et al., 2004; Tahvanainen et al., 2005). Feline and human PKD have similar clinical features (Eaton et al., 1997; Pedersen et al., 2003). Over 95% of cats with PKD will develop renal cysts by 8 months of age, which can be accurately determined by ultrasound. Depending on disease severity, cats can live a normal lifespan, 10–14 years, with PKD, or succumb within a few years of onset; similar disease variation and progressions are also seen in humans. Ultrasonographic identification of renal cysts allowed the accurate ascertainment of large pedigrees of Persian cats with PKD. In humans, two genes are responsible for a majority of PKD, *PKD1* and *PKD2*. Although *PKD1* may present more severely, the clinical presentations are not distinctive and can only be determined by mutation analyses. Approximately 85% of cases are caused by mutations related to the *PKD1* gene, while the remaining 15% are attributed to the *PKD2* gene (Peters et al., 1993). The two major genes produce very large transcripts and because limited domestic cat genetic sequence was available, a linkage analysis was first conducted to implicate a candidate gene for feline PKD prior to a candidate gene approach (Young et al., 2005). Once a cat microsatellite marker showed significant linkage to the region of the cat genome with *PKD1*, a gene scan identified a C10063A transversion that changes an argenine to an OPA stop codon, which should disrupt approximately 25% of the polycystin-1 protein (1).

Twenty-one mutations have been identified in the cat that cause diseases, four coat color loci represent an additional seven mutations, and a blood group mutation produces the Type B cat. Cats have even been shown to lack a "sweet tooth" due to the pseudogenization of *Tas1r2* sweet receptor gene (Li et al., 2006; Li et al., 2005). Veterinary medicine has identified over 277 traits that could be heritable conditions in the cat and many have been well documented. The identification of several additional mutations should be on the horizon as active linkage studies have implicated genes or genetic regions for white spotting, tabby, orange, progressive retinal atrophy in the Persian and the Abyssinian, and a craniofacial defect in the Burmese. Improved genetic resources will expedite mutation discoveries for the single gene traits of the cat.

3 Feline Genetic Resources

Early chromosome banding studies of the domestic cat revealed an easily distinguishable karyotype consisting of 18 autosomal and the XY sex chromosome pair, resulting in a $2N$ complement of 38 cat chromosomes (Wurster-Hill and Gray, 1973). The traditional groupings of chromosomes into alphabetic groups based on size and centromeric position has only relatively recently been renamed to more standard nomenclature for the cat (Cho et al., 1997). Basic light microscopy and giemsa banding also showed that domestic cats have a chromosomal architecture that is highly representative for all felids and even carnivores (Modi and O'Brien, 1988). Only minor chromosomal rearrangements are noted amongst the 36 extant felids, most noticeably a robertsonian fusion in the South American ocelot lineage leads to a reduced complement, $2N = 36$ (Wurster-Hill and Gray, 1973). The variation of chromosomal sizes allowed for the easy development of chromosome paints by flow sorting (Wienberg et al., 1997). Chromosome painting techniques supported early somatic cell hybrid maps in that the cat appeared to be highly conserved in chromosomal arrangement to humans, specifically as compared to mice (Stanyon et al., 1999). Hence, chromosome painting gave an excellent overview of cat genome organization, which greatly facilitates candidate gene approaches since the location of particular genes can be anticipated in cats from comparison with the genetic map of humans.

The low-resolution genetic comparisons provided by chromosomal studies have been augmented and complimented by medium resolution genetic and radiation hybrid maps of the cat. The current 5,000 Rad radiation hybrid map of the cat consists of 1784 markers (Menotti-Raymond et al., 2003a; Murphy et al., 2006, 1999, 2000) and the interspecies hybrid-based linkage map contains approximately 250 microsatellite markers (Menotti-Raymond et al., 2003b, 1999) that are affective for the initiation of linkage studies in families segregating for phenotypic traits. The linkage map is due for updating, which is likely in the coming year; however, already available linkage maps have assisted targeted candidate gene approaches, as seen for *PKD* (Young et al., 2005), linkage analyses for *Tabby* (Lyons et al., 2006), *white spotting* (Cooper et al., 2006) and *Orange* (Grahn et al., 2005), and the genetic map has led to the first disease gene isolated by positional cloning, *LIX1*, which causes spinal muscular atrophy in the Norwegian forest cat (Fyfe et al., 2006; He et al., 2005). Updated maps, deeper genomic sequence, and the RPCI-85 cat BAC library (http://bacpac.chori.org/) should provide sufficient and highly efficient genetic resources for future genetic studies in felids.

4 Reproductive Technologies

Even with limited genetic resources, the domestic cat has become the focus of research technologies that use reproductive methods to manipulate the cat genome. Although gene knock-outs for cat allergens have only been proposed, the currently available "hypoallergenic" cats have been developed by natural breeding methods (http://www.allerca.com/). However, cloning of the cat by either nuclear or chromatin transfer has been successful and even commercialized. To date, several domestic cats and the African wildcat have been successfully cloned (Gomez et al., 2004; Shin et al., 2002). At least one company has used client-provided materials to clone a pet as a commercial adventure (http://www.savingsandclone.com/). Clients are now asking veterinarians to collect viable cells as sample from their cats and send them to cloning companies for the eventual use of cloning the donor cat. Cat cloning has exemplified important aspects regarding this highly controversial topic. The first cloned cat, CC, was cloned from a calico cat, a female cat (Rainbow) that had white spotting and different alleles at the *Orange* locus on the X chromosome. The cloned cat, CC, did not express the orange allelic variants;

hence, X-inactivation did not get reprogrammed during the cloning process. This event highlights two important concepts: (1) researchers still do not recognize or understand all the normal biological processes during embryo development and (2) cloned animals can definitely look different that the original, donor individual. The successful cloning of the African wildcat potentially opens avenues for the reintroduction of the genetics of individuals that have been lost to conservation breeding programs, such as animals that have been sterilized or are too old or unhealthy to breed. However, domestic cat eggs were used for the cloning of the African wildcats, hence their mtDNA is not representative of the endangered species. But overall, cloning may provide unexpected opportunities in conservation, or perhaps in the replication of individuals with certain phenotypes of interest in research. Representing the ultimate of an inbred species, a series of clones from the same individual could foster interesting studies of nature versus nurture.

5 Future of Cat Genetics

Once causative mutations for heritable conditions are identified in cats, felines become a more important asset to human health. Several inborn errors of metabolism in cats are already exploring gene therapy approaches. Many murine models exist for the study of cystogenesis, the hallmark of PKD; however, each rodent model has its shortcomings. The cat is similar to human autosomal dominant PKD in several important aspects, including (1) a causative mutation in *PKD1*, (2) a similar type of mutation that causes a similar protein disruption, (3) similar variability in disease progression, (4) cystogenesis in other organs, including the liver and pancreas, (5) homozygotes for the mutation are lethal, and (6) cats are intermediate in regard to genetic variation, mimicking human populations and ethnic groups more closely than inbred strains of mice. None of the rodent models have a mutation in *PKD1* and many lack the other noted aspects that are important to human drug and gene therapy trials.

Thus, cats can be an alternative and supportive animal model to rodent models for many heritable conditions because (1) they provide a balance between cost and efficiency, (2) drug dosages are more easily translated between cats and humans, (3) the longer life span of the cat allows repeated therapy trials and longer term studies, (4) cats have strong conservation of biology, anatomy, and physiology to humans, (5) cats provide a second animal for validation and efficacy, (6) the larger size of the cat and its organs are more amenable to therapies, and (7) cats are intermediate in regard to genetic variation, mimicking human populations and ethnic groups more closely than inbred strains of mice.

The deeper sequencing of the cat genome and the investigation of variation by resequencing in different cat breeds will allow the cat to leap forward in genetics, from the analysis of single gene traits to the investigation of more complex traits, such as asthma, diabetes, and obesity. Cats are notorious for viral infections, hence the identification of susceptibility and resistance genes for acquired diseases should become a mainstay in cat research. The determination of linkage disequilibrium may assist disease studies in many breeds by potentially lowering the number of markers, likely SNP-based, that are required for full genome scans in complex trait analyses. However, many of the common diseases that plague humans, which are found in cats, will likely be examined in the outbred populations of non-purebred housecats, as only 10–15% of cats in the US are representatives by a fancy breed, a proportion that is higher than most other nations (Louwerens et al., 2005).

The available genetic resources for the cat now places the research bottleneck for felines back to the acquisition of appropriate patients with sufficient cases and controls. Hence, the primary care veterinarians, the veterinary specialists, and veterinary researchers need to join forces to properly characterize diseases and routinely collect research materials so that patients are not lost to important studies and health investigations. The development of the DNA tests for parentage and identification (http://www.isag.org.uk/), coat colors, and the prominent diseases, such as PKD and HCM, has encouraged cat breeders to explore genetic research in a more positive manner and has encouraged their participation in research studies. Forensic applications (Menotti-Raymond et al., 1997a, 1997b, 2005) and mass disasters have brought to the forefront the DNA profiling of loved ones, including our pets. For these reasons, more cat breeders are banking DNA sources on their animals or providing DNA to service and research laboratories. Many veterinary hospitals and large clinical conglomerates are developing electronic database systems that could facilitate the identification of proper patients, cases, and controls. Combined with DNA banking and specialty health care, the veterinary world stands to enhance the possibilities of complex disease research in the cat by leaps and bounds. Even though feline origins remain a mystery and "domesticated" may not be the most appropriate term for the domestic cat, the cat is unlocking its genetic secrets that explain the cat's form and function.

References

Auclair, D., Hein, L.K., Hopwood, J.J., and Byers, S., 2006, Intra-articular enzyme administration for joint disease in feline mucopolysaccharidosis VI: enzyme dose and interval, *Pediatr. Res.* **59**:538–543.

Barrs, V.R., Gunew, M., Foster, S.F., Beatty, J.A., and Malik, R., 2001, Prevalence of autosomal dominant polycystic kidney disease in Persian cats and related-breeds in Sydney and Brisbane, *Aust. Vet. J.* **79**:257–259.

Barthez, P.Y., Rivier, P., and Begon, D., 2003, Prevalence of polycystic kidney disease in Persian and Persian related cats in France, *J. Feline Med. Surg.* **5**:345–347.

Biller, D.S., Chew, D.J., and DiBartola, S.P., 1990, Polycystic kidney disease in a family of Persian cats, *J. Am. Vet. Med. Assoc.* **196**:1288–1290.

Biller, D.S., DiBartola, S.P., Eaton, K.A., Pflueger, S., Wellman, M.L., and Radin, M.J., 1996, Inheritance of polycystic kidney disease in Persian cats, *J. Hered.* **87**:1–5.

Brooks, S.A., and Bailey, E., 2005, Exon skipping in the KIT gene causes a Sabino spotting pattern in horses, *Mamm. Genome* **16**:893–902.

Cannon, M.J., MacKay, A.D., Barr, F.J., Rudorf, H., Bradley, K.J., and Gruffydd-Jones, T.J., 2001, Prevalence of polycystic kidney disease in Persian cats in the United Kingdom, *Vet. Rec.* **149**:409–411.

Casal, M., and Haskins, M., 2006, Large animal models and gene therapy, *Eur. J. Hum. Genet.* **14**:266–272.

CFA 2004, The Cat Fanciers' Association Complete Cat Book, HarperCollins Publishers, New York.

Cho, K.W., Youn, H.Y., Watari, T., Tsujimoto, H., Hasegawa, A., and Satoh, H., 1997, A proposed nomenclature of the domestic cat karyotype, *Cytogenet. Cell Genet.* **79**:71–78.

Clark, J.M., 1975, The effects of selection and human preference on coat colour gene frequencies in urban cats, *Heredity* **35**:195–210.

Cooper, M.P., Fretwell, N., Bailey, S.J., and Lyons, L.A., 2006, White spotting in the domestic cat (*Felis catus*) maps near KIT on feline chromosome B1, *Anim. Genet.* **37**:163–165.

Crawley, A.C., Yogalingam, G., Muller, V.J., and Hopwood, J.J., 1998, Two mutations within a feline mucopolysaccharidosis type VI colony cause three different clinical phenotypes, *J. Clin. Invest.* **101**:109–119.

Davis, B.K., and Davis, B.P., 1977, Allele frequencies in a cat population in Budapest, *J. Hered.* **68**:31–34.

Desnick, R.J., McGovern, M.M., Schuchman, E.H., and Haskins, M.E., 1982, Animal analogues of human inherited metabolic diseases: molecular pathology and therapeutic studies, *Prog. Clin. Biol. Res.* **94**:27–65.

Dial, S.M., Byrne, T., Haskins, M., Gasper, P.W., Rose, B., Wenger, D.A., and Thrall, M.A., 1997, Urine glycosaminoglycan concentrations in mucopolysaccharidosis VI-affected cats following bone marrow transplantation or leukocyte infusion, *Clin. Chim. Acta* **263**:1–14.

Eaton, K.A., Biller, D.S., DiBartola, S.P., Radin, M.J., and Wellman, M.L., 1997, Autosomal dominant polycystic kidney disease in Persian and Persian-cross cats, *Vet. Pathol.* **34**:117–126.

Eizirik, E., Yuhki, N., Johnson, W.E., Menotti-Raymond, M., Hannah, S.S., and O'Brien, S.J., 2003, Molecular genetics and evolution of melanism in the cat family, *Curr. Biol.* **13**: 448–453.

Fyfe, J.C., Menotti-Raymond, M., David, V.A., Brichta, L., Schaffer, A.A., Agarwala, R., Murphy, W.J., Wedemeyer, W.J., Gregory, B.L., Buzzell, B.G., Drummond, M.C., Wirth, B., and O'Brien, S.J., 2006, An approximately 140-kb deletion associated with feline spinal muscular atrophy implies an essential LIX1 function for motor neuron survival, *Genome Res.* **16**: 1084–1090.

Gomez, M.C., Pope, C.E., Giraldo, A., Lyons, L.A., Harris, R.F., King, A.L., Cole, A., Godke, R.A., and Dresser, B.L., 2004, Birth of African Wildcat cloned kittens born from domestic cats, *Cloning Stem Cells* **6**:247–258.

Grahn, R.A., Lemesch, B.M., Millon, L.V., Matise, T., Rogers, Q.R., Morris, J.G., Fretwell, N., Bailey, S.J., Batt, R.M., and Lyons, L.A., 2005, Localizing the X-linked orange colour phenotype using feline resource families, *Anim. Genet.* **36**:67–70.

Haskins, M., 1996, Bone marrow transplantation therapy for metabolic disease: animal models as predictors of success and in utero approaches, *Bone Marrow Transpl* **18**:S25–S27.

Haskins, M., Casal, M., Ellinwood, N.M., Melniczek, J., Mazrier, H., and Giger, U., 2002, Animal models for mucopolysaccharidoses and their clinical relevance, *Acta Paediatr. Suppl.* **91**:88–97.

Haskins, M.E., Wortman, J.A., Wilson, S., and Wolfe, J.H., 1984, Bone marrow transplantation in the cat, *Transplantation* **37**:634–636.

He Q., Lowrie, C., Shelton, G.D., Castellani, R.J., Menotti-Raymond, M., Murphy, W., O'Brien, S.J., Swanson, W.F., and Fyfe, J.C., 2005, Inherited motor neuron disease in domestic cats: a model of spinal muscular atrophy, *Pediatr. Res.* **57**:324–330.

Hemmer, H., 1972, Variations in brain size in the species *Felis silvestris*, *Experientia* **28**: 271–272.

Hemmer, H., 1976, Man's strategy in domestication – a synthesis of new research trends, *Experientia* **32**:663–666.

Hemmer, H., 1978, The evolutionary systematics of living Felidae: present status and current problems, *Carnivore* B:71–79.

Hemmer, H., 1990, "Domestication: The decline of environmental appreciation," Cambridge University Press, Cambridge.

Housley, D.J., and Venta, P.J. 2006, The long and the short of it: evidence that FGF5 is a major determinant of canine 'hair'-itability, *Anim. Genet.* **37**:309–315.

Ibsen, H.L., 1916, Tricolor inheritance. III. Tortoiseshell cats, *Genetics* **1**:377–386.

Imes, D.L., Geary, L.A., Grahn, R.A., and Lyons, L.A., 2006, Albinism in the domestic cat (*Felis catus*) is associated with a tyrosinase (TYR) mutation, *Anim. Genet.* **37**:175–178.

Ishida, Y., David, V.A., Eizirik, E., Schaffer, A.A., Neelam, B.A., Roelke, M.E., Hannah, S.S., O'Brien, S.J., and Menotti-Raymond, M., 2006, A homozygous single-base deletion in MLPH causes the dilute coat color phenotype in the domestic cat, *Genomics* **88**:698–705. Epub.

Johansson Moller, M., Chaudhary, R., Hellmen, E., Hoyheim, B., Chowdhary, B., and Andersson, L., 1996, Pigs with the dominant white coat color phenotype carry a duplication of the KIT gene encoding the mast/stem cell growth factor receptor, *Mamm. Genome* **7**:822–830.

Johnson, W.E., Eizirik, E., Pecon-Slattery, J., Murphy, W.J., Antunes, A., Teeling, E., and O'Brien, S.J., 2006, The late Miocene radiation of modern Felidae: a genetic assessment, *Science* **311**:73–77.

Kitchener, A., 1991, "The natural history of wild cats," Cornell University Press, New York.

Kittleson, M.D., Meurs, K.M., Munro, M.J., Kittleson, J.A., Liu, S.K., Pion, P.D., and Towbin, J.A., 1999, Familial hypertrophic cardiomyopathy in maine coon cats: an animal model of human disease, *Circulation* **99**:3172–3180.

Kukekova, A.V., Acland, G.M., Oskina, I.N., Kharlamova, A.V., Trut, L.N., Chase, K., Lark, K.G., Erb, H.N., and Aguirre, G.D., 2006, The genetics of domesticated behavior in Canids: what can dogs and Silver foxes tell us about each other? In E.A. Ostrander, U. Giger, and K. Lindblad-Tioh, Eds., "The Dog and Its Genome", pp. 515–537, Cold Spring Harbor Laboratory Press, Cold Spring Harbor.

Li X., Li, W., Wang, H., Cao, J., Maehashi, K., Huang, L., Bachmanov, A.A., Reed, D.R., Legrand-Defretin, V., Beauchamp, G.K., and Brand, J.G., 2005, Pseudogenization of a sweet-receptor gene accounts for cats' indifference toward sugar, *PLoS Genet.* **1**:27–35.

Li, X., Li, W., Wang, H., Bayley, D.L., Cao, J., Reed, D.R., Bachmanov, A.A., Huang, L., Legrand-Defretin, V., Beauchamp, G.K., and Brand, J.G., 2006, Cats lack a sweet taste receptor, *J. Nutr.* **136**:1932S–1934S.

Louwerens, M., London, C.A., Pedersen, N.C., and Lyons, L.A., 2005, Feline lymphoma in the post-feline leukemia virus era, *J. Vet. Intern. Med.* **19**:329–335.

Lyons, L.A., Bailey, S.J., Baysac, K.C., Byrns, G., Erdman, C.A., Fretwell, N., Froenicke, L., Gazlay, K.W., Geary, L.A., Grahn, J.C., Grahn, R.A., Karere, G.M., Lipinski, M.J., Rah, H., Ruhe, M.T., and Bach, L.H., 2006, The Tabby cat locus maps to feline chromosome B1, *Anim. Genet.* **37**:383–386.

Lyons, L.A., Biller, D.S., Erdman, C.A., Lipinski, M.J., Young, A.E., Roe, B.A., Qin, B., and Grahn, R.A., 2004, Feline polycystic kidney disease mutation identified in PKD1, *J. Am. Soc. Nephrol.* **15**:2548–2555.

Lyons, L.A., Foe, I.T., Rah, H., and Grahn, R.A., 2005a, Chocolate coated cat: *TYRP1* mutations for brown color in domestic cats, *Mamm. Genome* **16**:356–366.

Lyons, L.A., Imes, D.L., Rah, H.C., and Grahn, R.A., 2005b, Tyrosinase mutations associated with Siamese and Burmese patterns in the domestic cat (*Felis catus*), *Anim. Genet.* **36**:119–126.

Marklund, S., Kijas, J., Rodriguez-Martinez, H., Ronnstrand, L., Funa, K., Moller, M., Lange, D., Edfors-Lilja, I., and Andersson, L., 1998, Molecular basis for the dominant white phenotype in the domestic pig, *Genome Res.* **8**:826–833.

Menotti-Raymond, M., David, V.A., Agarwala, R., Schaffer, A.A., Stephens, R., O'Brien, S.J., and Murphy, W.J., 2003a, Radiation hybrid mapping of 304 novel microsatellites in the domestic cat genome, *Cytogenet. Genome Res.* **102**:272–276.

Menotti-Raymond, M., David, V.A., Chen, Z.Q., Menotti, K.A., Sun, S., Schaffer, A.A., Agarwala, R., Tomlin, J.F., O'Brien, S.J., and Murphy, W.J., 2003b, Second-generation integrated genetic linkage/radiation hybrid maps of the domestic cat (*Felis catus*), *J. Hered.* **94**:95–106.

Menotti-Raymond, M., David, V.A., Lyons, L.A., Schaffer, A.A., Tomlin, J.F., Hutton, M.K., and O'Brien, S.J., 1999, A genetic linkage map of microsatellites in the domestic cat (*Felis catus*), *Genomics* **57**:9–23.

Menotti-Raymond, M., David, V.A., Stephens, J.C., Lyons, L.A., and O'Brien, S.J., 1997a, Genetic individualization of domestic cats using feline STR loci for forensic applications, *J. Forensic Sci.* **42**:1039–1051.

Menotti-Raymond, M.A., David, V.A., and O'Brien, S.J., 1997b, Pet cat hair implicates murder suspect, *Nature* **386**:774.

Menotti-Raymond, M.A., David, V.A., Wachter, L.L., Butler, J.M., and O'Brien, S.J., 2005, An STR forensic typing system for genetic individualization of domestic cat (*Felis catus*) samples, *J. Forensic Sci.* **50**:1061–1070.

Meurs, K.M., Sanchez, X., David, R.M., Bowles, N.E., Towbin, J.A., Reiser, P.J., Kittleson, J.A., Munro, M.J., Dryburgh, K., Macdonald, K.A., and Kittleson, M.D., 2005, A cardiac myosin binding protein C mutation in the Maine Coon cat with familial hypertrophic cardiomyopathy, *Hum. Mol. Genet.* **14**:3587–3593.

Modi, W.S., and O'Brien, S.J., 1988, "Quantitative cladistic analysis of chromosomal banding data among species in three orders of mammals: Hominoid primates, felids, and arvicolid rodents," Plenum Publishing, New York.

Morris, D., 1999a, "Cat breeds of the world," Penguin Books, New York.

Morris, D., 1999b, "Cat Breeds of the World: A Complete Illustrated Encyclopedia," Viking Penguin, New York.

Muldoon, L.L., Neuwelt, E.A., Pagel, M.A., and Weiss, D.L., 1994, Characterization of the molecular defect in a feline model for type II GM2-gangliosidosis (Sandhoff disease), *Am. J. Pathol.* **144**:1109–1118.

Murphy, W.J., Davis, B., David, V.A., Agarwala, R., Schaffer, A.A., Pearks Wilkerson, A.J., Neelam, B., O'Brien, S.J., and Menotti-Raymond, M., 2006, A 1.5-Mb-resolution radiation hybrid map of the cat genome and comparative analysis with the canine and human genomes, *Genomics* **89**:189–196.

Murphy, W.J., Menotti-Raymond, M., Lyons, L.A., Thompson, M.A., and O'Brien, S.J., 1999, Development of a feline whole genome radiation hybrid panel and comparative mapping of human chromosome 12 and 22 loci, *Genomics* **57**:1–8.

Murphy, W.J., Sun, S., Chen, Z., Yuhki, N., Hirschmann, D., Menotti-Raymond, M., and O'Brien, S.J., 2000, A radiation hybrid map of the cat genome: implications for comparative mapping, *Genome Res.* **10**:691–702.

O'Brien, S.J., Haskins, M.E., Winkler, C.A., Nash, W.G., and Patterson, D.F., 1986, Chromosomal mapping of beta-globin and albino loci in the domestic cat. A conserved mammalian chromosome group, *J. Hered.* **77**:374–378.

Pedersen, K.M., Pedersen, H.D., Haggstrom, J., Koch, J., and Ersboll, A.K., 2003, Increased mean arterial pressure and aldosterone-to-renin ratio in Persian cats with polycystic kidney disease, *J. Vet. Intern. Med.* **17**:21–27.

Peters, D.J., Spruit, L., Saris, J.J., Ravine, D., Sandkuijl, L.A., Fossdal, R., Boersma, J., van Eijk, R., Norby, S., Constantinou-Deltas, C.D., and et al. 1993, Chromosome 4 localization of a second gene for autosomal dominant polycystic kidney disease, *Nat. Genet.* **5**:359–362.

Robinson, R., 1991, "Genetics for Cat Breeders," Pergamon Press, Oxford.

Seidensticker, J., and Lumpkin, S., "Great Cats: majestic creatures of the wild," Rodale Press, Enormaus, PA.

Sessa, A., Righetti, M., and Battini, G., 2004, Autosomal recessive and dominant polycystic kidney diseases, *Minerva Urol. Nefrol.* **56**:329–338.

Shin, T., Kraemer, D., Pryor, J., Liu, L., Rugila, J., Howe, L., Buck, S., Murphy, K., Lyons, L., and Westhusin, M., 2002, A cat cloned by nuclear transplantation, *Nature* **415**:859.

Simonaro, C.M., Haskins, M.E., Abkowitz, J.L., Brooks, D.A., Hopwood, J.J., Zhang, J., and Schuchman, E.H., 1999, Autologous transplantation of retrovirally transduced bone marrow or neonatal blood cells into cats can lead to long-term engraftment in the absence of myeloablation, *Gene Ther.* **6**:107–113.

Stanyon, R., Yang, F., Cavagna, P., O'Brien, P.C., Bagga, M., Ferguson-Smith, M.A., and Wienberg, J., 1999, Reciprocal chromosome painting shows that genomic rearrangement between rat and mouse proceeds ten times faster than between humans and cats, *Cytogenet. Cell Genet.* **84**:150–155.

Sundberg, J.P., Rourk, M H., Boggess, D., Hogan, M.E., Sundberg, B.A., and Bertolino, A.P., 1997, Angora mouse mutation: altered hair cycle, follicular dystrophy, phenotypic maintenance of skin grafts, and changes in keratin expression, *Vet. Pathol.* **34**:171–179.

Sunquist, M., and Sunquist, F., 2002, "Wild Cats of the World," University of Chicago Press, Chicago.

Tahvanainen, E., Tahvanainen, P., Kaariainen, H., and Hockerstedt, K. 2005, Polycystic liver and kidney diseases, *Ann. Med.* **37**:546–555.

Terry, R.R., Bailey, E., Bernoco, D., and Cothran, E.G., 2001, Linked markers exclude KIT as the gene responsible for appaloosa coat colour spotting patterns in horses, *Anim. Genet.* **32**: 98–101.

Todd, N.B., and Lloyd, A.T., 1984, Mutant allele frequencies in domestic cats of Portugal and the Azores, *J. Hered.* 75:495–497.

Todd, N.B., Robinson, R., and Clark, J.M., 1974, Gene frequencies in domestic cats of Greece, *J. Hered.* **65**:227–231.

Todd, N.E., Fagen, R.M., and Fagen, K., 1975, Gene frequencies in Icelandic cats, *Heredity* **35**:173–183.

Vella, C.M., Shelton, L.M., McGonagle, J.J., and Stanglein, T.W., 1999, "Robinson's Genetics for Cat Breeders and Veterinarians," Butterworth Heinemann, Boston.

Vite, C.H., McGowan, J.C., Niogi, S.N., Passini, M.A., Drobatz, K.J., Haskins, M.E., and Wolfe, J.H., 2005, Effective gene therapy for an inherited CNS disease in a large animal model, *Ann. Neurol.* **57**:355–364.

Wagner, A., and Wolsan, M., 1987, Pelage mutant allele frequencies in domestic cat populations of Poland, *J. Hered.* **78**:197–200.

Wastlhuber, J., 1991, History of domestic cats and cat breeds, In N.C. Pedersen, Ed., "Feline Husbandry: Diseases and management in the multiple-cat environment", pp. 1–59, American Veterinary Publications, Inc., Goleta, CA.

Wayne, R.K., 1986, Cranial morphology of domestic and wild canids: the influence of development on morphological change, *Evolution* **40**:243–261.

Wienberg, J., Stanyon, R., Nash, W.G., O'Brien, P.C., Yang, F., O'Brien, S.J., and Ferguson-Smith, M.A. 1997, Conservation of human vs. feline genome organization revealed by reciprocal chromosome painting, *Cytogenet. Cell Genet.* **77**:211–217.

Winand, N.J., Edwardsm M., Pradhanm D., Berianm C.A., and Cooper, B.J., 1994, Deletion of the dystrophin muscle promoter in feline muscular dystrophy, *Neuromuscul. Disord.* **4**:433–445.

Wurster-Hill, D.H., and Gray, C.W., 1973, Giemsa banding patterns in the chromosomes of twelve species of cats (Felidae), *Cytogenet. Cell Genet.* **12**:388–397.

Young, A.E., Biller, D.S., Herrgesell, E.J., Roberts, H.R., and Lyons, L.A., 2005, Feline polycystic kidney disease is linked to the PKD1 region, *Mamm. Genome* **16**:59–65.

APPENDIX: Table references

1. Berg, T., Tollersrud, O.K., Walkley, S.U., Siegel, D., and Nilssen, O., 1997, Purification of feline lysosomal alpha-mannosidase, determination of its cDNA sequence and identification of a mutation causing alpha-mannosidosis in Persian cats, *Biochem. J.* **328**:863–870.

2. Bighignoli, B., Grahn, R.A., Millon, L.V., Longeri, M., Polli, M., and Lyons, L.A., 2007, Genetic mutations for the feline AB blood group identified in *CMAH* mutations associated with the domestic cat AB blood group, *MBC Genet.* **8**:27.

3. De Maria, R., Divari, S., Bo, S., Sonnio, S., Lotti, D., Capucchio, M.T., and Castagnaro, M., 1998, Beta-galactosidase deficiency in a Korat cat: a new form of feline GM1-gangliosidosis, *Acta Neuropathol. (Berl)* **96**:307–314.

4. Eizirik, E., Yuhki, N., Johnson, W.E., Menotti-Raymond, M., Hannah, S.S., O'Brien, and S.J., 2003, Molecular genetics and evolution of melanism in the cat family, *Curr. Biol.* **13**:448–453.

5. Fyfe, J.C., Kurzhals, R.L., Lassaline, M.E., Henthorn, P.S., Alur, P.R., Wang, P., Wolfe, J.H., Giger, U., Haskins, M.E., Patterson, D.F., Sun, H., Jain, S., and Yuhki, N., 1999, Molecular basis of feline beta-glucuronidase deficiency: an animal model of mucopolysaccharidosis VII, *Genomics* **58**:121–128.

6. Fyfe, J.C., Menotti-Raymond, M., David, V.A., Brichta, L, Schaffer, A.A., Agarwala, R., Murphy, W.J., Wedemeyer, W.J., Gregory, B.L., Buzzell, B.G., Drummond, M.C., Wirth, B., and O'Brien, S.J., 2006, An approximately 140-kb deletion associated with feline spinal muscular atrophy implies an essential LIX1 function for motor neuron survival, *Genome Res.* **16**: 1084–1090.

7. Giger, U., Tcherneva, E., Caverly, J., Seng, A., Huff, A. M., Cullen, K., Van Hoeven, M., Mazrier, H., and Haskins, M. E., 2006, A missense point mutation in N-acetylglucosamine-1-phospotrans-ferase causes mucolipidosis II in domestic shorthair cats, *J. Vet. Intern. Med.* **20**:781.

8. Ginzinger, D.G., Lewis, M.E., Ma, Y., Jones, B.R., Liu, G., and Jones, S.D., 1996, A mutation in the lipoprotein lipase gene is the molecular basis of chylomicronemia in a colony of domestic cats, *J. Clin. Invest.* **97**:1257–1266.

9. Goree, M., Catalfamo, J.L., Aber, S., and Boudreaux, M.K., 2005, Characterization of the mutations causing hemophilia B in 2 domestic cats, *J. Vet. Intern. Med.* **19**:200–204.

10. Haskins, M., Jezyk, P., and Giger, U. 2005, Diagnostic tests for mucopolysaccharidosis. *J. Am. Vet. Med. Assoc.* **226**:1047.

11. He, X., Li, C.M., Simonaro, C.M., Wan, Q., Haskins, M.E., Desnick, R.J., and Schuchman, E.H., 1999, Identification and characterization of the molecular lesion causing mucopolysaccharidosis type I in cats, *Mol. Genet. Metab.* **67**:106–112.

12. Imes, D.L., Geary, L.A., Grahn, R.A., and Lyons, L.A., 2006, Albinism in the domestic cat (*Felis catus*) is associated with a tyrosinase (TYR) mutation, *Anim. Genet.* **37**:175–178.

13. Ishida, Y., David, V.A., Eizirik, E., Schaffer, A.A., Neelam, B.A., Roelke, M.E., Hannah, S.S., O'Brien, S.J., and Menotti-Raymond, M., 2006, A homozygous single-base deletion in MLPH causes the dilute coat color phenotype in the domestic cat, *Genomics* **88**:698–705.

14. Kanae, Y., Endoh, D., Yamato, O., Hayashi, D., Matsunaga, S., Ogawa, H., Maede, Y., and Hayashi, M., 2007, Nonsense mutation of feline beta-hexosaminidase beta-subunit (HEXB) gene causing Sandhoff disease in a family of Japanese domestic cats, *Res. Vet. Sci.* **82**:54–60.

15. Lyons, L.A., Foe, I.T., Rah, H.C., and Grahn, R.A., 2005, Chocolate coated cats: TYRP1 mutations for brown color in domestic cats, *Mamm. Genome.* **16**:356–366.

16. Lyons, L.A., Imes, D.L., Rah, H.C., and Grahn, R.A., 2005, Tyrosinase mutations associated with Siamese and Burmese patterns in the domestic cat (*Felis catus*), *Anim. Genet.* **36**: 119–126.

17. Lyons, L.A., Biller, D.S., Erdman, C.A., Lipinski, M.J., Young, A.E., Roe, B.A., Qin, B., and Grahn, R.A., 2004, Feline polycystic kidney disease mutation identified in PKD1, *J. Am. Soc. Nephrol.* **15**:2548–2555.

18. Martin, D.R., Krum, B.K., Varadarajan, G.S., Hathcock, T.L., Smith, B.F., and Baker, H.J., 2004, An inversion of 25 base pairs causes feline GM2 gangliosidosis variant, *Exp. Neurol.* **187**: 30–37.

19. Martin, D.R., Cox, N.R., Morrison, N.E., Kennamer, D.M., Peck, S.L., Dodson, A.N., Gentry, A.S., Griffin, B., Rolsma, M.D., and Baker, H.J., 2005, Mutation of the GM2 activator protein in a feline model of GM2 gangliosidosis, *Acta Neuropathol. (Berl.)* **110**:443–450.

20. Meurs, K., Sanchez, X., David, R., Bowles, N.E., Towbin, J.A., Reiser, P.J., Kittleson, J.A., Munro, M.J., Dryburgh, K., Boyer, M., Mathur, D., MacDonald, K.A., and Kittleson, M.D., 2005, Identification of a missense mutation in the cardiac myosin binding protein C gene in

a family of Maine Coon cats with hypertrophic cardiomyopathy, *Am. Col. Vet. Intern. Med. Forum* **14**:3587–3593.

21. Muldoon, L.L., Neuwelt, E.A., Pagel, M.A., and Weiss, D.L., 1994, Characterization of the molecular defect in a feline model for type II GM2-gangliosidosis (Sandhoff disease), *Am. J. Pathol.* **144**:1109–1118.

22. Somers, K.L., Royals, M.A., Carstea, E.D., Rafi, M.A., Wenger, D.A., and Thrall, M.A., 2003, Mutation analysis of feline Niemann-Pick C1 disease, *Mol Genet Metab.* **79**:99–103.

23. Winand, N.J., Edwards, M., Pradhan, D., Berian, C.A., and Cooper, B.J., 1994, Deletion of the dystrophin muscle promoter in feline muscular dystrophy, *Neuromuscul. Disord.* **4**:433–445.

24. Yogalingam, G., Litjens, T., Bielicki, J., Crawley, A.C., Muller, V., Anson, D.S., and Hopwood, J.J., 1996, Feline mucopolysaccharidosis type VI. Characterization of recombinant N-acetylgalactosamine 4-sulfatase and identification of a mutation causing the disease, *J. Biol. Chem.* **271**:27259–27265.

25. Yogalingam G, Hopwood J. J., Crawley A., and Anson, D. S., 2004, Mild feline mucopolysaccharidosis type VI. Identification of an N-acetylgalactosamine-4-sulfatase mutation causing instability and increased specific activity, *Hum. Mutat.* **24**:199–207.

Variation in Chicken Gene Structure and Expression Associated with Food-Safety Pathogen Resistance: Integrated Approaches to *Salmonella* Resistance

S.J. Lamont

Abstract The use of genetics to enhance immune response and microbial resistance in poultry is an environmentally sound approach to incorporate into comprehensive health programs. Many research strategies can be used to investigate the relationship of host genetics with immune response and disease resistance. Gene discovery to enhance poultry health and food safety should build upon well-defined genetic populations, cell lines, gene identification, genome maps, comparative genomics, and analysis of gene expression. Because each investigative approach has its own shortfalls, the strongest level of confidence comes from the convergence of evidence from an integrated approach of several independent experimental designs, such as whole-genome scans, candidate gene analyses, and functional genomics studies, all supporting the relationship of a specific gene with a resistance or immunity trait. Defining the causal genes, including genomic location and organization, epistatic and pleiotropic effects, and the encoded protein function, opens the door for genetic selection to improve health and also for enhancement of vaccine efficacy and innate immunity. This chapter reviews the rationale and strategies for uncovering genetic resistance to food-safety pathogens in poultry and summarizes successes in elucidating the genetic control of host resistance to *Salmonella*.

1 Rationale and Strategies for Uncovering Genetic Resistance to Food-Safety Pathogens in Poultry

Comprehensive programs of disease management should include a genetic enhancement approach, along with appropriate use of vaccination and approved therapeutic antibiotics, and strong biosecurity and sanitation programs. Genetic improvements are permanent and cumulative, thus they are a long-term, cost-effective, and environmentally friendly solution to maintaining bird health (Lamont, 1998; Lamont, 2003). Genetic enhancement of immunity can decrease disease and the need for pharmaceutical treatments, thus reducing drug residues in food. Genetic selection

S.J. Lamont
Department of Animal Science, Iowa State University, 2255 Kildee Hall, Ames, IA 50011, USA,
e-mail: sjlamont@iastate.edu

J.P. Gustafson et al. (eds.), *Genomics of Disease*,
© Springer Science+Business Media, LLC, 2008

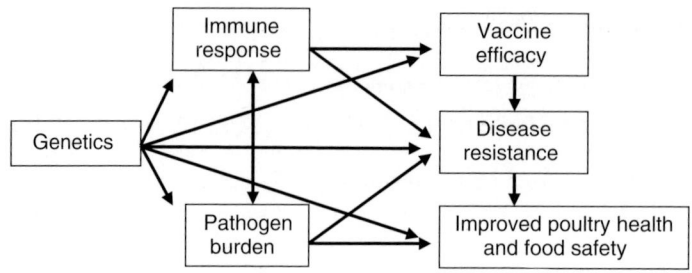

Fig. 1 Use of genetics to improve poultry health and food safety

for resistance to pathogens also often increases the efficacy of vaccine protection (Fig. 1). More virulent pathogens appear to be emerging and re-emerging, and spreading globally in animal production systems. Special challenges in infectious disease control are presented by the concentration of poultry in large production units or, conversely, the inability to eradicate pathogens from free-range environments. Enhancing disease resistance by genetic improvements will be beneficial in a wide range of environments and management situations.

Disease has negative impacts on the economics and welfare of poultry production. Of the potential profit in poultry production, 10–15% is estimated to be lost because of disease (Biggs, 1982). The biological impact of bearing a microbial burden reduces growth and reproductive performance (Klasing and Korver, 1997). This reduced performance can occur even in typical production environments in which microbes are present but are not causing clinical disease symptoms.

Zoonotic pathogens, such as *E. coli*, *Salmonella*, and *Campylobacter*, which cause disease in both animals and humans, present food-safety risks from consumption of contaminated, improperly prepared poultry products. Bacteria with asymptomatic presence in poultry may still be food-borne pathogens that cause disease in humans (Doyle and Erickson, 2006). Therefore, reduction of microbial burden in poultry is an important step in pre-harvest food safety. Consumers and the medical community are increasingly concerned about non-therapeutic use of antibiotics in food animals and the potential of introducing antibiotic-resistant pathogens into the food chain. To meet consumer demands and/or regulations, the future use of antibiotics in animal production is expected to decrease substantially. The identification of biomarkers associated with immune function can be effectively used to enhance host resistance to pathogens and thereby reduce disease and pharmaceutical treatments in the live bird and microbial contamination of poultry products (Lamont, 1998).

Because the immunophysiological networks that control resistance or susceptibility to most diseases are usually complex, resistance is a polygenic trait controlled by many quantitative trait loci (QTL). With disease resistance being, in essence, a measure of the bird's response to perturbations (pathogenic organisms) in its environment, the environmental component of disease phenotypes can be very high.

Heritabilities for overall resistance may be low to moderate, but dividing resistance into its component properties can identify immune phenotypes that each have moderate to high heritabilities that are amenable to genetic selection.

The complexity of disease resistance genetics has led to many different historical approaches being taken to improve health in commercial poultry stock. These have included genetic selection of breeders after pathogenic challenge and assessment of their relatives (sibs or progeny) for resistance traits, selection on positively correlated immune traits such as antibody level to vaccines or other antigens such as sheep red blood cells (Lamont et al., 2003), and indirect selection on protein markers (e.g., B blood group antigen for Marek's disease resistance).

The feasibility of using whole genome scan approaches to identify genomic regions controlling important biological traits in chickens, including traits related to disease resistance, has been amply demonstrated (for reviews, see Hocking, 2005; Abasht et al., 2006). The complete draft sequence of the chicken genome was recently released (Hillier et al., 2004), along with a companion map of 2.8 million single nucleotide polymorphisms (SNP) in the genome (Wong et al., 2004). Knowing the specific points of genetic variation in the chicken genome will accelerate identification of the genes that control health in poultry, by studying associations of the genomic variation with traits of host resistance to disease (Burt, 2005). Recent advances in genome sequencing and high-throughput genotyping technology have enabled genome-wide screening for QTL using high-throughput genotyping technologies.

Molecular markers associated with health traits must be identified in order to apply marker-assisted selection in breeding programs for enhanced health. These markers are two general types: within genes or linked to genes. Using linked markers has the advantages of not needing to know the specific identity or sequence of any gene; the goal is to localize the positions of resistance QTL in the genome. Markers closely linked to the causative variation can be effectively used in breeding programs until the linkage is lost to recombination, and can serve as the foundation to positionally identify the causative gene and mutation. Whole genome scans with microsatellites have been successfully used to identify QTL for disease resistance (e.g., McElroy et al., 2005). The current availability of the dense SNP map in now shifting investigations from medium-density microsatellite-based genome scans to high-density SNP-based studies.

Whole-genome scans can be complemented with studies of specific candidate genes that control resistance traits. These genes may be positional candidates because of their genomic position (mapping to a QTL location revealed through a genome scan), or they may be biological candidate genes because of a reported association with resistance or immunity traits in chickens or other species. Gene structural variation assessed by SNP has successfully been associated with disease resistance traits (e.g., Lamont et al., 2002).

Research approaches that evaluate DNA structural variation are powerful but lack the ability to characterize differences in gene expression, which for many traits may be the major genetic mechanism that modulates biological performance. Studying gene expression on a large scale has recently become a reality for poultry. The complementary characterization of both genome structural variation and gene expression differences is a very effective tool to identify genes affecting important traits and the biological networks involved. Chicken microarrays can be used to assess changes in gene expression that accompany different infection or resistance states (Morgan et al., 2001). The genes that show large expression differences are likely to be involved in some aspect of the relevant pathways, although it is not possible from these studies to determine whether the expression differences are a cause of, or a result of, the phenotypic differences in resistance. The convergence of associations of both structure and expression of the same genes with disease resistance phenotypes provides additional evidence of an important role of the identified genes (Liu et al., 2001).

Proteomics, which is currently in early stages of use in poultry, can help to identify the genes underlying resistance and immunity by profiling the proteins that exist in specific tissues. Molecular genetics also provides other approaches to improve poultry health. Genes that encode immune-system proteins, such as cytokines, can be transferred to expression systems that produce large quantities of pure recombinant protein. These chicken proteins may then be used as natural immune enhancers or vaccine adjuvants. Production of transgenic poultry that are completely resistant to specific pathogens offers possibilities for those diseases with single-gene control (such as viral receptors), but only if there is an increase in the general efficiency of techniques used to produce transgenic animals and public acceptance of transgenic animals for food.

The complicated genetic architecture of disease resistance is shaped by variation in both structure and expression of many genes. Thus, a comprehensive picture of the genetics of poultry health is obtained only by integrating approaches that combine genome-wide and gene-specific studies of structural and expressional variation. Each of the individual research approaches to identify markers has special strengths, as well as limitations, in the type of information that it can yield.

Some genes are expected to control responses that are very specific to the individual pathogen. However, other genes are likely to function at critical control points in host defense mechanisms and may have more general beneficial effects, such as aiding resistance to a whole group of similar pathogens (e.g., intracellular bacteria). Identifying biochemical pathways that have general effects on resistance properties will be important, especially in the cases of pathogens of emerging importance such as *Campylobacter*, or pathogens that frequently mutate and appear in new antigenic combinations, such as avian influenza. The major diseases studied to define host genetic resistance in poultry include Marek's disease, a viral disease (Liu et al., 2001; Morgan et al., 2001; Yonash et al., 1999; McElroy et al., 2005), Salmonellosis, a bacterial disease (Hu et al., 1997; Mariani et al., 2001; Kramer et al., 2003; Liu and Lamont, 2003; Malek et al., 2004), and coccidiosis, a parasitic disease (Lillehoj et al., 1989). As an example illustrating the principles of eludicating the genetic control of complex resistance traits in poultry, this chapter reviews the research to determine the molecular genetic control of host response to *Salmonella*, a food-safety pathogen.

2 Genetic Control of *Salmonella* Resistance in Poultry

Poultry salmonellosis is a food-safety concern and *Salmonella enteritidis* is a major food-borne pathogen, accounting for 82% of food-borne salmonellosis (Kramer et al., 1998). Contaminated eggs are a leading cause of *Salmonella* food poisoning and poultry flocks with high levels of *S. enteritidis* and contaminated manure pose the largest threat of egg contamination (Kramer et al., 1998). Understanding genetic mechanisms controlling bacterial and cellular interactions to decrease bacterial colonization and fecal shedding and manure contamination would significantly decrease this microbial contamination problem.

Some species of *Salmonella* bacteria are highly pathogenic in poultry, while other species cause little response in the host birds, which can then become asymptomatic carriers. Birds with subclinical salmonellosis can transmit the zoonotic bacteria into the human food chain. Gast and Holt (1998) showed that chicks infected with *Salmonella* immediately after hatching can be persistently colonized to maturity, when the bacteria are shed vertically to infect table or hatching eggs, or horizontally to infect other hens. Thus, control of salmonellosis presents special challenges to the poultry industry.

To establish the feasibility of host genetic approaches to help improve resistance to salmonellosis, it was first necessary to establish that there was, indeed, a measure of genetic control over these traits. The heritability of chick mortality after *Salmonella* challenge was estimated at 0.14 and 0.62 for sire and dam components, respectively (Beaumont et al. 1999). Estimated heritability of resistance to cecal carrier state, measured by enrichment culture, in laying hens was 0.20 (Berthelot et al., 1998). Heritabilities of the number of bacteria persisting in internal organs ranged from 0.02 to 0.29 (Girard-Santosuosso et al., 2002). Estimated heritability of antibody response to *Salmonella* ranged widely, from 0.03 to 0.26 (Kaiser et al., 1997; Beaumont et al., 1999). Measuring antibody response is more economical and less laborious than measuring bacterial colonization and because vaccine antibody has a negative genetic correlation with cecal colonization, vaccine antibody is a useful biomarker to improve resistance to colonization (Kaiser et al., 2002). Heritability estimates of various parameters of *Salmonella* response, therefore, indicate that genetic selection to improve resistance to salmonellosis is feasible.

As a polygenic trait, many genes are associated with genetic resistance to *Salmonella* species. Most genes have an individual effect on the disease phenotypic variation that is relatively small, often only 3–5% of the total variation, which is consistent with complex control of the disease by many genes. Factors such as genetic line, population structure, definition of host response (antibody, mortality, systemic, or enteric colonization), and serovar of *Salmonella* may result in some variation in results. However, a generally consistent picture of the existence of genetic control of host resistance to *Salmonella* has emerged.

Comparative genomics strategies have been very effectively used to identify genes controlling salmonellosis resistance in chickens, because response to *Salmonella* was previously studied in detail in the mouse as a model organism. Chicken homologs of major loci controlling natural resistance of mice to infection with *S. typhimurium* were examined as candidate genes. Variation in NRAMP1 (natural resistance-associated macrophage protein 1, now called SLC11A1) and TNC were shown to account for 33% of the differential resistance in *Salmonella*-induced mortality in a backcross population of inbred lines (Hu et al., 1997). The TNC locus is closely linked to LPS, now known as TLR4, which binds lipopolysaccharide – a major component of Gram-negative bacteria membranes. The NRAMP1 association was additionally confirmed for many different specific traits of host response to *Salmonella* in other chicken populations (Beaumont et al., 2003; Kramer et al., 2003; Liu et al., 2003), as was the TLR4 association (Beaumont et al., 2003; Leveque et al., 2003; Malek et al., 2004).

Success in the comparative genomics approach provided a foundation to select additional candidate genes to test. Some genes were positional candidates based upon genomic position, such as *CD28* and *VIL1* in the NRAMP1 region (Girard-Santosuosso et al., 1997). The *CD28* gene was found to be associated with enteric *Salmonella* infection (Malek et al., 2004), and *VIL1* with visceral infection (Girard-Santosuosso et al., 2002).

Involvement in pathways that are hypothesized to be important in host response to *Salmonella* has been a source of information for selection of other candidate genes to study. The product of the *MD2* gene interacts with the TLR4 receptor on the cell surface, and SNPs in *MD2* established to be associated with persistence of *Salmonella* colonization in the cecum (Malek et al., 2004). The *MHC*, because of its crucial role in antigen processing and presentation, has been investigated. In a series of 12 B-complex congenic lines, line differences occurred in morbidity and mortality after *Salmonella* challenge (Cotter et al., 1998). Variation in the *MHC class I* gene was associated with resistance to *Salmonella* colonization in the spleen in an experimental cross (Liu et al. 2002). Genes in apoptotic pathways include *CASP1* and *IAP1*. A CASP1 SNP was associated with *Salmonella* persistence in the spleen and cecum in an experimental cross (Liu and Lamont, 2003) and in the liver and cecum in commercial broilers (Kramer et al., 2003). The *IAP1* gene was associated with spleen (Liu and Lamont, 2003) and cecum bacterial load (Kramer et al., 2003). The genomic region on chromosome 3 that contains a cluster of 13 gallinacin genes, which encode a newly identified family of antimicrobial peptides, was associated with *Salmonella* vaccine antibody response (Hasenstein et al., 2006).

Most candidate gene experiments, because of linkage disequilibrium, cannot exclude the possibility that the causal gene could be a nearby gene rather than the specific one studied. Supporting lines of evidence, such as confirmation in independent populations, with QTL scans, by gene expression data, or from comparative genomics, adds confidence in the detected gene-resistance associations. Gene expression studies in *Salmonella* challenged versus unchallenged, or resistant versus susceptible animals, are currently revealing differential expression in genes that may be active in pathways controlling resistance (Cheeseman et al., 2007; Kaiser et al., 2006; Zhou and Lamont, 2007). Combining data on *Salmonella*-response associations with variation in gene structure and in gene expression, all conducted on the same resource populations in the Lamont lab, has identified some predominant

Table 1 Gene families/pathways associated with resistance to *Salmonella* in poultry, as collectively determined from multiple candidate gene and functional genomic studies in the Iowa Salmonella Response resource population F_1 and/or F_8 generation

Gene family/pathway	Representative genes
Toll-like receptors	MD-2, TLR1, TLR2, TLR4, TLR5
Cytokines	IL2, IL4, IL6, IL8, IL10, IL18, IFNG, K60, MIP1B, RANTES, SOCS3, TGFB2, TGFB3, TGFB4, ah294
Cell-surface antigens	CD3, CD40, MHC2A, MHC2B
Apoptosis	Bcl-x, CASP1, Fas, IAP, TRAIL, TNF-R1
Antimicrobial peptides	GAL2, GAL3, GAL5, GAL6, GAL11, GAL13

families or pathways of genes that are highly likely to be involved in resistance to *Salmonella* in poultry (Table 1).

Use of whole-genome scans have identified QTL regions controlling *Salmonella* resistance. Using a backcross population of resistant and susceptible parental inbred lines (Lines 6_1 and 15I, respectively), Mariani et al. (2001) genotyped birds with extreme bacterial counts in spleen and found a significant linkage between markers on chromosome 5 and the colonization trait (Mariani et al., 2001). Fine-mapping of the chromosomal location of the QTL found a very strong effect in the region near the *CKB* and *DNCH1* genes, accounting for 50% of the parental variation difference. The specific gene associated with resistance is yet positively identified, and the *Salmonella* resistance QTL in this region, SAL1, is considered a novel locus. The distance of almost 50 cM between SAL1 and the *Salmonella* resistance QTL linked to marker ADL0298 (Kaiser and Lamont, 2002) makes them unlikely to represent the same locus, although both influenced the resistance trait of bacterial colonization in the spleen.

Tilquin et al. (2005) performed a genome scan for *Salmonella* resistance QTL using F_2 and BC populations formed from Line N and 6_1, and assessment of cloacal and cecal carrier state. This study confirmed an association of the SAL1 region with enteric carrier state, and new QTL regions were identified at significant or suggestive levels on chromosomes 1, 2, 5, and 16, with some having effects as large as 37.5% of the phenotypic variance. The QTL on chromosome 16 lies in the MHC, although the two inbred lines are considered to have the same MHC haplotype. The QTL on chromosome 5 maps near TGFB3, which was shown in studies of other populations to be associated with spleen bacterial colonization (Kramer et al., 2003) and bacterial level in the cecal contents (Malek and Lamont, 2003).

3 Conclusions

Because disease resistance traits are difficult and expensive to measure, they represent a strong opportunity for application of molecular, genetic marker–assisted selection in commercial breeding programs. Poultry breeders, however, must already account for dozens of important traits in their selection program and the addition of every new criterion will lower the selection intensity on other criteria.

Therefore, the use of new selection tools, such as molecular markers, needs to be carefully balanced so that the health improvements will complement, but not compromise, current successes in poultry breeding.

Multiple approaches can be taken to identify genes associated with important biological traits such as resistance to *Salmonella*. The integration of multiple, complementary approaches helps to resolve the shortfalls inherent in each approach. Genome scans and candidate gene studies can reveal associations with disease resistance, and techniques to study gene expression are also powerful tools in disease genetics research. The newly available chicken genome sequence will accelerate the discovery process. Because many of these resources, however, have only recently come on line, a limited number of specific genes or regions of the genome associated with poultry health have yet been identified.

Genetic selection for improved immunity and resistance traits, which can be accomplished after sufficient genetic markers are defined, can improve poultry health and production efficiency, as well as reducing the potential for the introduction of food-borne pathogens into the human food chain. Improving poultry health by increasing genetic resistance to disease is essential to meet the increasing emphasis of consumers and the industry on animal welfare, food safety, environmental concerns, and efficiency of production.

Acknowledgments I gratefully acknowledge the many colleagues and research group members who have contributed to this discovery program in the genetics of host resistance to *Salmonella*. The work was funded by the Iowa Agriculture and Home Economics Experiment Station, Ames, Iowa; Animal Health, Hatch Act, State of Iowa and Iowa State University Center for Integrated Animal Genomics Funds; Midwest Poultry Research Program, National Research Initiative Grant no. 2004-35205-14234 from the USDA Cooperative State Research, Education, and Extension Service; and Research Grant US-3408-03 from BARD, the Binational Agriculture Research and Development fund.

References

Abasht, B., Dekkers, J.C.M., and Lamont, S.J., 2006, Review of quantitative trait loci identified in the chicken, *Poultry Sci.* **85**:2079–2096.

Beaumont, C., Protais, J., Guillot, J. F., Colin, P., Proux, K., Millet, N., and Pardon, P., 1999, Genetic resistance to mortality of day-old chicks and carrier-state of hens after inoculation with *Salmonella enteritidis*, *Avian Pathol.* **28**:131–135.

Beaumont, C., Protais, J., Pitel, F., Leveque, G., Malo, D., Lantier, F., Plisson-Petit, F., Colin, P., Protais, M., Le Roy, P., Elsen, J. M., Milan, D., Lantier, I., Neau, A., Salvat, G., and Vignal, A., 2003, Effect of two candidate genes on the *Salmonella* carrier state in fowl, *Poultry Sci.* **82**: 721–726.

Berthelot, F., Beaumont, C., Mompart, F., Girard-Santosuosso, O., Pardon, P., and Duchet-Suchaux, M., 1998, Estimated heritability of the resistance to cecal carrier state of *Salmonella enteritidis* in chickens, *Poultry Sci.* **77**:797–801.

Biggs, P.M., 1982, The world of poultry diseases, *Avian Path.* **11**:281–300.

Burt, D.W., 2005, Chicken genome: Current status and future opportunities, *Genome Res.* **15**: 1692–1698.

Cheeseman, J.H., Kaiser, M.G., Ciraci, C., Kaiser, P., and Lamont, S.J., 2007, Breed effect on early cytokine mRNA expression in spleen and cecum of chickens with and without *Salmonella enteritidis* infection, *Dev. Comp. Immunol.* **31**:52–60.

Cotter, P.F., Taylor, R.L., Jr., and Abplanalp, H., 1998, B-complex associated immunity to *Salmonella enteritidis* challenge in congenic chickens, *Poultry Sci.* **77**:1846–1851.

Doyle, M., and Erickson, M.C., 2006, Reducing the carriage of foodborne pathogens in livestock and poultry, *Poultry Sci.* **85**:960–973.

Gast, R.K. and Holt, P.S., 1998, Persistence of *Salmonella enteritidis* from one day of age until maturity in experimentally infected layer chickens, *Poultry Sci.* **77**:1759–1762.

Girard-Santosuosso, O., Bumstead, N., Lantier, I., Protais, J., Colin, P., Guillot, J.F., Beaumont, C., Malo, D., and Lantier, F., 1997, Partial conservation of the mammalian NRAMP1 syntenic group on chicken chromosome 7, *Mamm. Genome* **8**:614–616.

Girard-Santosuosso, O., Lantier, F., Lantier, I., Bumstead, N., Elsen, J. M., and Beaumont, C., 2002, Heritability of susceptibility to *Salmonella enteritidis* infection in fowls and test of the role of the chromosome carrying the NRAMP1 gene, *Genet. Sel. Evol.* **34**:211–219.

Hasenstein, J.R., Zhang, G., and Lamont, S.J., 2006, Analyses of five gallinacin genes and the *Salmonella enterica* serovar Enteritidis response in poultry, *Infect. Immun.* **74**:3375–3380.

Hillier, L.W., Miller, W., Birney, E., Warren, W., Hardison, R.C., et al., 2004, Sequence and comparative analysis of the chicken genome provide unique perspectives on vertebrate evolution, *Nature* **432**:695–716.

Hocking, P.M., 2005, Review on QTL mapping results in chickens, *World's Poult. Sci. J.* **61**: 215–226.

Hu, J., Bumstead, N., Barrow, P., Sebastiani, G., Olien, L., Morgan, K., and Malo, D., 1997, Resistance to salmonellosis in the chicken is linked to NRAMP1 and TNC, *Genome Res.* **7**:693–704.

Kaiser, M.G., Cheeseman, J.H., Kaiser, P., and Lamont, S.J 2006, Cytokine expression in chicken peripheral blood mononuclear cells after in vitro exposure to *Salmonella enterica* serovar Enteritidis, *Poultry Sci.* **85**:1907–1911.

Kaiser, M.G., Lakshmanan, N., Wing, T., and Lamont, S.J., 2002, *Salmonella enterica* serovar enteritidis burden in broiler breeder chicks genetically associated with vaccine antibody response, *Avian Dis.* **46**:25–31.

Kaiser, M.G., and Lamont, S.J., 2002, Microsatellites linked to *Salmonella enterica* serovar Enteritidis burden in spleen and cecal content of young F1 broiler-cross chicks, *Poultry Sci.* **81**: 657–663.

Klasing, K.C., and Korver, D.R., 1997, Leukocytic cytokines regulate growth rate and composition following activation of the immune system, *J. Anim. Sci.* **75**:58–67.

Kaiser, M.G., Wing, T., Cahaner, A., and Lamont, S.J., 1997, Aviagen, 12th International Symposium on Current Problems in Avian Genetics, Prague, Czech Republic.

Kramer, J., Malek, M., and Lamont, S.J., 2003, Association of twelve candidate gene polymorphisms and response to challenge with *Salmonella enteritidis* in poultry, *Anim. Genet.* **34**: 339–348.

Kramer, T.T., Reinke, C.R., and James, M., 1998, Reduction of fecal shedding and egg contamination of *Salmonella enteritidis* by increasing the number of heterophil adaptations, *Avian Dis.* **42**:585–588.

Lamont, S.J., 1998, Impact of genetics on disease resistance, *Poult Sci.* **77**:1111–1118.

Lamont, S.J., Kaiser, M.G., and Liu, W., 2002, Candidate genes for resistance to *Salmonella enteritidis* colonization in chickens as detected in a novel genetic cross, *Vet. Immunol. Immunop.* **87**:423–428.

Lamont, S.J., Pinard-van der Laan, M.-H., Cahaner, A., van der Poel, J.J., and Parmentier, H.K., 2003, Selection for disease resistance: direct selection on the immune response, in W.M. Muir and S.E. Aggrey, eds., Poultry Genetics, Breeding and Biotechnology, CAB International, Oxon UK, pp. 399–418.

Leveque, G., Forgetta, V., Morroll, S., Smith, A.L., Bumstead, N., Barrow, P., Loredo-Osti, J.C., Morgan, K., and Malo, D., 2003, Allelic variation in TLR4 is linked to susceptibility to *Salmonella enterica* serovar Typhimurium infection in chickens, *Infect. Immun.* **71**:1116–1124.

Lillehoj, H.S., Ruff, M.D., Bacon, L.D., Lamont, S.J., and Jeffers, T.K., 1989, Genetic control of immunity to *Eimeria tenella*. Interaction of MHC genes and non-MHC linked genes influences levels of disease susceptibility in chickens, *Vet. Immunol. Immunop.* **20**:135–148.

Liu, H.C., Cheng, H.H., Tirunagaru, V., Sofer, L., and Burnside, J., 2001, A strategy to identify positional candidate genes conferring Marek's disease resistance by integrating DNA microarrays and genetic mapping, *Anim. Genet.* **32**:351–359.

Liu, W., Kaiser, M.G., and Lamont, S.J., 2003, Natural resistance-associated macrophage protein 1 gene polymorphisms and response to vaccine against or challenge with *Salmonella enteritidis* in young chicks, *Poultry Sci.* **82**:259–266.

Liu, W., and Lamont, S.J., 2003, Candidate gene approach: potential association of caspase-1, inhibitor of apoptosis protein-1, and prosaposin gene polymorphisms with response to *Salmonella enteritidis* challenge or vaccination in young chicks, *Anim. Biotechnol.* **14**:61–76.

Liu, W., Miller, M. M., and Lamont, S.J., 2002, Association of MHC class I and class II gene polymorphisms with vaccine or challenge response to *Salmonella enteritidis* in young chicks, *Immunogenetics* **54**:582–590.

Malek, M., Hasenstein, J.R., and Lamont, S.J., 2004, Analysis of chicken TLR4, CD28, MIF, MD-2, and LITAF genes in a *Salmonella enteritidis* resource population, *Poultry Sci.* **83**: 544–549.

Malek, M., and Lamont, S.J., 2003, Association of INOS, TRAIL, TGF-beta2, TGF-beta3, and IgL genes with response to *Salmonella enteritidis* in poultry, *Genet. Sel. Evol.* **35** (Suppl 1): S99–S111.

Mariani, P., Barrow, P.A., Cheng, H.H., Groenen, M.M., Negrini, R., and Bumstead, N., 2001, Localization to chicken chromosome 5 of a novel locus determining salmonellosis resistance, *Immunogenetics* **53**:786–791.

McElroy, J.P., Dekkers, J.C.M., Fulton, J.E., O'Sullivan, N.P., Soller, M., Lipkin, E., Zhang, W., Koehler, K.J., Lamont, S.J., and Cheng, H.H., 2005, Microsatellite markers associated with resistance to Marek's disease in commercial layer chickens, *Poultry Sci.* **84**:1678–1688.

Morgan, R.W., Sofer, L., Anderson, A.S., Bernberg, E.L., Cui, J., and Burnside, J., 2001, Induction of host gene expression following infection of chicken embryo fibroblasts with oncogenic Marek's disease virus, *J. Virol.* **75**:533–539.

Tilquin, P., Barrow, P.A., Marly, J., Pitel, F., Plisson-Petit, F., Velge, P., Vignal, A., Baret, P.V., Bumstead, N., and Beaumont, C., 2005, A genome scan for quantitative trait loci affecting the *Salmonella* carrier-state in the chicken, *Genet. Sel. Evol.* **37**:539–561.

Wong, G.K., Liu, B., Wang, J., Zhang, Y., Yang, X., et al., 2004, A genetic variation map for chicken with 2.8 million single-nucleotide polymorphisms, *Nature* **432**:717–722.

Yonash, N., Bacon, L.D., Witter, R.L., and Cheng, H.H., 1999, High resolution mapping and identification of new quantitative trait loci (QTL) affecting susceptibility to Marek's disease, *Anim. Genet.* **30**:126–135.

Zhou, H., and Lamont, S.J., 2007, Global gene expression profile after *Salmonella enterica* Serovar enteritidis challenge in two F8 advanced intercross chicken lines. Cytogenet, *Genome Res.* **117**:131–138.

Functional Genomics and Bioinformatics of the *Phytophthora sojae* Soybean Interaction

Brett M. Tyler, Rays H.Y. Jiang, Lecong Zhou, Sucheta Tripathy,
Daolong Dou, Trudy Torto-Alalibo, Hua Li, Yongcai Mao, Bing Liu,
Miguel Vega-Sanchez, Santiago X. Mideros, Regina Hanlon, Brian M. Smith,
Konstantinos Krampis, Keying Ye, Steven St. Martin, Anne E. Dorrance,
Ina Hoeschele, and M.A. Saghai Maroof

Abstract Oomycete plant pathogens such as *Phytophthora* species and downy mildews cause destructive diseases in an enormous variety of crop plant species as well as forests and native ecosystems. These pathogens are most closely related to algae in the kingdom Stramenopiles, and hence have evolved plant pathogenicity independently of other plant pathogens such as fungi. We have used bioinformatic analysis of genome sequences and EST collections, together with functional genomics to identify plant and pathogen genes that may be key players in the interaction between the soybean pathogen *Phytophthora sojae* and its host. In *P. sojae*, we have identified many rapidly diversifying gene families that encode potential pathogenicity factors including protein toxins, and a class of proteins (avirulence or effector proteins) that appear to have the ability to penetrate plant cells. Transcriptomic analysis of quantitative or multigenic resistance against *P. sojae* in soybean has revealed that there are widespread adjustments in host gene expression in response to infection, and that some responses are unique to particular resistant cultivars. These observations lay the foundation for dissecting the interplay between pathogen and host genes during infection at a whole-genome level.

1 Introduction

Plant pathogens from the genus *Phytophthora* cause destructive diseases in an enormous variety of crop plant species as well as forests and native ecosystems (Erwin and Ribiero, 1996). The soybean pathogen *Phytophthora sojae* is also causing serious losses to the United States soybean crop on the order of US$ 100–200 million annually (Grau et al., 2004). The newly emerged *Phytophthora* species, *Phytophthora ramorum*, is attacking trees and shrubs of coastal oak forests

B.M. Tyler
Virginia Bioinformatics Institute, Virginia Polytechnic Institute and State University, Blacksburg, VA 24061, USA

J.P. Gustafson et al. (eds.), *Genomics of Disease*,
© Springer Science+Business Media, LLC, 2008

in California, including the keystone live oak species (sudden oak death disease) (Rizzo et al., 2002). Taxonomically, *Phytophthora* pathogens are oomycetes, which are organisms that resemble fungi but belong to a kingdom of life called Stramenopiles that are most closely related to algae such as kelp and diatoms. Hence conventional fungal control measures often fail against these pathogens. Oomycetes include many other destructive plant pathogens in addition to *Phytophthora*, in particular the downy mildews and more than 100 species of the genus *Pythium*. The downy mildew of *Arabidopsis thaliana*, caused by *Hyaloperonospora parasitica*, is an important model system for understanding plant responses to infection (Slusarenko and Schlaich, 2003). *P. sojae* and the potato and tomato pathogen *Phytophthora infestans* have been used for many basic studies of *Phytophthora* because these species are easy to genetically manipulate (Judelson, 1997; Tyler, 2007). Protection of soybean against *P. sojae* infection has traditionally relied on major resistance genes (R genes) (Schmitthenner, 1985). At least 14 R genes that protect against *P. sojae* infection (*Rps* genes) have been identified in soybean (Burnham et al., 2003b). At least 12 avirulence genes have been identified genetically in *P. sojae* that interact in a gene-for-gene manner with *Rps* genes (Gijzen et al., 1996; May et al., 2002; Tyler et al., 1995; Tyler, 2002; Whisson et al., 1994, 1995). One avirulence gene, *Avr1b-1*, has been cloned and extensively characterized (Shan et al., 2004). Cloning of *Avr1a* (MacGregor et al., 2002) and *Avr4b* (Whisson et al., 2004) is also close to completion.

Although major gene resistance has historically been effective against *P. sojae*, changes in pathogen populations over the last 10–15 years have progressively eroded the effectiveness of these genes (Dorrance et al., 2003b; Schmitthenner, 1985; Schmitthenner et al., 1994). In particular, sexual reproduction appears to have greatly increased the complexity of the *P. sojae* population in the US and Canada, resulting in genotypes of the pathogen able to overcome most combinations of *Rps* genes (Dorrance et al., 2003b; Fšrster et al., 1994; Schmitthenner et al., 1994). As a result, there has been increased interest in quantitative resistance against *P. sojae* (also called partial resistance, multigenic resistance, field resistance, rate-reducing resistance, general resistance, or tolerance) (Dorrance and Schmitthenner, 2000; Dorrance et al., 2003a; Tooley and Grau, 1982, 1984). Quantitative resistance has also been the main focus of breeding efforts against *P. infestans* in tomato and potato (Colon et al., 1995), where R gene resistance has been much more rapidly overcome (Fry and Goodwin, 1997). This form of resistance is more durable against changes in pathogen populations because it relies on small contributions from multiple genes and does not depend on the presence of specific genes in the pathogen for effectiveness. The principal mechanism of quantitative resistance in the soybean lines characterized to date is the ability to reduce the rate of lesion expansion following infection (Mideros et al., 2007; Tooley and Grau, 1982, 1984). Quantitative resistance in soybean is highly heritable, but the multigenic nature of the resistance makes it difficult to introduce by conventional breeding. Although there has been some molecular characterization of resistant lines (Burnham et al., 2003a; Vega-Sanchez et al., 2005), the multigenic nature of the resistance as well as the small contributions of each gene also impede the development of a better understanding of the molecular mechanisms of quantitative resistance.

2 Sequencing of Oomycete Genomes

The complete genome sequence of an organism is an excellent starting point for identifying genes involved in pathogenicity or any other process of interest. The genome sequence can also aid substantially in developing genetic tools for detecting and tracking the pathogen. The draft genome sequences of *P. sojae* and *P. ramorum* have been completed (Tyler et al., 2006). The 95-Mb genome of *P. sojae* was sequenced to a depth of ×9 while the 65-Mb genome of *P. ramorum* was sequenced to a depth of ×7, both by whole-genome shotgun sequencing. Automated gene calling software has identified 19,027 predicted genes (called gene models) in the genome of *P. sojae* and 15,743 in the genome of *P. ramorum*. Analysis and visualization of gene predictions and annotations of the *P. sojae* and *P. ramorum* genomes are available at the VBI Microbial Database (VMD) (phytophthora.vbi.vt.edu) and the JGI Genome Portals (www.jgi.doe.gov/genomes). VMD stores not only genome sequence data, but also Expressed Sequence Tag (EST) and microarray data and includes a Genome Community Annotation Tool (GCAT) (Tripathy et al., 2006). Genome sequencing of four other ošmycetes, *P. infestans*, *Phytophthora capsici*, *Hyaloperonospora parasitica*, and *Pythium ultimum*, is also under way (Govers and Gijzen 2006).

One of the principal interests in the genomes is to identify genes that potentially contribute to the ability of *P. sojae* and *P. ramorum* to infect their plant hosts. The availability of the two genome sequences for comparison enables the identification of rapidly evolving genes. Such genes are candidates for being involved in infection because the co-evolutionary battle between the infection mechanism of the pathogens and the defense systems of the plants is expected to accelerate the evolution of these genes. The different host ranges of the two pathogens also predict extensive variation in the genes that control host specificity. *P. sojae* is principally confined to soybean while *P. ramorum* infects a wide range of woody trees and shrubs.

Of the predicted genes in *P. sojae* and *P. ramorum*, 9,768 are similar enough in sequence between the two species that they are predicted to have the same function. These genes, often called orthologs, are arranged in very similar orders along the chromosomes (i.e., exhibit synteny) (Tyler et al., 2006; Jiang et al., 2006). However, a large proportion of the predicted genes (49% in *P. sojae*; 38% in *P. ramorum*) have diverged so much that they have no matching gene in the other species. Of these divergent genes, 1,755 genes are unique to *P. sojae* and 624 are unique to *P. ramorum*. The remainder of the predicted genes represent gene families that are present in both species but whose members do not match one another precisely. Proteins that are secreted by the pathogen during infection and are also divergent between *P. sojae* and *P. ramorum* are excellent candidates for proteins that function to promote the infection process. An example is the explosive expansion and divergence of one secreted protein family, NPP1, which encodes a phytotoxin (Fellbrich et al., 2002; Qutob et al., 2002; Tyler et al., 2006). Although the genomes of some fungal and bacterial species contain up to four *NPP1* genes, *P. sojae* and *P. ramorum* have 29 and 40 copies, respectively, and *H. parasitica* also has multiple *NPP1* genes.

Comparisons of the mechanisms of pathogenicity in ošmycetes to mechanisms in other plant pathogens such as bacteria fungi and nematodes is also of considerable interest (Latijnhouwers et al., 2003). The pathogens in these diverse kingdoms of life have presumably acquired the ability to attack plants by a process of convergent evolution. Identifying common mechanisms among these pathogens may advance our understanding of the fundamental mechanisms required by all plant pathogens, and from there we may be able to better understand what are the essential components of the plant defense mechanisms. To facilitate the identification of common mechanisms of pathogenesis, the Plant Associated Microbe Gene Ontology Project (pamgo.vbi.vt.edu) was initiated in 2005 to add standardized terms to The Gene Ontology (The Gene Ontology Consortium 2000) for annotating the contributions of genes to interactions between microbes and hosts. This multi-institution collaboration, coordinated by B. Tyler, recently added 472 new terms to the Gene Ontology and is currently engaged in using the terms to annotate genomes of nine pathogens from diverse kingdoms including *P. sojae* and *P. ramorum*.

3 Effector Genes in Oomycete Genomes

Effector proteins are produced by pathogens to modify the physiology of the host cells to facilitate infection. Effector proteins have been characterized extensively in bacteria (Chang et al., 2004). Many bacterial effector proteins were initially discovered as the products of avirulence genes, which are genes whose products are detected by receptors encoded by plant major resistance genes (R genes). Almost all bacterial effector proteins are introduced into the cytoplasm of the host plant cell by means of a specialized type III secretion system (Chang et al., 2004), where many of them interfere with the signal transduction pathways responsible for triggering plants' defense responses to infection (see other chapters in this book).

R genes also protect plants against oomycetes (Tyler, 2002). Several oomycete avirulence genes have been identified that interact genetically with these R genes (Tyler, 2002), including 10 in *P. sojae*. In addition to the *P. sojae* avirulence gene *Avr1b*-1 (Shan et al., 2004) mentioned above, a number of other oomycete avirulence genes have been cloned including *Atr1* (Rehmany et al., 2005) and *Atr13* (Allen et al., 2004) from *H. parasitica* and *Avr3a* from *P. infestans* (Armstrong et al., 2005). All four of these cloned oomycete avirulence genes encodes a small secreted protein and in each case the cognate resistance gene encodes an intracellular NBS-LRR protein (Allen et al., 2004; Armstrong et al., 2005; Bhattacharyya et al., 2005; Rehmany et al., 2005; Shan et al., 2004). Particle bombardment experiments with *Avr1b*-1 (N. Bruce, S. Kale, B. Tyler, unpublished), *Avr3a*

(Armstrong et al., 2005), *Atr13* (Allen et al., 2004) and *Atr1* (Rehmany et al., 2005) have confirmed that the pathogen proteins are recognized inside the plant cell. By inference, therefore, there must be a mechanism for those proteins to enter the plant cells. However, the nature of that mechanism is presently unknown. Since many bacterial avirulence genes encode effector proteins that enter the cytoplasm of plant cells via a type III secretion pathway to increase the susceptibility of the plant tissue to infection (Chang et al., 2004), there is a strong likelihood that oomycete avirulence gene products also act to increase susceptibility of the host tissue to infection. Consistent with this hypothesis, transient expression of the *P. infestans Avr3a* gene in *Nicotiana benthamiana* suppressed the hypersensitive defense response triggered by the ubiquitous *Phytophthora* protein elicitin (Bos et al., 2006). *Avr1b*-1 also encodes an effector protein that contributes positively to virulence, since *P. sojae* transformants that overexpress the *Avr1b-1* gene have increased virulence on some soybean cultivars (D. Dou and B. Tyler, unpublished).

The sequencing of the *P. sojae* and *P. ramorum* genomes revealed a very large, very diverse superfamily of genes with similarity to the four cloned ošmycete avirulence genes. There are more than 350 members of this superfamily in each of the *P. sojae* and *P. ramorum* genomes (Tyler et al., 2006) (Fig. 1), and large numbers of these genes can also be detected in the draft genome sequences of *P. infestans*, *P. capsici*, and *H. parasitica*. The members of the superfamily share two conserved motifs, RXLR and dEER, a short distance from the end of the secretory leader (Birch et al., 2006; Rehmany et al., 2005; Tyler et al., 2006). Since the four characterized oomycete avirulence proteins are inferred to have the ability to enter the plant cell, it was hypothesized that the RXLR motif, with or without the dEER motif, is involved in the ability of these proteins to enter the plant cell (Birch et al., 2006; Rehmany et al., 2005; Tyler et al., 2006). In support of this hypothesis, mutations in the RXLR motif of *Avr1b*-1 block the ability of the protein to interact with the plant when the protein is secreted into the extracellular space, but not when the protein is synthesized inside the plant cell (D. Dou, S. Kale and B. Tyler, unpublished). The mechanism by which the RXLR motif is utilized to transport oomycete proteins into plant cells is currently not known.

A motif similar to the RXLR motif is found in proteins secreted by malaria parasites (*Plasmodium* species) that have the ability to cross the membrane enclosing the parasite (the parasitiphorous vacuolar membrane) into the lumen of the red blood cell. The *Plasmodium* motif, called Pexel (Marti et al., 2004) or VTF (Hiller et al., 2004), is also located near the secretory leader (Fig. 2). The similarity between the RXLR and the Pexel/VTF motifs suggests that the two pathogens use a similar mechanism to introduce virulence proteins into the cytoplasm of their host cells (Fig. 2). Indeed, Bhattacharjee et al. (2006) demonstrated that RXLR motifs from *P. infestans* effector proteins could replace the VTF motif in transporting proteins across the parasitiphorous vacuolar membrane of the red blood cell.

The physical and functional resemblance between the *Plasmodium* and the oomycete RXLR motifs is fascinating because the two groups of pathogens belong to different kingdoms of life (Alveolata and Stramenopila respectively), while the same is also true for their host species (animalia and planta respectively)

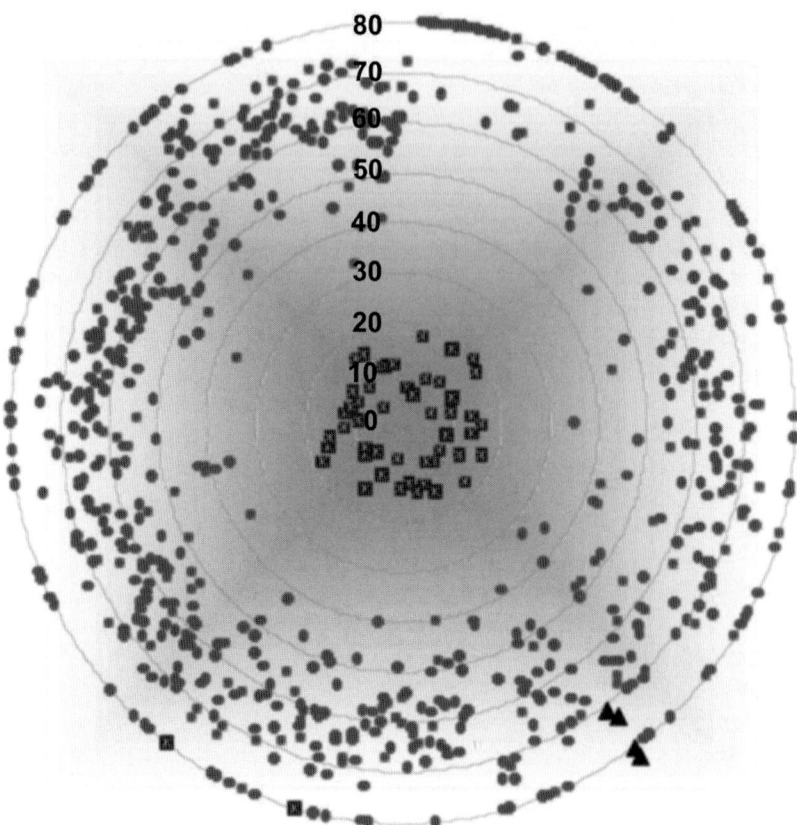

Fig. 1 Diversity of the effector gene families in *P. sojae* and *P. ramorum*. The percentage amino acid sequence divergence of each predicted RXLR protein in *P. sojae* and *P. ramorum* (Tyler et al., 2006) from its most similar homolog in the other species is shown. *Filled circles* represent RXLR genes. *Triangles* represent the small sub-family that contains *P. sojae* Avr1b-1 and *P. infestans* Avr3a. *Crosses* represent a set of randomly chosen *P. infestans* genes that have been experimentally characterized. The *outer rim* represents 80% sequence divergence (i.e., 20% identify) while the *center* represents 0% divergence

(Baldauf et al., 2000). This dispersed evolutionary distribution suggests that the host molecules targeted by the motif are very ancient. The target appears to be widely distributed within the land plants since the hosts, known to be targeted by these proteins, include legumes, crucifers, and solanaceae. On the pathogen side, the motif might have an ancient common origin or may have arisen by convergent evolution. If the host target molecule is indeed so ancient and widely distributed, the question arises as to its natural function within the hosts since it is surely not present simply to benefit attacking pathogens. The function must be sufficiently important that selection pressure from pathogens utilizing the target has not resulted in its loss from the host genomes. One possible function for plant proteins that transit the plasma membrane could be in signaling. Plant and animal genomes do not contain readily

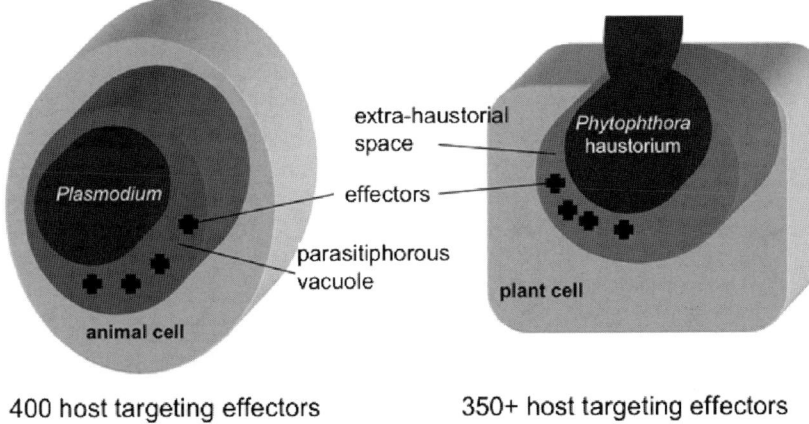

400 host targeting effectors   350+ host targeting effectors

Fig. 2 Comparison of effector release between *Plasmodium* and *Phytophthora*

identifiable RXLR proteins, but it is possible that the putative signaling proteins contain some currently unrecognized variant of the motif.

4 Counter-Play of Plant and Pathogen Genes During *Phytophthora* Infection of Soybean

As described above, the evolutionary "arms race" between plants and oomycete pathogens has resulted in large numbers of rapidly evolving pathogen genes with the potential to contribute to virulence. The same process also results in large numbers of plant genes that contribute to resistance to pathogens such as oomycete. Multigenic disease resistance (also called partial or quantitative resistance) has proven valuable to crop breeders who have found that improving this form of resistance gives much longer protection to crops than single resistance genes, which are quickly overcome by new strains of pathogens. However, this kind of resistance is much harder to improve by conventional breeding because multigenic resistance is the result of many genes making small contributions (Young, 1996). It is also much harder to study the complex molecular mechanisms by which the genes act.

In this section, we summarize progress in a project that is focused on characterizing mechanisms of quantitative resistance in soybean against *P. sojae*, using a combination of transcriptional profiling and quantitative trait locus (QTL) mapping. These technologies are well suited to dissecting a phenotype with a complex genetic basis. The goals of the project are to characterize gene expression differences among a selection of soybean cultivars with different levels of quantitative resistance, and then to use genetic segregation in a large (300 line) recombinant inbred population to dissect the genetic regulatory networks that are associated with resistance.

An initial analysis of the transcriptional profiles of eight cultivars with varying levels of quantitative resistance was recently completed. The profiles were determined in both mock and *P. sojae*-inoculated roots, 3 or 5 days after inoculation. The transcriptional profiles were measured using an Affymetrix GeneChip that contains probes for 37,590 soybean genes and 15,800 *P. sojae* genes. The experimental design, summarized in Fig. 3, included the pooling of samples from 60 plants for each measurement, and four overall replications of the experiment. The data were normalized using GC-RMA (Wu et al. 2004), and analyzed by linear mixed model analysis. Genes showing significant variation in response to each modeled factor were identified using the two-step false discovery rate (TST-FDR) estimation (Benjamini et al., 2006).

Table 1 shows the number of genes that show significant variation in response to the various factors, such as cultivar differences, the presence of the pathogen and the time of sampling, as well as interactions among those factors. A major finding of this analysis is the very large numbers of genes showing statistically significant variation in each category. For example, 97% of all the genes with detectable mRNAs showed significant variation among the eight cultivars, irrespective of the infection status or

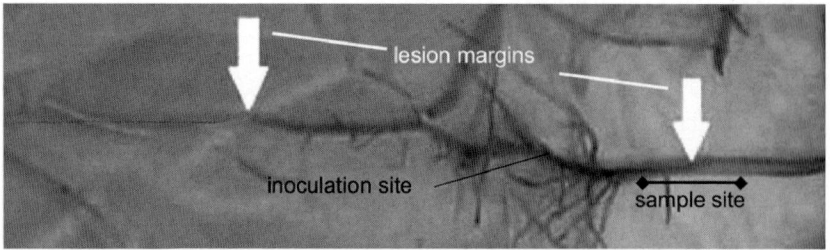

Four experimental replicates (128 GeneChips)			
	Sampling Time	Mock inoculation	*P. sojae* inoculation
Eight cultivars 4 resistant	3 days post inoculation	RNA extracted from 2 sets of 30 plants each and pooled	
2 moderate 2 susceptible	5 days post inoculation		

Fig. 3 Experimental design for expression profiling of infection responses of soybean cultivars differing in quantitative resistance. Mock or infected tissue samples were collected 3 or 5 days after inoculation from a region spanning 7.5 mm each side of the upper edge of the visible lesion. RNA was extracted from powdered frozen tissue from pools of 30 seedlings using Qiagen RNeasy Mini Plant Kits and submitted to the Virginia Bioinformatics Institute Core Laboratory Facility for Affymetrix GeneChip hybridization. Soybean seedlings were inoculated using the slant board method (McBlain et al., 1991; Olah and Schmitthenner, 1985) after growth in vermiculite in the greenhouse for 7 days. A 5 mm scrape wound on the root 2 cm below the stem was inoculated with 0.5 ml of macerated lima bean agar which was either sterile (mock inoculation) or infested with *P. sojae*. Four resistant cultivars (V71-370, Conrad, General, and Athow), two moderately resistant cultivars (Williams and PI), and two susceptible cultivars (Sloan and OX20-8) were tested

Table 1 Linear mixed-model analysis (LMMA) of soybean gene expression changes in the eight cultivar experiment. LMMA model: y = Cul + Inf + Time + Cul × Inf + Cul × Time + Inf × Time + Cul × Inf × Time + Exp + Exp × Cul + Exp × Inf + Error, where Cul (cultivar), Inf (infection/mock), and Time were fixed factors while Exp (experimental replication) was random. GC-RMA background subtraction and quantile normalization were performed prior to LMMA. LMMA was carried out using Proc Mixed of SAS v9. *P*-values for each of the three fixed factors and their interactions were computed based on normality assumption and adjusted for multiple testing using TST-FDR

LMMA category	Biological interpretation	Number of soybean genes
Genes with significant	Genes which respond to	
cultivar effect	differences across all conditions	29102 (97%)
infection effect	common to all cultivars and times	24983 (83%)
time effect	irrespective of cultivar or infection	18704 (62%)
cultivar × infection interaction	cultivar-sensitive infection effects	23446 (78%)
cultivar × time interaction	time effects vary among cultivars	4884 (16%)
infection × time interaction	time-specific infection effects	12827 (43%)
cultivar × infection × time interaction	infection effects specific to cultivars AND time	3178 (11%)
All detectable mRNAs		29953 (100%)
All genes on chip		37590

time of sampling. In the category of most interest for this study, 78% of detectable genes showed significant variation as a result of cultivar × infection interaction, i.e., showed infection responses that were significantly affected by which cultivar was assayed. The reason that such a large percentage of the soybean transcriptome is showing significant variation is that the experimental design enables low levels of transcriptional variations to be detected efficiently with statistical significance. For example, 25% of genes with significant variation as a result of cultivar × infection interaction show no more than 40% variation in any specific contrast between two cultivars. By comparison, 30% of the genes showed a maximum contrast of more than 2-fold. Thus the sensitivity afforded by the experimental design enables us to detect a large number of small variations in gene expression levels. Our biological interpretation of these large numbers of small changes is that they reflect a large number of small adjustments that occur in the transcriptional program of the sampled tissues as a result of the treatments, in concert with the more major responses that may occur.

Acknowledgments We thank Felipe Arredondo, Xia Wang, Daryoush Nabati, Scott McIntyre, and Sue Ann Berry for technical assistance, Clive Evans of the Virginia Bioinformatics Institute Core Laboratory Facility for Affymetrix GeneChip hybridizations, and Carol Volker for assistance with manuscript preparation. This work was supported by grants from the National Research Initiative of the United States Department of Agriculture (USDA) Cooperative State

Research, Education and Extension Service, numbers 2001-35319-14251, 2002-35600-12747, 2004-35600-15055, and 2005-35600-16370, from the US National Science Foundation, numbers MCB-0242131, DBI-0211863, EF-0412213 and EF-0523736, and by funds from the US Department of Energy Joint Genome Institute and the Virginia Bioinformatics Institute.

References

Allen, R.L., Bittner-Eddy, P.D., Grenville-Briggs, L.J., Meitz, J.C., Rehmany, A.P., Rose, L.E., and Beynon, J.L., 2004, Host-parasite coevolutionary conflict between *Arabidopsis* and downy mildew, *Science* **306**:1957–1960.

Armstrong, M.R., Whisson, S.C., Pritchard, L., Bos, J.I., Venter, E., Avrova, A.O., Rehmany, A.P., Bohme, U., Brooks, K., Cherevach, I., Hamlin, N., White, B., Fraser, A., Lord, A., Quail, M.A., Churcher, C., Hall, N., Berriman, M., Huang, S., Kamoun, S., Beynon, J.L., and Birch, P.R., 2005, An ancestral oomycete locus contains late blight avirulence gene *Avr3a*, encoding a protein that is recognized in the host cytoplasm, *Proc. Natl. Acad. Sci. USA* **102**:7766–7771.

Baldauf, S.L., Roger, A.J., Wenk-Siefert, I., and Doolittle, W.F., 2000, A kingdom-level phylogeny of eukaryotes based on combined protein data, *Science* **290**:972–977.

Benjamini, Y., Krieger, A.M., and Yekutieli, D., 2006, Adaptive linear step-up procedures that control the false discovery rate, *Biometrika* **93**:491–507.

Bhattacharjee, S., Hiller, N.L., Liolios, K., Win, J., Kanneganti, T.D., Young, C., Kamoun, S., and Haldar, K., 2006, The malarial host-targeting signal is conserved in the Irish potato famine pathogen, *PLoS Pathog.* **2**:e50.

Bhattacharyya, M.K., Narayanan, N.N., Gao, H., Santra, D.K., Salimath, S.S., Kasuga, T., Liu, Y., Espinosa, B., Ellison, L., Marek, L., Shoemaker, R., Gijzen, M., and Buzzell, R.I., 2005, Identification of a large cluster of coiled coil-nucleotide binding site—leucine rich repeat-type genes from the Rps1 region containing *Phytophthora* resistance genes in soybean, *Theor. Appl. Genet.* **111**:75–86.

Birch, P.R., Rehmany, A.P., Pritchard, L., Kamoun, S., and Beynon, J.L., 2006, Trafficking arms: oomycete effectors enter host plant cells, *Trends Microbiol.* **14**:8–11.

Bos, J.I.B., Kanneganti, T.-D., Young, C., Cakir, C., Huitema, E., Win, J., Armstrong, M., Birch, P.R.J., and Kamoun, S., 2006, The C-terminal half of *Phytophthora infestans* RXLR effector AVR3a is sufficient to trigger R3a-mediated hypersensitivity and suppress INF1induced cell death in *Nicotiana benthamiana*, *Plant J.* **48**:165–176.

Burnham, K.D., Dorrance, A.E., VanToai, T.T., and St. Martin, S.K., 2003a, Quantitative trait loci for partial resistance to *Phytophthora sojae* in soybean, *Crop Sci.* **43**:1610–1617.

Burnham, K.D., Dorrance, A.E., Francis, D.M., Fioritto, R.J., and St-Martin, S.K., 2003b, Rps8, A New Locus in Soybean for Resistance to *Phytophthora sojae*, *Crop Sci.* **43**:101–105.

Chang, J.H., Goel, A.K., Grant, S.R., and Dangl, J.L. 2004. Wake of the flood: ascribing functions to the wave of type III effector proteins of phytopathogenic bacteria, *Curr. Opin. Microbiol.* **7**:11–18.

Colon, L.T., Budding, D.J., Keizer, L.C.P., and Pieters, M.M.J., 1995, Components of resistance to late blight (*Phytophthora infestans*) in eight South American *Solanum* species, *Eur. J. Plant Pathol.* **101**:441–456.

Dorrance, A.E., and Schmitthenner, A.F., 2000, New sources of resistance to *Phytophthora sojae* in the soybean plant introductions, *Plant Dis.* **84**:1303–1308.

Dorrance, A.E., McClure, S.A., and St. Martin, S.K., 2003a, Effect of partial resistance on *Phytophthora* stem rot incidence and yield of soybean in Ohio, *Plant Dis.* **87**:308–312.

Dorrance, A.E., McClure, S.A., and DeSilva, A., 2003b, Pathogenic diversity of *Phytophthora sojae* in Ohio soybean fields, *Plant Dis.* **87**:139–146.

Erwin, D.C., and Ribiero, O.K., 1996, *Phytophthora* Diseases Worldwide, APS Press, St. Paul, Minnesota.

Fellbrich, G., Romanski, A., Varet, A., Blume, B., Brunner, F., Engelhardt, S., Felix, G., Kemmerling, B., Krzymowska, M., and Nurnberger, T., 2002, NPP1, A *Phytophthora* associated trigger of plant defense in parsley and *Arabidopsis*, *Plant J.* **32**:375–390.

Fšrster, H., Tyler, B.M., and Coffey, M.D., 1994, *Phytophthora sojae* races have arisen by clonal evolution and by rare outcrosses, *Mol. Plant Microbe In.* **7**:780–791.

Fry, W.E., and Goodwin, S.B., 1997, Re-emergence of potato and tomato late blight in the United States, *Plant Dis.* **81**:1349–1357.

Gijzen, M., Forster, H., Coffey, M.D., and Tyler, B.M., 1996, Cosegregation of *Avr4* and *Avr6* in *Phytophthora sojae*, *Can. J. Bot.* **74**:800–802.

Govers, F., and Gijzen, M., 2006, *Phytophthora* genomics: the plant destroyers' genome decoded, *Mol. Plant Microbe In.* **19**:1295–1301.

Grau, C.R., Dorrance, A.E., Bond, J., and Russin, J.S., 2004, Chapter 14. Fungal diseases, in H.R. Boerma and J.E. Specht, eds., Soybeans: Improvement, Production, and Uses, Amer. Soc. Agron., Madison, Wisconsin.

Hiller, N.L., Bhattacharjee, S., van-Ooij, C., Liolios, K., Harrison, T., Lopez-Estra–o, C., and Haldar, K., 2004, A host-targeting signal in virulence proteins reveals a secretome in malarial infection, *Science* **306**:1934–1937.

Jiang, R.H., Tyler, B.M., and Govers, F., 2006, Comparative analysis of *Phytophthora* genes encoding secreted proteins reveals conserved synteny and lineage-specific gene duplications and deletions, *Mol. Plant Microbe Interact.* **19**:1311–1321.

Judelson, H.S., 1997, The genetics and biology of *Phytophthora infestans*: modern approaches to a historical challenge, *Fungal Genet. Biol.* **22**:65–76.

Latijnhouwers, M., de Wit, P.J., and Govers, F., 2003, Oomycetes and fungi: similar weaponry to attack plants, *Trends Microbiol.* **11**:462–469.

MacGregor, T., Bhattacharyya, M., Tyler, B.M., Bhat, R., Schmitthenner, A.F., and Gijzen, M., 2002, Genetic and physical mapping of *Avr1a* in *Phytophthora sojae*, *Genetics* **160**:949–959.

Marti, M., Good, R.T., Rug, M., Knuepfer, E., and Cowman, A.F., 2004, Targeting malaria virulence and remodeling proteins to the host erythrocyte, *Science* **306**:1930–1933.

May, K.J., Whisson, S.C., Zwart, R.S., Searle, I.R., Irwin, J.A.G., Maclean, D.J., Carroll, B.J., and Drenth, A., 2002, Inheritance and mapping of eleven avirulence genes in *Phytophthora sojae*, *Fungal Genet. Biol.* **37**:1–12.

McBlain, B.A., Zimmerly, M.M., Schmitthenner, A.F., and Hacker, J.K., 1991, Tolerance to *phytophthora* rot in soybean: I. Studies of the cross Ripley × Harper, *Crop Sci.* **31**:1405–1411.

Mideros, S., Nita, M., and Dorrance, A.E., 2007, Characterization of components of partial resistance, Rps2, and root resistance to *Phytophthora sojae* in soybean, *Phytopath.* **97**:655–662.

Olah, A.F., and Schmitthenner, A.F., 1985, A growth chamber test for measuring *phytophthora* root rot tolerance in soybean [Glycine max] seedlings. *Phytopath.* **75**:546–548.

Qutob, D., Kamoun, S., and Gijzen, M., 2002, Expression of a *Phytophthora sojae* necrosis-inducing protein occurs during transition from biotrophy to necrotrophy, *Plant J.* **32**:361–373.

Rehmany, A.P., Gordon, A., Rose, L.E., Allen, R.L., Armstrong, M.R., Whisson, S.C., Kamoun, S., Tyler, B.M., Birch, P.R., and Beynon, J.L., 2005, Differential recognition of highly divergent downy mildew avirulence gene alleles by RPP1 resistance genes from two *Arabidopsis* lines, *Plant Cell* **17**:1839–1850.

Rizzo, D.M., Garbelotto, M., Davidson, J.M., Slaughter, G.W., and Koike, S.T., 2002, *Phytophthora ramorum* as the cause of extensive mortality of *Quercus* spp. and *Lithocarpus densiflorus* in California, *Plant Dis.* **86**:205–214.

Schmitthenner, A.F., 1985, Problems and progress in controlling *Phytophthora* root rot of soybean, *Plant Dis.* **69**:362–368.

Schmitthenner, A.F., Hobe, M., and Bhat, R.G. 1994. *Phytophthora sojae* races in Ohio over a 10-year interval. *Plant Dis.* **78**: 269–276.

Shan, W., Cao, M., Leung, D., and Tyler, B.M., 2004, The *Avr1b* locus of *Phytophthora sojae* encodes an elicitor and a regulator required for avirulence on soybean plants carrying resistance gene *Rps1b*, *Mol. Plant Microbe In.*. **17**:394–403.

Slusarenko, A.J., and Schlaich, N.L., 2003, Pathogen profile. Downy mildew of *Arabidopsis thaliana* caused by *Hyaloperonospora parasitica* (formerly *Peronospora parasitica*), *Mol. Plant Path.* **4**:159–170.

The Gene Ontology Consortium, 2000, Gene ontology: tool for the unification of biology, *Nat. Genet.* **25**:25–29.

Tooley, P.W., and Grau, C.R., 1982, Identification and quantitative characterization of rate-reducing resistance to *Phytophthora megasperma* f.sp. *glycinea* in soybean seedlings, *Phytopath.* **72**:727–733.

Tooley, P.W., and Grau, C.R., 1984, Field characterization of rate-reducing resistance to *Phytophthora megasperma* f.sp. *glycinea* in soybean, *Phytopath.* **74**:1201–1208.

Tripathy, S., Pandey, V.N., Fang, B., Salas, F., and Tyler, B.M., 2006, VMD: A community annotation database for microbial genomes, *Nucl. Acids Res.* **34**:D379–D381.

Tyler, B., Forster, H., and Coffey, M.D., 1995, Inheritance of avirulence factors and restriction fragment length polymorphism markers in outcrosses of the oomycete *Phytophthora sojae*, *Mol. Plant Microbe In.* **8**:515–523.

Tyler, B.M., 2002, Molecular basis of recognition between *Phytophthora* species and their hosts, *Annu. Rev. Phytopath.* **40**:137–167.

Tyler, B.M., 2007, *Phytophthora sojae*: root rot pathogen of soybean and model oomycete, *Mol. Plant Path.* **8**:1–8.

Tyler, B.M., Tripathy, S., Zhang, X., Dehal, P., Jiang, R.H.Y., and et al., 2006, *Phytophthora* genome sequences uncover evolutionary origins and mechanisms of pathogenesis, *Science* **313**:1261–1266.

Vega-Sanchez, M.E., Redinbaugh, M.G., Costanzo, S., and Dorrance, A.E., 2005, Spatial and temporal expression analysis of defense-related genes in soybean cultivars with different levels of partial resistance to *Phytophthora sojae*, *Physiol. Mol. Plant Path.* **66**:175–182.

Whisson, S.C., Drenth, A., Maclean, D.J., and Irwin, J.A.G., 1994, Evidence for outcrossing in *Phytophthora sojae* and linkage of a DNA marker to two avirulence genes, *Curr. Genet.* **27**:77–82.

Whisson, S.C., Drenth, A., Maclean, D.J., and Irwin, J.A.G., 1995, *Phytophthora sojae* avirulence genes, RAPD and RFLP markers used to construct a detailed genetic linkage map, *Mol. Plant Microbe In.* **8**:988–995.

Whisson, S.C., Basnayake, S., Maclean, D.J., Irwin, J.A., and Drenth, A., 2004, *Phytophthora sojae* avirulence genes *Avr4* and *Avr6* are located in a 24kb, recombination-rich region of genomic DNA, *Fungal Genet. Biol.* **41**:62–74.

Wu, Z., Irizarry, R.A., Gentleman, R., Martinez-Murillo, F., and Spencer, F., 2004, A model-based background adjustment for oligonucleotide expression arrays, *J. Am. Stat. Assoc.* **99**:909–917.

Young, N.D., 1996, QTL mapping and quantitative disease resistance in plants, *Ann. Rev. Phytopathol.* **34**:479–501.

Canine SINEs and Their Effects on Phenotypes of the Domestic Dog

Leigh Anne Clark, Jacquelyn M. Wahl, Christine A. Rees, George M. Strain, Edward J. Cargill, Sharon L. Vanderlip, and Keith E. Murphy

Abstract Short interspersed elements (SINEs) are mobile elements that contribute to genomic diversity through the addition of genetic material. Recent genomic analyses have vastly augmented our knowledge of both human- and canine-specific SINEs. SINEC_Cf is a major SINE of the canid family that has undergone recent expansion and is thought to be present in half of all genes. To date, only three phenotypes of the domestic dog have been attributed to a SINE. One of these is merle, a coat pattern characterized by patches of full color on a diluted background and associated with ocular and auditory anomalies. A SINEC_Cf in the *SILV* gene causes merle patterning by altering the cDNA transcript and has unique characteristics that are likely responsible for the random nature of the phenotype.

1 Short Interspersed Elements

Short interspersed elements (SINEs) are non-autonomous transposons of approximately 100–400 bp. SINEs have an internal promoter and are transcribed by RNA polymerase III but rely on long interspersed elements (LINEs) for reverse transcription and integration into the genome (Dewannieux et al., 2003). A typical SINE is characterized by three features: the head, which contains the promoter necessary for initiation of transcription; the body, which is similar at the 3' end to LINE RNA and is necessary for reverse transcription; and the tail, which is usually A- or AT-rich but can vary in sequence and length and may be involved in transcription termination, RNA delivery, and/or RNA stability (Kramerov and Vassetzky, 2005).

Our knowledge of SINEs has increased greatly in recent years, largely due to the assembly of whole genome sequences. Analyses of the human sequence revealed that SINEs comprise 13% of the genome (Lander et al., 2001). In humans, the dominant SINE is the *Alu* element, which is also the only *active* SINE family in the genome (Lander et al., 2001). The promoter region of *Alu* elements is derived from 7SL RNA sequences. This is different from most SINEs in that the majority

K.E. Murphy

Canine Genetics Laboratory, Department of Pathobiology, College of Veterinary Medicine, Texas A&M University, College Station, TX 77843-4467

of SINEs have promoters derived from tRNA sequences (Ullu and Tschudi, 1984; Lander et al., 2001). It is estimated that 75% of human genes contain *Alu* elements, but many of these were inserted prior to divergence and thus are fixed in the population (Wang and Kirkness, 2005).

In 1992, Minnick et al. described a family of SINEs specific to canids. These SINEs, which are tRNA-Lys derived, are characterized by a (TC) repeat and an oligo-dA rich tail (Minnick et al., 1992). A subfamily of these SINEs is SINEC_Cf. It is estimated that there are 170,000 SINEC_Cf elements in the genome and that half of all genes contain a SINEC_Cf insertion (Wang and Kirkness, 2005). Analyses of the first available canine genome sequence, which provides ×1.5 coverage (a Standard Poodle was used), suggest that SINEC_Cf elements have experienced a large expansion in recent history and that this subfamily is highly conserved, exhibiting only 4.8% divergence from the consensus sequence (Kirkness et al., 2003). When this sequence was compared to the ×7.5 coverage sequence (from a Boxer), 10,000 insertion sites were found to be bimorphic (differing by the presence or absence of a SINE), indicating that there are a substantial number of bimorphic sites present in the entire canine population (Lindblad-Toh et al., 2005).

The first published report of a disease-causing retrotransposon in the dog came in 1999 when Lin et al. discovered that narcolepsy in the Doberman Pinscher results from a 226 bp SINEC_Cf insertion in intron 3 of *Hcrtr2* (Lin et al., 1999). The SINE, located 35 bp upstream of the exon 4 splice acceptor, causes skipping of exon 4. Exon skipping as a result of an intronic SINE is also reported in humans, causing autoimmune lymphoproliferative syndrome (Tighe et al., 2002) and hemophilia A (Ganguly et al., 2003). In both organisms, this phenomenon presumably results from displacement of the lariat branch point sequence.

A second canine disorder was recently attributed to a retrotransposon. Pele et al. (2005) showed that exon 2 of *PTPLA* is disrupted by a 236 bp SINEC_Cf element in Labrador retrievers having autosomal recessive centronuclear myopathy. The result of the insertion is variable; seven different cDNA transcripts are identified. Exon-skipping and exonization (the incorporation of SINE sequence into mRNA) are observed in the mutant transcripts. Interestingly, one of the seven transcripts is wild-type, indicating that the SINE is spliced out.

2 Merle Patterning

While the most devastating impact of SINE insertions is disease such as the two mentioned above, SINEs may also have a dramatic effect on physical appearance of dogs, without necessarily causing overt clinical disease. An example of this is a SINEC_Cf insertion in the gene *SILV* that causes merle patterning in several breeds of dog (Clark et al., 2006). In a single dose, the mutation causes dilution of the base fur color, which is determined by a separate locus, and can also cause blue eye color. The predominant characteristic of merling is random patches of full color distributed in various sizes and patterns across the coat. In some breeds (e.g., Australian

Fig. 1 Three red (*e/e*) Australian Shepherd Dogs. From *left* to *right*: a homozygous merle, a non-merle, and a heterozygous merle

Shepherd Dog, Catahoula Leopard Dog) this coat pattern is extremely popular and, according to breed standard, is the preferred phenotype.

Unfortunately, merle is problematic when two mutant copies of *SILV* are inherited (dogs may be referred to as double merles or double dilutes). Double merles have very little pigmentation and are afflicted with a variety of auditory and ocular defects (Fig. 1). The most common abnormality in double merles is deafness. One study concluded that 54.6% of double merle and 36.8% of single merle dogs have mild to severe deafness (Reetz et al., 1977). Also, merle dogs exhibit greater frequencies of ocular abnormalities than do non-merle dogs. These include increased ocular pressure, ametropic eyes, microphthalmia, and colobomas (Klinkmann et al., 1987; Klinkmann and Wegner, 1987; Sorsby and Davey, 1954; Gelatt and McGill, 1973; Dausch et al. 1977). Skeletal defects and sterility have also been reported in double merle dogs (Sponenburg and Bowling, 1985; Treu et al., 1976).

A linkage disequilibrium approach using 32 non-merle and 9 merle Shetland sheepdogs was used to identify the merle locus (Clark et al., 2006). *SILV* (also known as *Pmel17*; gp100) was selected as a candidate gene for its location and its role in pigmentation of the mouse and chicken (Kwon et al. 1995; Kerje et al. 2004). Although *SILV* is clearly critical for pigmentation, its precise function remains controversial (Theos et al., 2005). Studies suggest that the SILV protein is necessary for the formation of the fibril matrix upon which melanin intermediates are deposited (Theos et al., 2005).

Sequencing of *SILV* revealed a SINEC_Cf insertion at the intron 10/exon 11 boundary in merle dogs. The SINE is flanked by a 15 bp target duplication site that includes the exon 11 splice acceptor, making it impossible to determine if the insertion is exonic or intronic. Initial sequencing of the SINE was carried out using DNA from a single dog and the total insertion size was determined to be 262 bp.

cDNA transcripts from double merle dogs have a portion of the SINE incorporated between exons 10 and 11 (unpublished). This exonization is possible because the SINE is situated in reverse orientation and the reverse complement sequence of the SINEC_Cf element has an intron splice acceptor site (Fig. 2) (Kirkness, 2006). Immediately following the splice acceptor is the characteristic GA tandem repeat, which is subject to strand slippage and thus is variable in length. Two alleles have been identified to date. One allele, with seven GA repeats, maintains the reading frame. The *SILV* gene is transcribed in full with the mutant protein having a 52 amino acid insertion (Fig. 3). A second allele has only six GA repeats and the

```
Wild-type  CTTGTCCATTGCTAATCAGTTTCTCCTTTATTCTCCCAATGT------------------
Merle      CTTGTCCATTGCTAATCAGTTTCTCCTTTATTCTCCCAATGTTAGGGGAAGACCTCTTTT


Wild-type  ----------------------------------------------------------
Merle      TTTTTTTTTTTTTTTTTTTTTTTTTTTTTTTTTTTTTTTTTTTTTTTTTTTTTTTTTTTTT

Wild-type  ----------------------------------------------------------
Merle      TTTTTTTTTTAAATTTTTATTTATTTATGATAGTCACAGAGAGAGAGAGAGGCGCAGAGA

Wild-type  ----------------------------------------------------------
Merle      CACAGGCAGAGGGAGAAGCAGGCTCCATGCACCGGGAGCCCGACGTGGGATTCGATCCCG

Wild-type  ----------------------------------------------------------
Merle      GGTCTCCAGGATCGCGCCCTGGGCCAAAGGCAGGCGCCAAACCGCTGCGCCACCCAGGGA

Wild-type  ----TAGGCGAAGACTTCTGAAGCAGGGCTCAGCTCTCCCCCTTCCCCAGCTACCACGTG
Merle      TCCCTAGGCGAAGACTTCTGAAGCAGGGCTCAGCTCTCCCCCTTCCCCAGCTACCACGTG

Wild-type  GTAGCACCCACTGGCTACGCCTGCCCCAGGTCTTCCGCTCTTGCCCCATTGGTGAGAACA
Merle      GTAGCACCCACTGGCTACGCCTGCCCCAGGTCTTCCGCTCTTGCCCCATTGGTGAGAACA

Wild-type  GACCCCTCCTCAATGGGCAGCAGCAGGTCTGAGGACTCTCATGT
Merle      GACCCCTCCTCAATGGGCAGCAGCAGGTCTGAGGACTCTCATGT
```

Fig. 2 Genomic sequence of the 3′ end of *SILV* with SINEC_Cf insertion. The putative lariat branch point is boxed. The 15 bp target duplication site is underlined and contains the exon 11 3′ splice site (in *bold*). The cryptic splice acceptor within the SINE is in *bold* and the *arrow* depicts where transcription begins in merle dogs

Wild-type MNLVPRKCLLHVAVMGVLLAVGATEGPRDQDWLGVPRQLTTKAWNRQLYPEWTETQRPDC 60
Merle(GA)$_7$ MNLVPRKCLLHVAVMGVLLAVGATEGPRDQDWLGVPRQLTTKAWNRQLYPEWTETQRPDC 60
Merle(GA)$_6$ MNLVPRKCLLHVAVMGVLLAVGATEGPRDQDWLGVPRQLTTKAWNRQLYPEWTETQRPDC 60

Wild-type WRGGQVSLKVSNDGPTLVGANASFSIALHFPESQKVLPDGQVVWANNTIIDG SQVWGGQP 120
Merle(GA)$_7$ WRGGQVSLKVSNDGPTLVGANASFSIALHFPESQKVLPDGQVVWANNTIIDGSQVWGGQP 120
Merle(GA)$_6$ WRGGQVSLKVSNDGPTLVGANASFSIALHFPESQKVLPDGQVVWANNTIIDGSQVWGGQP 120

Wild-type VYPQVLDDACIFPDGRACPSGPWSQTRSFVYVWKTWGQYWQVLGGPVSGLSIVTGKAVLG 180
Merle(GA)$_7$ VYPQVLDDACIFPDGRACPSGPWSQTRSFVYVWKTWGQYWQVLGGPVSGLSIVTGKAVLG 180
Merle(GA)$_6$ VYPQVLDDACIFPDGRACPSGPWSQTRSFVYVWKTWGQYWQVLGGPVSGLSIVTGKAVLG 180

Wild-type THTMEVTVYHRRESQSYVPLAHSCSAFTITDQVPFSVSVSQLQALDGGNKHFLRNHPLTF 240
Merle(GA)$_7$ THTMEVTVYHRRESQSYVPLAIISCSAFTITDQVPFSVSVSQLQALDGGNKHFLRNHPLTF 240
Merle(GA)$_6$ THTMEVTVYHRRESQSYVPLAHSCSAFTITDQVPFSVSVSQLQALDGGNKHFLRNHPLTF 240

Wild-type ALRLHDPSGYLSGADLSYTWDFGDHTGTLISRALVVTHTYLESGPITAQVVLQAAIPLTS 300
Merle(GA)$_7$ ALRLHDPSGYLSGADLSYTWDFGDHTGTLISRALVVTHTYLESGPITAQVVLQAAIPLTS 300
Merle(GA)$_6$ ALRLHDPSGYLSGADLSYTWDFGDHTGTLISRALVVTHTYLESGPITAQVVLQAAIPLTS 300

Wild-type CGSSPVPVTTDGHAPTAEIPGTTAGRVPTAEVISTTPGQVPTAEPSGATAVQMTTTEVTG 360
Merle(GA)$_7$ CGSSPVPVTTDGHAPTAEIPGTTAGRVPTAEVISTTPGQVPTAEPSGATAVQMTTTEVTG 360
Merle(GA)$_6$ CGSSPVPVTTDGHAPTAEIPGTTAGRVPTAEVISTTPGQVPTAEPSGATAVQMTTTEVTG 360

Wild-type TTLAQMPTTEGIGTTPEQVPTSEVISTTLAETTGTTPEGSTAEPSGTTGEQVTTKESVEP 420
Merle(GA)$_7$ TTLAQMPTTEGIGTTPEQVPTSEVISTTLAETTGTTPEGSTAEPSGTTGEQVTTKESVEP 420
Merle(GA)$_6$ TTLAQMPTTEGIGTTPEQVPTSEVISTTLAETTGTTPEGSTAEPSGTTGEQVTTKESVEP 420

Wild-type TAGEGPTPETKGPDTNLFVPTEGITGSQSALLDGTATLILAKRETPLDCVLYRYGSFSLT 480
Merle(GA)$_7$ TAGEGPTPETKGPDTNLFVPTEGITGSQSALLDGTATLILAKRETPLDCVLYRYGSFSLT 480
Merle(GA)$_6$ TAGEGPTPETKGPDTNLFVPTEGITGSQSALLDGTATLILAKRETPLDCVLYRYGSFSLT 480

Wild-type LDIVRGIENAEILQAVPSSEGDAFELTVSCQGGLPKEACMDISSPGCQPPAQRLCQPVPP 540
Merle(GA)$_7$ LDIVRGIENAEILQAVPSSEGDAFELTVSCQGGLPKEACMDISSPGCQPPAQRLCQPVPP 540
Merle(GA)$_6$ LDIVRGIENAEILQAVPSSEGDAFELTVSCQGGLPKEACMDISSPGCQPPAQRLCQPVPP 540

Wild-type SPACQLVLHQVLKGGSGTYCLNVSLADANSLAMVSTQLVMPGQEAGVGQAPLFMGILLVL 600
Merle(GA)$_7$ SPACQLVLHQVLKGGSGTYCLNVSLADANSLAMVSTQLVMPGQEAGVGQAPLFMGILLVL 600
Merle(GA)$_6$ SPACQLVLHQVLKGGSGTYCLNVSLADANSLAMVSTQLVMPGQEAGVGQAPLFMGILLVL 600

Wild-type LAMVLVSLIY-- 610
Merle(GA)$_7$ LAMVLVSLIYSHRERERGAETQAEGEAGSMHREPDVGFDPGSPGSRPGPKAGAKPLRHPG 660
Merle(GA)$_6$ LAMVLVSLIYSHRERERRRDTGRGRSRLHAPGARRGIRSRVSRIAPWAKGRRQTAAPPRD 660

↑

Wild-type --RRRLLKQGSALPLPQLPRGSTHWLRLPQVFRSCPIGENRPLLNGQQQV* 658
Merle(GA)$_7$ IPRRRLLKQGSALPLPQLPRGSTHWLRLPQVFRSCPIGENRPLLNGQQQV* 710
Merle(GA)$_6$ P*   661

Fig. 3 Wild-type and mutant SILV protein sequences. The *arrow* denotes where the truncated GA repeat alters the reading frame

insertion disrupts the reading frame. Fifty-one amino acids are incorporated after exon 10, and a premature stop codon occurs near the end of the insertion (Fig. 3).

3 A-Tails Are Important

Gel eletrophoresis analysis of the *SILV* SINE showed that the insertion size varies from dog to dog. Further sequencing revealed that the variability in the SINE is found in the poly(A) tail. A-tail length is an important factor in retrotransposition. Roy-Engel et al. (2002) analyzed A-tail length in *Alu* elements and found that insertions that result in disease (many are *de novo* events) have a mean length nearly twofold longer than other insertions. They also observed that overall, younger *Alu* elements have longer A-tails than older *Alu* elements (Roy-Engel et al., 2002). These data suggest that the A-tail is evolutionarily unstable and subject to mutation and degradation over time. This phenomenon may exist in part because A-tails are subject to strand slippage during replication and unequal crossing over (Roy-Engel et al., 2002).

The SINE element that disrupts the *SILV* gene is longer than other SINEC_Cf elements that have been described in the dog (Pele et al., 2005; Fletcher et al., 2001; Jeoung et al., 2000). The variation in length is again found in the A-tail, which we define as the region after the (TC) repeat, between the last G and the duplicated sequence. The A-tail of the SINEs in the *PTPLA*, *dystrophin*, and *D2 dopamine receptor* genes are 64, 50, and 46 bp long, respectively (Pele et al., 2005; Fletcher et al., 2001; Jeoung et al., 2000). In merle dogs, the tail length of the *SILV* SINE insertion ranges from 91 to 101 bp.

Along with changes in length, A-tails accumulate non-A bases that disrupt the pure A stretches. These interruptions reduce the likelihood of strand slippage, resulting in greater tail stability. The *SILV* SINE has fewer A-tail interruptions than do the *PTPLA* and *dystrophin* SINEs. Consequently, the former SINE has pure A stretches extending as long as 83 bp, while the latter SINEs have maximum pure A stretches of 14 and 15 bp, respectively. The length and purity of the A-tail of the *SILV* insertion suggest that it is a young SINEC_Cf and is subject to greater levels of instability.

A surprising find is the presence of the *SILV* SINE in dogs that do not have the merle phenotype (Clark et al., 2006). In these dogs, the poly(A)-tail is shortened, ranging from 54 to 65 bp. This finding suggests that the SINE insertion is necessary, but not sufficient for the merle pattern (Cordaux and Batzer, 2006). Although the threshold has not been determined, it is apparent that a long A-tail is requisite to produce the phenotype. A smaller insertion size may bring the lariat branch point to a more reasonable distance with the true splice site, encouraging proper splicing.

These data offer a possible explanation for merle patterning. During development, melanoblasts migrate from the neural crest and differentiate into the pigment-producing melanocytes (Steingrimsson et al., 2004). Instability of the poly(A)-tail during this migration could result in cell populations with varying tail lengths. Melanocytes having a larger *SILV* SINE insertion would produce diluted pigment, while those with a truncated A-tail would produce full pigment. A similar mechanism is found in pigs: black spotting results from somatic reversions of a C expansion in the *MC1R* gene (Kijas et al., 2001).

The randomness of merling may also provide an explanation for the variability of the abnormalities associated with the double merle phenotype. In 2006, a hearing test known as the brainstem auditory evoked response (BAER), which detects electrical activity in the cochlea and auditory pathways in the brain (Wilson and Mills, 2005), was performed on 70 young to middle-aged merles representing five breeds (unpublished). Twenty-two double merles were examined: eight were bilaterally deaf and two were unilaterally deaf. Although Reetz et al. (1977) suggested that about one-third of heterozygous merles have deafness, only one of the 48 heterozygous merles in this study was deaf (unilateral). This dog, a Great Dane, was also piebald and consequently the deafness cannot be positively attributed to the *SILV* mutation (Strain, 1999).

Fifteen of the double merles tested were Catahoula Leopard Dogs and of these only four were deaf. Although the sample size is small, this study suggests that only about 26% of double merle Catahoula Leopard Dogs are deaf while roughly 85% of double merles from other breeds tested (Australian Shepherd, Collie, and Shetland Sheepdog) are deaf. This finding is not entirely surprising because double merle Catahoula Leopard Dogs exhibit larger amounts of pigmentation than do other breeds. These phenotypic differences may result from a shorter average poly(A)-tail length in the breed or from modifying genes.

4 Summary

Mobile elements promote genomic diversity through the addition, and occasional deletion, of genetic material. Those mutations, which occur in or near coding regions may also influence phenotypic diversity by altering gene expression and function. These changes may manifest themselves as detrimental genetic disorders or as simple physical traits. In addition, variability within phenotypes may result from instability of the causative element. This is exemplified by merle patterning in the dog in which random spotting is determined by the total size of the SINE insertion.

Acknowledgments This research was supported in part by the National Organization for Hearing Research Foundation. We are grateful to Don Abney, Kerry Kirtley, Sherry Lindsey, and Susan Schroeder for providing tissue samples.

References

Clark, L.A., Wahl, J.M., Rees, C.A., and Murphy, K.E., 2006, Retrotransposon insertion in SILV is responsible for merle patterning of the domestic dog, *Proc. Natl. Acad. Sci. USA* **103**: 1376–1381.

Cordaux, R., and Batzer, M.A., 2006, Teaching an old dog new tricks: SINEs of canine genomic diversity, *Proc. Natl. Acad. Sci. USA* **103**:1157–1158.

Dausch, D., Wegner, W., Michaelis, M., and Reetz, I., 1977, Ophthalmologische befunde in einer merlezucht. *Dtsch. Tierarztl. Wschr.* **84**:453–492.

Dewannieux M., Esnault, C., and Heidmann, T., 2003, LINE-mediated retrotransposition of marked Alu sequences. *Nat. Genet.* **35**:41–48.

Fletcher, S., Carville, K.S., Howell, J.M., Mann, C.J., and Wilton, S.D., 2001, Evaluation of a short interspersed nucleotide element in the 3' untranslated region of the defective dystrophin gene of dogs with muscular dystrophy, *Am. J. Vet. Res.* **62**:1964–1968.

Ganguly, A., Dunbar, T., Chen, P., Godmilow, L., and Ganguly, T., 2003, Exon skipping caused by an intronic insertion of a young Alu Yb9 element leads to severe hemophilia A, *Hum. Genet.* **113**:348–352.

Gelatt, K.N., and McGill, L.D., 1973, Clinical characteristics of microphthalmia with colobomas of the Australian shepherd dog, *J. Am. Vet. Med. Assoc.* **162**:393–396.

Jeoung, D., Myeong, H., Lee, H., Ha, J., Galibert, F., Hitte, C., and Park, C., 2000, A SINE element in the canine D2 dopamine receptor gene and its chromosomal location, *Anim. Genet.* **31**: 333–346.

Kerje, S., Sharma, P., Gunnarsson, U., Kim, H., Bagchi, S., Fredriksson, R., Schultz, K., Jensen, P., von Heijne, G., Okimoto, R., and Andersson, L., 2004, The *dominant white*, *dun* and *smokey* color variants in chickens are associated with insertion/deletion polymorphisms in the *PMEL17* gene, *Genetics* **168**:1507–1518.

Kijas, J.M.H., Moller, M., Plastow, G., and Anderson, L., 2001, A frameshift mutation in *MCR1* and a high frequency of somatic reversions cause black spotting in pigs, *Genetics* **158**: 779–785.

Kirkness, E.F., 2006, SINEs of canine genomic diversity, in Elaine A. Ostrander, Urs Giger, Kerstin Lindblad-Toh, eds., The Dog and Its Genome, Cold Spring Harbor Laboratory Press, Woodbury, New York, pp. 209–219.

Kirkness, E.F., Bafna, V., Halpern, A.L., Levy, S., Remington, K., Rusch, D.B., Delcher, A.L., Pop, M., Wang, W., Fraser, C.M., and Venter, J.C., 2003, The dog genome: survey sequencing and comparative analysis, *Science* **301**:1898–1903.

Klinkmann, G., Koniszewski, G., and Wegner, W., 1987, Lichtmikroskopische untersuchungen an den corneae von merle-Dachshunden, *Dtsch. Tierarztl. Wschr.* **94**:338–341.

Klinkmann, G., and Wegner, W., 1987, Tonometry in merle dogs, *Dtsch. Tierarztl. Wschr.* **94**: 337–338.

Kramerov, D.A., and Vassetzky, N.S., 2005, Short retroposons in eukaryotic genomes, *Int. Rev. Cytol.* **247**:165–221.

Kwon, B.S., Halaban, R., Ponnazhagan, S., Kim, K., Chintamaneni, C., Bennett, D., and Pickard, R.T., 1995, Mouse *silver* mutation is caused by a single base insertion in the putative cytoplasmic domain of Pmel 17, *Nucleic Acids Res.* **23**:154–158.

Lander, E.S., and the International Human Genome Sequencing Consortium, 2001, Initial sequencing and analysis of the human genome, *Nature* **409**:860–921.

Lin, L., Faraco, J., Li, R., Kadotani, H., Rogers, W., Lin, X., Qiu, X., de Jong, P.J., Nishino, S., and Mignot, E., 1999, The sleep disorder canine narcolepsy is caused by a mutation in the *Hypocretin (Orexin) Receptor 2* gene, *Cell* **98**:365–376.

Lindblad-Toh, K., Wade, C.M., Mikkelsen, T.S., Karlsson, E.K., Jaffe, D.B., Kamal, M., Clamp, M., Chang, J.L., Kulbokas III, E.J., Zody, M.C. et al., 2005, Genome sequence, comparative analysis and haplotype structure of the domestic dog, *Nature* **438**:803–819.

Minnick, M.F., Stillwell, L.C., Heineman, J.M., and Stiegler, G.L., 1992, A highly repetitive DNA sequence possibly unique to canids, *Gene* **110**:235–238.

Pele, M., Tiret, L., Kessler, J.L., Blot, S., and Panthier, J.J., 2005, SINE exonic insertion in the PTPLA gene leads to multiple splicing defects and segregates with the autosomal recessive centronuclear myopathy in dogs, *Hum. Mol. Genet.* **14**:1417–1427.

Reetz, I., Stecker, M., and Wegner, W., 1977, Audiometrische befunde in einer merlezucht, *Dtsch. Tierarztl. Wschr.* **84**:253–292.

Roy-Engel, A.M., Salem, A.H., Oyeniran, O.O., Deininger, L., Hedges, D.J., Kilroy, G.E., Batzer, M.A., and Deininger, P.L., 2002, Active Alu element "A-tails": size does matter, *Gen. Res.* **12**:1333–1344.

Sorsby, A., and Davey, J.B., 1954, Ocular accociations of dappling (or merling) in the coat color of dogs, *J. Genet.* **54**:425–440.

Sponenburg, D.P., and Bowling, A.T., 1985, Heritable syndrome of skeletal defects in a family of Australian shepherd dogs, *J. Hered.* **76**:393–394.

Steingrimsson, E., Copeland, N.G., and Jenkins, N.A., 2004, Melanocytes and the microphthalmia transcription factor network, *Annu. Rev. Genet.* **38**:365–411.

Strain, G.M., 1999, Congenital deafness and its recognition, *Vet. Clin. North Am. Small Anim. Pract.* **29**:895–907.

Theos, A.C., Truschel, S.T., Raposo, G., and Marks, M.S., 2005, The *Silver* locus product Pmel17/gp100/Silv/ME20: controversial in name and function, *Pigm. Cell Res.* **18**: 322–336.

Tighe, P.J., Stevens, S.E., Dempsey, S., Le Deist, F., Rieux-Laucat, F., and Edgar, J.D.M., 2002, Inactivation of the Fas gene by Alu insertion: retrotransposition in an intron causing splicing variation and autoimmune lymphoproliferative syndrome, *Genes Immun.* **3**:S66–S70.

Treu, H., Reetz, I., Wegner, W., Krause, D., 1976, Andrological findings in merled Dachshunds, *Zuchthygiene* **11**:49–61.

Ullu, E., and Tschudi, C., 1984, Alu sequences are processes 7SL RNA genes, *Nature* **312**: 171–172.

Wang, W., and Kirkness, E.F., 2005, Short interspersed elements (SINEs) are a major source of canine genomic diversity, *Gen. Res.* **15**:1798–1808.

Wilson, W.J., and Mills, P.C., 2005, Brainstem auditory-evoked response in dogs, *Am. J. Vet. Res.* **66**:2177–2187.

Ovine Disease Resistance: Integrating Comparative and Functional Genomics Approaches in a Genome Information-Poor Species

H.W. Raadsma, K.J. Fullard, N.M. Kingsford, E.T. Margawati,
E. Estuningsih, S. Widjayanti, Subandriyo, N. Clairoux, T.W. Spithill,
and D. Piedrafita

Abstract In combination with goats, sheep represent the two most numerous agricultural species for which no cultural or ethical restrictions apply in their use as a source for milk, fibre and red meat. Particularly, in the developing world these species often represent the sole asset base for small-holder livestock farmers. Despite their global significance, genomic tools and approaches in disease resistance have lagged behind the efforts in the economically more influential beef and dairy cattle industries. In particular, infectious diseases have a significant economic impact on livestock production systems worldwide. The most frequently investigated diseases in sheep have focused on the economically important burdens including gastrointestinal nematodes, dermatophilosis, footrot, myiases and fasciolosis. In this study we describe the use of Indonesian Thin Tail sheep (ITT) as a resource which has been shown to have innate and acquired resistance to tropical fasciolosis (*Fasciola gigantica*). Using the contrast between the resistant ITT and the highly susceptible Merino in a combined functional and comparative genomics approach, we have identified putative QTL (quantitative trait loci) for an extensive panel of parasite and immune response phenotypes and putative resistance pathways and effector molecules. On refinement of candidate gene analyses and effector mechanisms we propose to map these in the economically important target species, namely cattle and buffalo. In addition we exploit the relative susceptibility of ITT sheep to temperate fasciolosis (*Fasciola hepatica*) to contrast parasite–host interactions and identify parasite immune evasion strategies to boost the discovery of new vaccine candidates and effector pathways, which may be amenable to exogenous control. The study highlights the power and utility of direct gene discovery in ruminant model systems. To overcome the shortage of genomic tools required for such investigations, we have drawn on the development of integrated comparative maps and alignment

H.W. Raadsma
Reprogen-Centre for Advanced Technologies in Animal Genetics and Reproduction, Faculty of Veterinary Science, University of Sydney, Camden NSW, Australia

J.P. Gustafson et al. (eds.), *Genomics of Disease*,
© Springer Science+Business Media, LLC, 2008

to the genome information–rich species such as human, murine and recently cattle. Similarly the use of bovine transcriptome tools have shown cross utility in sheep. The only species-specific requirement is the development of genome-wide high resolution SNP mapping tools which are now under development.

1 Introduction

Diseases in livestock can broadly be divided into three classes: inherited disease, environmental/metabolic disease and infectious disease. Due to the complicated interactions between host, parasite and the environment, the latter has proved most refractory to amelioration. Given the global estimate of 8.74 billion head of cattle and 1.05 billion sheep estimates in 2004 (GLiPHA website in July 2007), it becomes obvious that the economic impact of infectious disease is substantial. It is therefore not surprising that ongoing research has focussed on this aspect, considering the potential benefits from successful preventative measures or integrating complementary strategies for disease control. While much of the work has been done on cattle, this chapter discusses the use of sheep as a model organism for ruminants.

In sheep, predominant diseases concern endo- and ecto-parasites. In particular gastro-intestinal nematode (GIN) infestation causes significant loss of productivity and death, through inefficiency of feed utilization, anaemia, loss of body condition, poor growth, reproductive fitness and lactation performance. Ovine footrot, flystrike and body strike (cutaneous myasis) are the three cutaneous diseases of greatest economic impact on sheep production after internal parasitism. Other diseases of relevance are parasitical (trypanosomes, ticks and trematodes), protozoal (toxoplasmosis, coccidiosis, giardia and cryptosporidia), bacterial (brucellosis, mastitis, paratuberculosis or Johne's disease, dermatophilosis and salmonellosis) and viral (foot and mouth, bovine leukaemia virus, sheep pox, scabby mouth and blue tongue), with prion diseases such as scrapie also possibly falling within the infectious disease category. Some of the diseases are present in both cattle and sheep and a comparative genomics approach will be relevant for both species.

To date, a combination of disease control strategies are utilised to minimise infection and maximise survival of flocks or herds, relying on vaccination, sanitation, therapeutic/pharmaceutical applications, quarantine/isolation, culling and eradication. All of these approaches have limitations including, variously, the cost of implementation, the need for farmer education, the difficulty of eradication practices where farmers depend on the existence of the animal, and the development of drug resistance to multiple classes of pharmaceutical control measures. In developing countries, different control measures to those employed in more affluent regions may be appropriate.

The selection of genetically resistant, tolerant or less susceptible hosts as a further arm of disease control strategies has been encouraged by the widespread evidence for host genetic variation for most diseases of interest in livestock production (as reviewed by Axford et al., 2000). Gibson and Bishop (2005) elegantly describe epidemiological models, which incorporate host resistance to disease, and suggest that success in selection for resistance to disease is strongly dependent on the transmission pathway, virulence and initial level of resistance in the host population. Diseases brought about by fast-replicating highly infectious pathogens, for instance foot and mouth, may be less amenable to genetic improvement. In contrast, for internal parasites with a slower generation time and an intermediate reservoir stage, elimination of the most susceptible sub-population may prove beneficial. In such cases, it is envisaged that conventional breeding strategies would be able to produce stock with improved levels of innate resistance to infection. The relatively seamless integration of "ready-made" resistant stock into existing breeding programmes would be especially beneficial to developing countries. However, conventional genetic improvement programmes aimed at increasing profitability have found it difficult to incorporate disease resistance in multi-trait breeding objectives.

One of the hurdles has been the problem of accurately identifying resistant breeding replacements, given the low to moderate heritabilities of such traits, the often low and sporadic expression of disease, and the frequent use of surrogate traits for actual infection (that are often the only convenient phenotypes to measure). More work is needed to ensure that the genetic correlation is high between the disease traits (such as parasite burden) and the surrogate trait such as faecal egg count and antibody titre. A distinction is also made between the ability to prevent or counteract initial infection (resistance) and the ability to survive and thrive despite infection (resilience).

A further difficulty with incorporating disease resistance in animal production systems is the need for selection for resistance to multiple diseases simultaneously, with evidence mounting that there is no "silver bullet" of a gene that will convey protection against multiple pathogens. In addition, selection pressures on pathogens may well force the evolution of parasite resistance in the same manner as for chemical prophylaxis, potentially necessitating the stacking of favourable alleles for multiple resistance mechanisms per pathogen. Finally for most disease resistance traits, there is a potentially negative impact of reduced selection efficiency on production traits.

Many of these problems could be solved by a better understanding of the mechanisms underpinning the resistance, the genes involved and the use of marker-assisted selection. Research would also help to identify multiple resistance pathways, to allow multiple breeding targets and suggest novel chemical prophylaxis or immunization strategies.

There are many tools that may be used to achieve gene discovery for disease resistance, thereby addressing the above points. Thus far, an important research tool has been the numerous quantitative trait loci (QTL) studies employed in livestock research (Brenig et al., 2004). QTL analyses employ statistical methods to correlate phenotypic variation with the correspondingly underlying genetic variation in the genome. Regions of the genome that segregate together with a particular phenotype across multiple meiotic events are likely to hold genes influencing that phenotype. Published QTL analyses almost invariably identify significant chromosomal regions. Unfortunately, as more studies are conducted, more regions are identified with a bewildering lack of repeatability and concordance between studies. Few research studies have moved beyond a primary report on significant but unfocussed QTL regions. To attain the level of biochemical understanding necessary to understand the genetic architecture of QTL for disease resistance, resolution must increase to the level of individual candidate genes and the functional mutations causing the difference in phenotype – or Quantitative Trait Nucleotide (or so called QTN). Some of the processes and the tools employed to move from QTL to QTN are discussed further.

2 Tools Used to Obtain Candidate Genes

2.1 Resource Flocks for QTL Analysis and Mapping

Many research flocks have been established with sheep (Table 1) to map QTL in a broad range of disease phenotypes. Despite this initial effort in primary QTL mapping, few resources have resolved functional and causative genes underlying such QTL.

2.2 Integrated Maps, Comparative Mapping and Meta-analysis

Integrated and Comparative Maps

The pooling of data from multiple sources is now providing dependable, high-resolution information for researchers. These large-scale projects are often under the control of major consortiums, such as the International Sheep Genomics Consortium (www.sheephapmap.org), which provides publicly available Web-based resources where the data is not commercially sensitive. Such consortiums have helped to establish high-resolution linkage maps generated for humans and mice medium to high resolution maps for the bovine genome and medium density maps for sheep For example, according to the NCBI UniSTS database there are 45,138 independent STS markers in human transcript map 99,16,754 independent STS markers in the Mouse Genome Database Genetic Map, 3,484 independent STS markers in the ILIX-2005 map and 1411 independent STS loci in the International Sheep Mapping Flock v.4.7 as of July 2007. Likewise, there are 5,689,286 validated human SNPs (dbSNP build 127), 6,447,366 validated mouse SNPs (dbSNP build 126), 14,371 validated bovine SNPs (dbSNP build 127) but only 66 ovine SNPs (dbSNP build 126) documented in NCBI dbSNP. There are also radiation hybrid maps for all species, physical maps obtained through fluorescence in situ hybridisation, restriction digestion maps and other means, and in the former three cases whole genome sequence is available. Much current work is involved in uniting all available maps within a species into a single "integrated" map, and overlaying this on sequence information.

Mammalian genomes are relatively highly conserved. Markers that have been designed to amplify cattle DNA, also often amplify sheep DNA. Once the position

Table 1 Flocks for QTL detection of disease resistance in sheep

Institute or working title	Location	Breeds	Disease traits	Major resource/selected references
GeneSheepSafety (EU)	France, UK, Italy and Spain	Churra, Sarda, Lacaune, Manech	Mastitis, GI nematodes, oestrus ovis	Barillet et al., 2003, 2005; Bayon et al., 2004; Moreno et al., 2006; Carta et al., 2002
French Region Centre, INRA 'FEOGA (EU)	France French West Indies, Guiana	West Indian Black Belly, INRA 401	GI nematodes	Moreno et al., 2006; Gruner et al. 2003
Bourges-la Sapinière, INRA	France	INRA401 (Romanov, Berichon du Cher)	Salmonellosis, PrP gene, mastitis	Elsen et al., 2006; Moreno et al. 2003
Indonesian Thin Tail – Sydney University	Australia, Indonesia	Garut and Sumatran Indonesian Thin Tail and Merino	GI nematodes, *F. gigantica*, *F. hepatica*	Raadsma et al.. 2002
SheepGenomics	Australia, New Zealand	Merino (Falkiner Flock)	GI nematodes	Oddy et al., 2005 personal communication www.sheepgenomics.com
CSIRO *T. colubriformis* flock	Australia	Peppin Merino	GI nematodes	Beh et al., 2002,
AgResearch/Ovita	New Zealand	Romney, Merino, Perendale	GI nematodes, Facial eczema	Crawford et al., 1997a, b; Crawford and McEwan, 1998; Diez-Tascon et al., 2002
Louisiana State University	US	Suffolk, Gulf Coast Native	GI nematodes	Li et al., 2001; Miller et al., 2002; Shay et al., 2002
Roslin/Glasgow	UK	Scottish Black Face	GI nematodes	Davies et al., 2006
Golden Ram – AGBU	Australia	Merino	GI nematodes	Marshall et al. 2005
ILRI/Wageningen University	Kenya	Maasai, Dorper	GI nematodes	Baker et al., 2003
Soay sheep	UK	Soay	GI nematodes	Coltman et al., 2001; Paterson et al., 1998

of matching markers or genes have been established in two genomes, it is possible to deduce the chromosomal rearrangements that have taken place between the two. Marker order by and large remains conserved in those regions between historical breakpoints, and so-called "comparative mapping" allows information to be transferred between the relevant genomes. Such comparisons have become vital for those species whose genomes have not been sequenced, such as the sheep. As of 15 November 2006, a virtual sequence has been established for the sheep genome, based entirely on the ordering of sequenced BAC ends relative to established human (bg17), cattle (Btau2.6) and dog (canFam2) frameworks (International Sheep Genomics Consortium). Information on virtual sheep chromosomes are found at www.livestockgenomics.csiro.au. For sheep researchers, this provides access to the far greater functional and positional knowledge available in other species. Likewise, knowledge generated in the sheep genome can be transferred to other species through the vehicle of integrated maps and Oxford grids, in which shared features between two species are used to align the two integrated maps in a grid format as is shown for sheep, cattle and human in Fig. 1 (Nicholas, 2005; Liao et al., 2007).

2.2.1 Meta-analysis

Once a number of QTL studies have been undertaken on a particular disease, meta-analyses may be employed to help eliminate type 2 errors and suggest critical genomic regions affecting a disease across multiple populations. Meta-analysis employs comparable data from all such studies in a joint statistical analysis. By combining the information across studies, refined confidence intervals of critical regions are also possible. In sheep, the possibility of collating the data gathered on gastro-intestinal nematodes is being considered (Jill Maddox, personal Communicaton, 2006). A summary of chromosomes with significant or suggestive QTL for GI nematode resistance/susceptibility is depicted in Fig. 2 and includes a crude cross species analysis, in which QTL identified on bovine chromosomes are translated to their corresponding ovine chromosomes according to the appropriate Oxford grid. Even this rudimentary depiction suggests a major gene for GI nematode resistance on OAR3 and possibly OAR1, OAR2, OAR6, OAR14 and OAR20. For a comprehensive statistical meta-analyses, Khatkar et al. (2004) detail the general procedures and requirements from the data, to provisionally locate QTL to best-bet locations and make distinctions between single QTL and multiple QTL in target regions.

2.3 Association Studies, SNP Chips and LD Mapping

Most primary QTL screens have confidence intervals that are typically in the order of 50–100 cM, thus encompassing potentially thousands of positional candidate genes. The first step towards refining the position of the actual QTL is to extract maximal information from such studies by adding more markers to the same linkage mapping population, and if possible adding more individuals. This approach is

Human

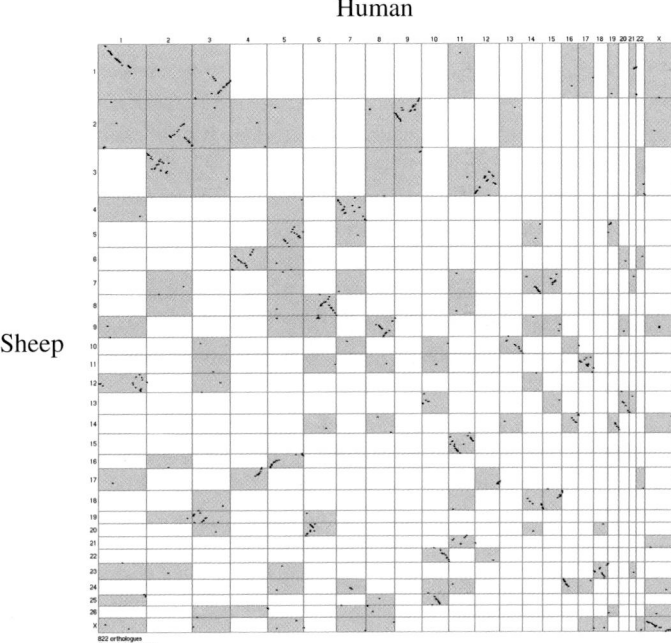

Cattle

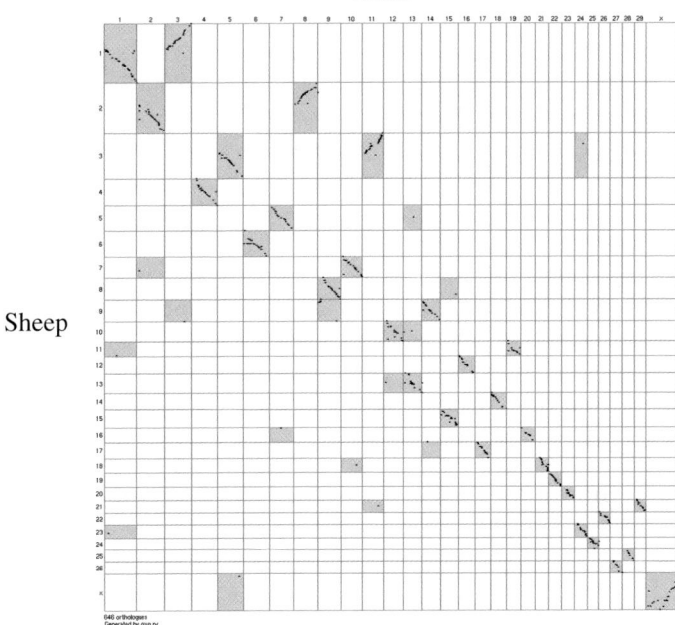

effective until no additional informative recombination is observed, and adding more animals is unfeasible. Typically, confidence intervals may have shrunk to a sub 50 cM region at this stage depending on the size of the population, but there may still be hundreds of potential candidate genes represented under the QTL peak. Without very strong supporting evidence of a lead positional candidate gene, further "fine mapping" is often required of such QTL.

To jump to the next level of resolution, the large number of historical recombination events within a population may be utilised, as opposed to the limited recombination generated within the QTL mapping pedigree. Within a population, very tightly linked markers tend to segregate together as a haplotype in 'linkage disequilibrium' (LD) with the mutation event that generated the QTN. Meiotic crossovers occurring since the coalescent ancestor would tend to break down the extent of markers in disequilibrium with the QTN. Thus, association between markers and phenotype within a random mating population would suggest that the markers may lie very close to the polymorphism causing the phenotypic variation. The population-wide average extent of long range LD in cattle appears to be in the order of 10 Mb, but LD generated by ancestral mutation and in strong (short range) LD occurs on a much smaller scale. LD blocks showing no signs of historical recombination had an average length of approximately 40–100 kb on BTA6 (Khatkar et al., 2006). Thus, markers in the order of 10 Mb might show evidence of an association with the QTN, but only once the marker was within a few kb of the QTN would there be a chance of a particular marker allele always being in phase with a trait "direction".

A pertinent example may be found in the directed and shotgun association studies carried out on OAR 3 for GIN resistance. As shown in Fig. 2, a number of genome-wide QTL studies on gastrointestinal nematodes in sheep identified OAR 3 as significant. Directed fine mapping of chromosome 3 occurred by adding markers at approximately 2cM intervals over approximately 15 cM, which narrowed down the QTL confidence interval to a 5 cM region. This target region encompassed the *interferon gamma* (*IFNG*) gene (Paterson et al., 2001). A number of polymorphisms were detected in linkage disequilibrium within the *IFNG* gene region between resistant and susceptible selection lines (Crawford and McEwan, 1998). This lead was followed by association studies across other populations conducted by other groups, which made use of the same markers (Fig. 3). Interestingly, while many of the studies reported significant association between markers predominantly near to the 5′ region of the gene and resistance, the phase of the resistant haplotype was not consistent (Paterson et al., 2001, Coltman et al., 2001, Sayers et al., 2005, Stear et al., 2006).

Fig. 1 Oxford grid contrasting the sheep linkage map v3.1 against the human and cattle genome sequence available at http://oxgrid.angis.org.au/. Each point represents a gene or marker that has been positioned in both genomes, either physically or through linkage. The length and direction of the two sides of the boxes are indicative of the relative length and directions of respective chromosomes

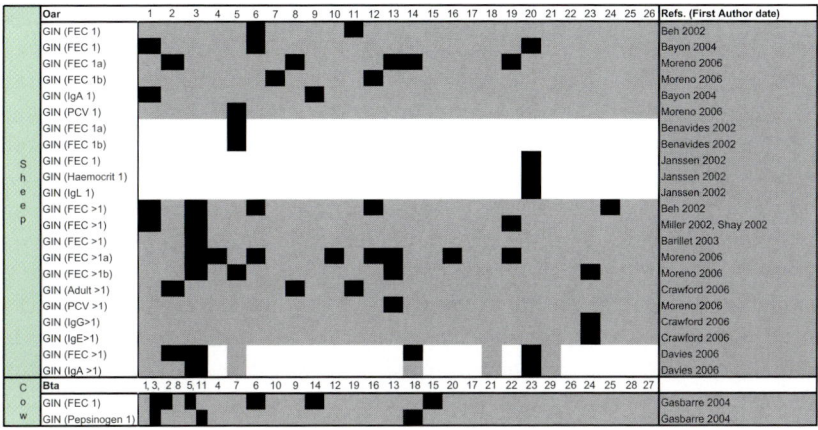

Fig. 2 QTL studies of nematode infection in sheep (Oar = *Ovis aries*) and cattle (Bta = *Bos taurus*). Cattle chromosomes are aligned with homologous sheep chromosomes. *Dark grey* = autosomal chromosomes reported as significant or indicative, *light grey* = non-significant chromosomes, *white* = unexamined. GIN = gastro intestinal nematode, FEC = faecal egg count, 1 = naïve infection, >1 = secondary infection, IgA/E/G = immunoglobulin A/E/G, Adult = adult worm count, Pepsinogen = serum pepsinogen level, PCV = packed cell volume, a/b = separate flocks within the same study

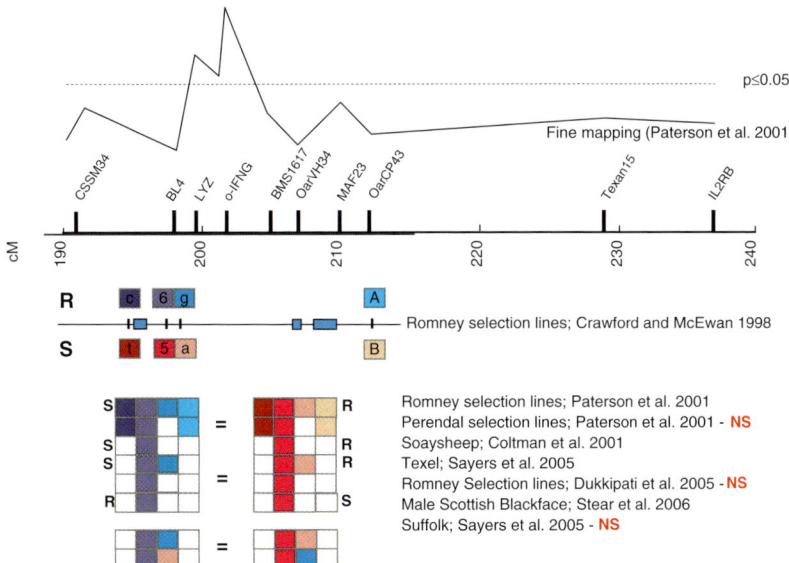

Fig. 3 Schematic representation of fine mapping and haplotype association studies of the Interferon Gamma gene. R haplotype = resistant and S haplotype = susceptible phenotypes, respectively. Coloured blocks indicate which of the four SNP polymorphisms were analysed in each study

In addition, three studies reported a lack of significant association. The first used Perendale selection lines for high and low levels of GIN resistance (Paterson et al., 2001). Romney lines selected for traits independent of parasite resistance also proved non-significant (Dukkipati et al., 2005). A proposed founder effect is unlikely to be the sole explanation for association between *IFNG* and GIN resistance, given the number of studies, and the broad base of breeds reporting association. The last was in the Suffolk breed (Sayers et al., 2005), which had four segregating haplotypes, instead of the two present in other breeds, representing an ancestral and potentially highly informative recombination between two closely linked markers. Taken together, the combined work by many laboratories suggests that the markers that have been used in association studies to date are not in sufficiently close linkage disequilibrium to be in perfect phase with the real resistance polymorphism.

To date, such fine mapping and association studies have primarily been conducted on a targeted basis across limited regions in the genome. However, using high-density SNP genotyping technologies, whole-genome scans of a similar level of resolution are now possible. Single nucleotide polymorphisms (SNPs) offer a plentiful supply of bi-allelic variability throughout the genome, occurring on average once in every 100 bp. Advances in technology now make it possible to take full advantage of the dense genome-wide information that these markers can provide, with mapping arrays of up to 1,000,000 markers now available in humans. With a 100,000 K chip, marker density on a map is approximately in the order of 1 in every 30 kb. At this level, population-wide levels of linkage disequilibrium may be exploited. The transition from targeted QTL mapping studies to genome wide selection (GWS) studies with sufficiently dense SNP arrays will allow markers significantly associated with a trait to be used directly as selection aids in breeding programmes. Development of a 60,000 ovine SNP array has already been proposed (www.sheephapmap.org).

2.4 Microarrays, SELDI-TOF MS and Other High Density Genomic or Proteomic Functional Tools

Functional genomic approaches offer other means to narrow down the approximately 30,000 genes found within a genome to a few candidates, which influence a particular trait. Micro-arrays, whether they be oligo-nucleotide-based chips or c-DNA arrays, seek to detect changes in expression levels between two states (i.e. resistant vs. susceptible, or infected vs. non-infected). The caveat to bear in mind is that genes demonstrating changed expression levels may be downstream in a biochemical pathway from influential "controller" genes with small changes in their expression or undetectable changes to their conformation state, or indeed may not be represented on the array and therefore undetectable in the target studies. Nevertheless, sophisticated algorithms are now available to group differentially expressed genes according to their response profiles or promoter sequences, thereby suggesting functional links to lead pathways. This information, in conjunction with knowledge of the pertinent biochemical pathways obtained in other species, may lead to a number of candidate genes near the top of cascades.

Only a single large ovine array has been established based on 27 different tissues, with 19,968 spots representing 9,238 RefSeqs, which has been used successfully to examine baseline differences between GIN resistant and susceptible Perendale sheep (Keane et al., 2006). Sheep researchers also have access to multiple large bovine arrays. The commercially available bovine Affymetrix array represents approximately 19,000 unigene clusters and has a 75% cross hybridisation to the ovine genome (Ross Tellam, personal communication). In addition, a 10,204 bovine cDNA array has successfully been utilised for cross species hybridisation with sheep, in which duodenal tissue from infected GIN resistant and susceptible Perendale sheep was compared (Diez-Tascon et al., 2005). A specialist ruminant innate immune response array has been constructed with 1480 characterised genes and 5376 cDNAs derived from subtracted and normalised libraries, which detected 94% of ovine transcripts tested (Donaldson et al., 2005). Other high throughput transcriptome techniques of a similar nature include serial analysis of gene expression (SAGE) used to investigate the response to trypanosome infection in cattle (Berthier et al., 2003) and massively parallel signature sequencing (MPSS).

A proteomic approach using SELDI-TOF mass spectroscopy, or similar, may also be used to identify differentially expressed genes based on their protein gene products as leads for further analyses. Again, the process interrogates samples from animals subjected to two different states and attempts to identify proteins that are differentially up or down regulated. This process has the benefit of being one step closer to the actual functional cellular mechanism, given that even though genes may be up or down regulated, there are still cellular controls that may prevent a corresponding up or down translation of the transcripts into proteins. Once again, differentially expressed proteins may not necessarily represent the most important genes controlling the trait, but may well contribute to a better-characterised pathway analysis. A major limitation to the proteomic approach has been the limited annotation of the many thousands of peptide fragments to genes and their function.

2.5 Positional Functional Integration

Despite large expenditure, most positional mapping studies are unsuccessful in refining the number of candidate genes to a manageable number, especially since many of the putative genes lying under the most likely positions remain uncharacterised. Further study would require exhaustive functional investigation of each in relation to the specific disease trait of interest. Likewise, candidates derived from functional transcriptomic, immunological or proteomic studies could number in the hundreds, without any clear idea of which ones, if any, were causative, merely reflecting a differentially expressed pathway, or even worse were spurious in often under powered micro-array studies.

However, a combination of the two approaches may give maximum power to identify true candidates. Transecting functional studies based on full transcriptome analyses with comprehensive micro-array and/or alternative techniques, such as

SAGE and MPSS, with QTL studies has the potential to yield functional information on relevant genes underlying QTL or so-called functional-positional candidate genes. Joint candidates obtained through both approaches may typically be screened in vitro and in vivo and sequenced to identify putative QTN.

Unequivocal proof that a specific mutation is causative of a QTL or a phenotype change remains elusive in livestock, with limited scope for gene substitution and targeted mutation induction to obtain such proof. A rare example of an appropriate expression study has used a murine NRAMP1-susceptible (now known as solute carrier family II member *A1* gene) macrophage cell line as a constant background to investigate the effect of a mutation in the bovine NRAMP gene under the control of the bovine NRAMP1 promoter, with concommittant differences in the levels of gene expression, as described in Barthel et al. (2001).

Functional transcriptome analyses and co-localisation studies with known QTL regions have been conducted for mastitis (Schwerin et al., 2003), trypanosome resistance using micro-array and SAGE approaches, respectively (Hill et al., 2005, Berthier et al., 2003) and for GIN in cattle (Gasbarre et al., 2004) and recently in sheep (Keane et al., 2006).

A technique that has long been proposed, but is still uncommon, is that of "microarray QTLs" or "e-QTLs" in which a suite of phenotypically variable animals are used and micro-array analyses of each individual animal used as the phenotype in QTL studies. Those genes in common between QTL location and expression profiles (*cis* acting) may lead to so-called master regulatory genes, whereas those not collocated with the QTL but differentially expressed or *trans* acting may lead to major pathway analyses in regulating gene expression. The difficulty of linking such e-QTL and their candidate genes to defined disease traits is still a complex issue, although once again meta-analyses of all functional and positional mapping studies should reveal functionally important genes for further analysis.

3 An Example: Mapping Genes for Ruminant Fasciolosis

An example of an integrated gene mapping and functional genomics programme to identify genes for resistance to fasciolosis has been described in full by Raadsma et al. (2005). In collaboration with many others, this example is informative in that multiple tools have been utilised in a targeted gene discovery programme. Two primary research arms are involved, a QTL mapping experiment to identify major genes for innate and acquired resistance to *F. gigantica* and an immunological characterisation of resistance mechanisms to fasciolosis to identify candidate gene(s) and pathways. The combination of information derived from both arms allowed greater refinement of a list of candidate genes than would have been possible using each approach alone.

Fasciolosis is a major parasitic disease of livestock with over 700 million production animals at risk of infection and worldwide economic losses estimated at >US\$3.2 billion per annum (Spithill et al., 1999). The parasite is especially prevalent in tropical regions of Asia and Africa, where it is considered the single most important helminth infection of cattle (Fabiyi, 1987; Schillhorn van Veen, 1980; Roy and Tandon, 1992; Pholpark and Srikitjakarn, 1989; Soesetya, 1975; Spithill et al., 1999). Infected animals show lower weight gain, anaemia, reduced fertility, reduced milk production, lower feed conversion efficiency and a diminished work capacity. In African and Southeast Asian developing countries, the latter trait significantly impacts on production, where ruminants provide 80% of the draught power (Spithill et al., 1999, Sukhapesna et al., 1994). In addition, there is a significant zoonotic infection rate in humans with the WHO recognising >2.4 million cases of infection.

Anthelminthics are currently utilised against fasciolosis, but they are expensive and require frequent administration, which prevents their application in many developing countries (Spithill et al., 1999). In addition, there is growing evidence of resistance to Triclabendazole, which has seriously compromised its use (Overend and Bowen, 1995; Spithill and Dalton, 1998). Vaccination is a second control mechanism that might be used and, indeed, experimental vaccination against *Fasciola* is well established (Spithill and Dalton, 1998). However, the vaccine-induced and natural acquired immune mechanisms linked to rejection of *Fasciola* infection in ruminants are not well defined (Piedrafita et al., 2004).

3.1 Resistance to Fasciola

At the outset of this programme observations had been made that the Indonesian Thin Tail (ITT) sheep breed expressed relative resistance to fasciolosis, whereas the Merino was shown to be highly susceptible (Wiedosari and Copeman, 1990; Roberts et al., 1997a, b, c). ITT sheep expressed resistance to *F. gigantica* within the prepatent period of a primary infection and acquired a higher level of resistance after exposure (Roberts et al., 1997a, c). The acquired resistance of ITT sheep, expressed within 3–4 weeks of infection, could be suppressed by dexamethasone, implying that this resistance was immunologically based and that the killing of many migrating parasites occurred within 2–4 weeks of infection (Roberts et al., 1997b; Spithill et al., 1999). Despite displaying resistance to *F. gigantica*, ITT sheep are fully susceptible to *F. hepatica* (Roberts et al., 1997a). These findings suggest that *F. hepatica* and *F. gigantica* differ in some fundamental biological trait(s), which renders *F. gigantica* more susceptible to immune effector mechanisms (Spithill et al., 1997; Piedrafita et al., 2004). The factors responsible for this apparent natural resistance to *Fasciola* infection in certain sheep breeds are currently unknown.

3.2 The Resource Flock for Mapping Fasciolosis Resistance

The resistance of ITT sheep represents a unique biological situation, which has been exploited in the construction of a QTL mapping resource flock. ITT sheep were backcrossed into susceptible Merino (Fig. 4) to generate 10 F1 ram families. Two strains of ITT sheep were used, namely Sumatra and Garut (West Java) to allow for possible strain variation in genetic resistance. Reciprocal matings were used in the creation of the F1 sires, to allow for recognition of sex-linked or imprinted genes. Four families were augmented in number by further matings of the F1 sire. In two cases, these augmented matings involved alternate backcross to ITT ewes and F2 intercross matings. Animals were maintained parasite free in holding pens above ground at the Indonesian Institute for Science, Bogor, Indonesia, under natural light conditions, and were fed ad libertum. A total of 694 animals from 10 linkage families were genotyped and phenotyped but to achieve this number, a total of 1435 animals were bought or born in the flock. As for all experimental flocks, a large investment is necessary to achieve a reasonable number of experimental animals.

Over 3 years, naïve experimental animals were subjected to two experimentally controlled *Haemonchus contortus* infections prior to a midyear challenge (average 373 days of age) with 300 metacercariae of *F. gigantica*. In the fourth year following the *H. contortus* trials, animals were sensitised with 50 metacercariae of *F. gigantica* (sensitization), drenched after 6 weeks, and then challenged with 250 metacercariae of *F. gigantica* (challenge at average 446 days of age). Challenges were slaughtered 12–15 weeks post-infection. Various phenotypes were measured including repeat measures of live weight (LW), packed cell volume (PCV), antibody levels, eosinophil, neutrophil and lymphocyte levels and liver enzymes. Flukes were recovered post slaughter (F_l), liver damage was assessed as a score (LvrSc) from 1 (least damage) to 5 (most damage) and liver weight (LvrWT) together with fluke biomass (WW) was obtained. In addition, use of this major resource flock was optimised by taking detailed carcass, growth and fleece production data to provide a better understanding of major genes affecting production traits.

3.3 Linkage and QTL Analysis for Fasciolosis

A low density genome screen of autosomal chromosomes was undertaken using 138 microsatellite markers selected on the basis of map location (Maddox et al., 2001) and polymorphic information content (PIC) scores from the Australian Sheep Gene Mapping Resource (http://rubens.its.unimelb.edu.au/~jillm/jill.htm). This is one of a few free Web-based resources with updated information on the International Mapping Flock sheep linkage map and associated tools such as CMap allowing comparative analysis to human, canine and bovine linkage and sequence information.

QTL analysis was undertaken for a range of parasite resistance and immune response traits, using the half-sib analysis capability of QTL Express (Seaton et al., 2002). This Web-based analysis tool stemming from collaborative work between the Roslin Institute and the University of Edinburgh is widely used and designed to allow user friendly regression-based QTL analysis of outbred populations. An initial analysis across all families for each trait included screens

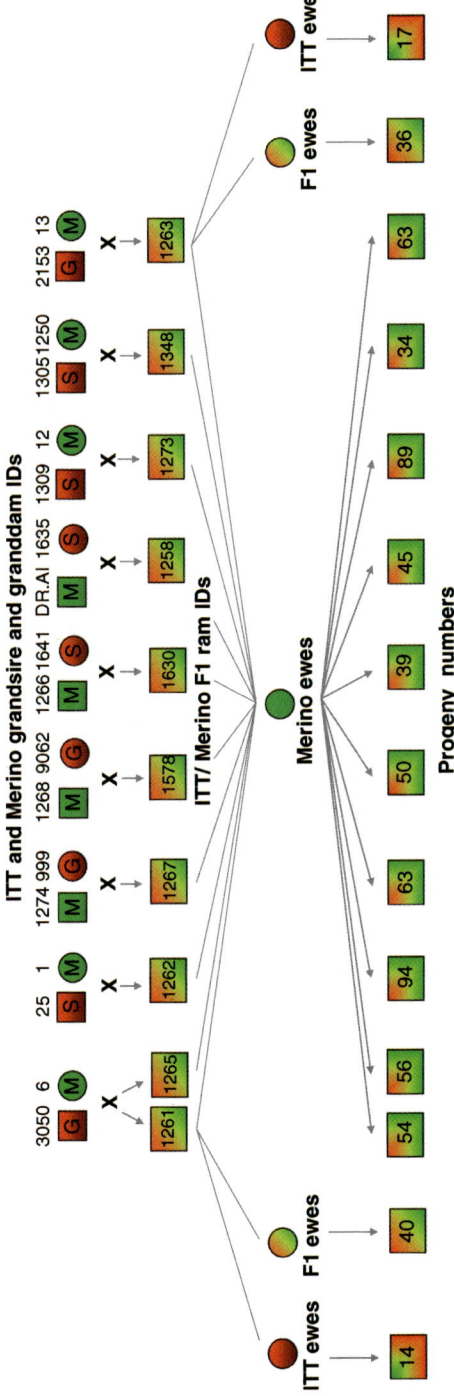

Fig. 4 Mating design of multi-family QTL mapping resource flock between Indonesian Thin Tail (*red*) (Garut and Sumatra) and Merino(*green*) parental breeds for mapping of disease resistance traits. Number of mapped progeny shown in progeny line

to determine the presence of potential QTL within selected families, on the basis of t-values. Chromosome-wide significance was empirically determined in QTL Express using 1000 permutations: significance to a level of $p \leq 0.05$ was considered suggestive and is summarised in Table 2.

As can be seen, the false discovery rate (FDR) suggests that approximately 1 in 2 of the QTL are potentially spurious, a relatively common (FDR) for QTL mapping studies. Lifting the significance threshold to improve the FDR may give rise to the problem of "throwing out the baby with the bathwater". Since it is unlikely that *F. gigantica* resistance is caused by a single major gene, a strong QTL signal is not expected, and increasing stringency may well cause an unacceptable level of type I error.

One means of weeding out the real QTL from the false is by examining multiple correlated traits. Where QTL are confirmed in two or more independently measured traits, with the same families exhibiting significant signal, more credence may be placed in their validity. For instance, OAR 17 displays a suggestive peak for both fluke number and fluke biomass. This peak appears mainly due to the influence of the first three cohorts, suggesting that it is likely to be caused by an innate response rather than an acquired immune response. In contrast, other QTL peaks appear to be due to the development of a protective immune response, and their validity is further informed by the parallel immunological work undertaken in this project.

Table 2 Summary of putative across-family QTL and significant individual-family t-values identified in a panel of parasite and immune response traits measured in 10 families after *F. gigantica* challenge. FDR = False discovery rate, calculated as observed/expected, where expected is 26×0.05 for number of genome-wide QTL, and $26 \times 10 \times 0.05$ for number of significant QTL observed in single families

Trait	Number putative QTL detected above chromosome-wide $p \leq 0.05$ threshold	Number QTL above $p \leq 0.05$ threshold in individual families	FDR: QTL genome wide, single family t-tests
Fluke count	4	36	3.08, 2.77
Fluke weight – biomass	3	33	2.31, 2.54
Fluke size	2	22	1.54, 1.69
Liver damage	2	26	1.54, 2.00
Liver weight	2	25	1.54, 1.92
Weight change	2	29	1.54, 2.23
PCV change	1	25	0.77, 1.92
Late average GLDH	2	17	1.54, 1.31
Late average GGT	3	24	2.31, 1.85
Late average AST	2	21	1.54, 1.62
Late average Albumin	4	30	3.08, 2.31
Late average Globulin	3	39	2.31, 3.00
Early Eosinophil volume	4	28	3.80, 2.15
Early Eosinophil count	1	20	0.77, 1.54
Early Neutrophil count	3	22	2.31, 1.69
Early Lymphocyte count	1	27	0.77, 2.08
IgA concentration Wk2	1	27	0.77, 2.08
Mean			1.81, 2.04

3.4 Mapping Fasciolosis QTL in Cattle and Buffalo

By exploiting the strong genomic similarity across ruminants, it is now feasible to use one species (i.e. sheep) as a model, allowing comparative biological mechanisms and gene mapping for the economically more important target species (i.e. buffalo and cattle). Based on Oxford grid analysis, the strong QTL identified on OAR17 maps almost exclusively to BTA17 which suggests that a putative QTL/gene with a role in fasciolosis resistance in cattle may be present on BTA17 (Fig. 5). While some ovine genes have been located on OAR17 through physical or linkage mapping, the location and function of the vast majority are inferred solely through comparative mapping with other species. Essentially, investigation of the gene content of the QTL region on ovine chromosome 17 has to occur in another species. Transferring the QTL peak and confidence intervals to homologous regions on BTA17 using integrated comparative map information suggests a list of a few hundred candidate genes found in the region.

In most cases, a few of these genes can be immediately targeted as potential candidates, given known functions in other species. For instance, the *IL2* gene and the *MIF* gene lie within the BTA17-transformed QTL region. The *IL2* gene has been shown to contain a GATA-3 transcription factor motif with a nucleotide polymorphism within the second intron (Luhken et al., 2005). As GATA-3 motifs appear to regulate the switch from a Th1 to a Th2 response in men and mice (Lee et al., 2000), its presence in the *IL2* gene may be of importance in the Th1 response to fasciola infection. On the other hand, MIF is a glutathione-s-transferase molecule that recruits immune cells after damage has been done. Allelic variation in the gene may affect the ability of the liver to regenerate.

However, immediate construction of a candidate gene shortlist is fraud with dangers. Given the very wide confidence intervals in most primary QTL analyses, there remain many genes of unknown function, or genes that have partially understood function, which may ultimately prove to harbour the QTN (evident in Fig. 5). In our project, the immunological characterisation of resistance to fasciolosis is providing some additional insight in short listing putative candidate genes.

3.5 Immunological Characterisation for Functional Positional Integration

The immune response(s) against invading *F. gigantica* parasites within the peritoneum is likely to involve complicated processes with numerous pathways and effector mechanisms. With such a complicated system, we needed to develop strategies to focus on key end-points (or "bottlenecks") of these pathways, mediating parasite death in the peritoneum. One such strategy used was to focus on which cellular responses occur at the time when killing of the parasite occurs, using high throughput and targeted tools. This strategy is based on the early finding from the literature, still as relevant today, that there are a relatively finite number of immune effector cells which can kill large extracellular parasites (Butterworth, 1984; Maizels et al.,

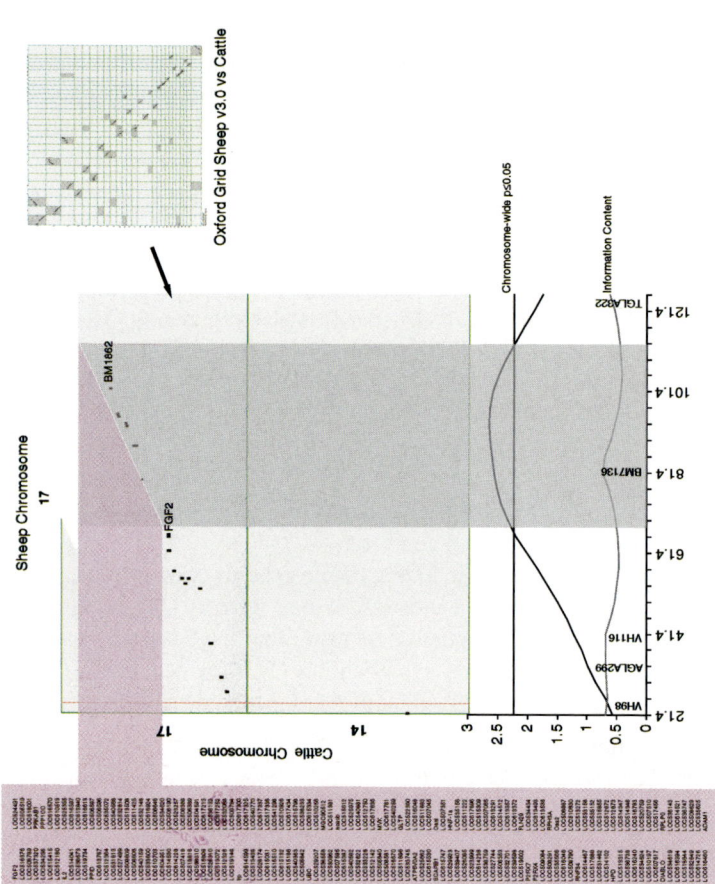

Fig. 5 QTL on sheep chromosome 17 with 95% confidence intervals highlighted in the *grey zone*, and the use of an Oxford Grid (*green block*) to infer bovine genes underlying the peak (*pink shading*). The *pink line* within the Oxford grid indicates a chromosome breakpoint, where the first part of sheep chromosome 17 maps to bovine chromosome 14, and the remainder maps to bovine chromosome 17

1993). The concept of the research was that by investigating their role in killing in vitro, and the toxic molecules produced which can kill the parasite, we would be able to predict the central pathway(s) directing the protective immune response(s) and identify the underlying resistance mechanism in ITT sheep. This would then allow identification of candidate genes, which control these pathways, to be compared with that generated through the QTL analyses.

High resistance of ITT sheep to *F. gigantica* appeared to follow a model whereby parasites were killed within the gut wall and/or peritoneum or shortly after reaching the liver, and that the newly excysted immature juvenile parasite (NEJ) was the primary target of an effective immune response in ITT sheep. We therefore decided to study the peritoneal cavity for immune mediators correlating with resistance in ITT sheep, using peritoneal lavage cells obtained from ITT and Merino sheep.

Our studies in this regard have shown that an antibody-dependent cell cyto-toxicity (ADCC) mechanism is expressed in ITT sheep (Piedrafita et al., 2007). NEJ of *F. gigantica* are susceptible to killing in vitro by peritoneal macrophages and eosinophils present in the peritoneal lavage cells. The mechanism involves superoxide radicals as evidenced by reversal of cytotoxicity following inclusion of superoxide dismutase (SOD) in the culture; SOD neutralizes immune cell–generated superoxide radicals. We suggest that this killing mechanism may be important in the resistance of ITT sheep to *F. gigantica* infection (Piedrafita et al., 2004; Meeusen and Piedrafita, 2003). In contrast, ITT sheep do not acquire resistance to *F. hepatica* infections (Roberts et al., 1997a) and the identified superoxide-mediated killing mechanisms effective against *F. gigantica* are ineffective against *F. hepatica* in vitro (Piedrafita et al., 2000, 2001, 2007).

3.6 High Density Proteomic and Genomic Functional Screening

In an attempt to unravel host responses expressed in sheep against *Fasciola* sp., we have recently begun to apply SELDI-TOF Mass Spectrometry (Issaq et al., 2002; Tang et al., 2004) to derive a protein profile of serum during infection of sheep with *F. hepatica*. Although these data are preliminary, the results suggest that serum biomarkers associated with *F. hepatica* infection may be readily discernible in sheep (Fig. 6). Sequence identification of these markers may reveal insights into the pathobiology of liver fluke infections in sheep. Comparative biomarker studies are underway in the resistant ITT and susceptible Merino after *F. gigantica* challenge. We have now commenced gene-specific expression profiles by RT-PCR analyses during different stages of infection in ITT and Merino backgrounds for a broad repertoire of cytokine profiles. In addition, we will be conducting transcriptome analyses using bovine micro-arrays and ruminant immune-response arrays.

3.7 Future Studies and Potential Applications

To our knowledge, this is the first genome screen conducted in ruminants for identification of resistance genes and their functional importance in fasciolosis. The

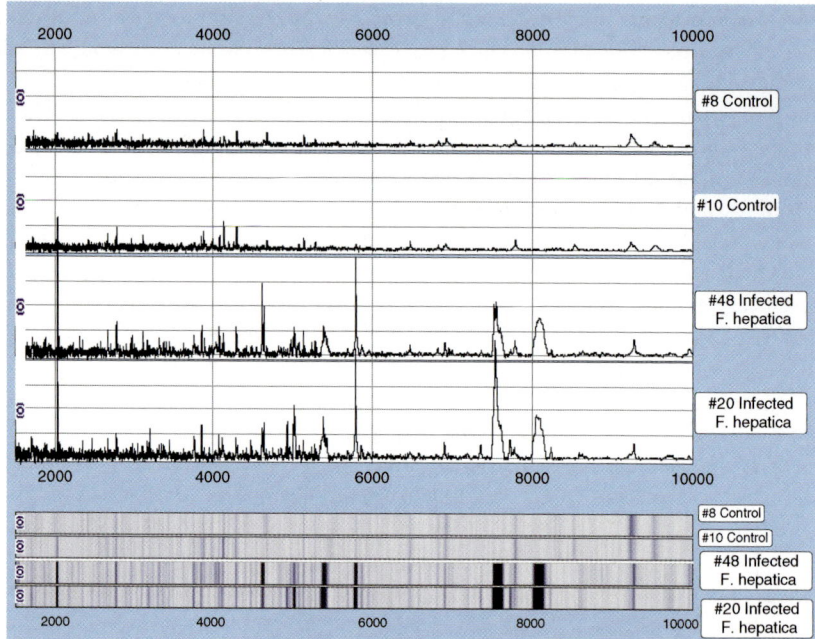

Fig. 6 SELDI-TOF MS analysis of serum biomarkers detected in sheep infected with *F. hepatica*. Panel A biomarkers detected in the pH 7 fraction in the range 1,500–10,000 Da. The *upper sections* show the trace views (mass spectrum) for two control sheep (#8, 10) and two infected sheep (# 48, 20). The *bottom sections* show the gel view conversion of the trace views to simplify comparisons of the profiles between groups (courtesy of Prof. Spithill)

combined approach to identify functional candidate genes by expression profiling of animals with defined genetic contrasting backgrounds and the use of positional candidate gene mapping provides an extremely powerful tool to dissect complex functional pathways in host responses to *F. gigantica*. The identification of candidate genes in ITT sheep will allow rapid identification of similar genes in cattle and buffalo due to the high degree of homology between the ovine and the bovine genomes (Cockett et al., 2001). Such genes could be used as DNA markers for resistance to identify elite resistant animals for selective breeding programmes, which would improve the productivity of cattle/buffalo herds in Indonesia as well as Southeast Asia, Asia and Africa. Identification of such gene(s) and understanding their mode of action can also lead to novel biological methods for parasite control, through the identification of new biological compounds. The model described here highlights the potential power and impact to discover target molecules, molecular pathways and genetic variation in genes regulating such pathways using a model animal system which is more closely aligned to the target species than the traditional mouse or human systems. Although the genome information–rich model organisms will continue to have high utility in early stages of discovery, ultimate validation studies have to be conducted in the target animal species. The use of appropriate ruminant models can accelerate this process.

Acknowledgments The research on fasciolosis reported in this chapter was strongly supported by funds from the Australian Centre for International Agricultural Research (ACIAR) in a bilateral research project between Australia and Indonesia. The input from Dr P. Thomson in the development of the statistical procedures for segregation analyses is gratefully acknowledged. The strong technical and managerial support from a very large team of highly dedicated technical staff both in Indonesia and in Australia has made this project possible, for which we are very grateful. Studies in Canada are supported by grants from the National Sciences and Engineering Research Council of Canada and the Canada Research Chair Program.

References

Axford, R.F.E., Bishop, S.C., Nicholas, F.W., and Owen, J.B., 2000, Genetics of susceptibility to production diseases in cattle and sheep, in R.F.E. Axford, S.C. Bishop, F.W. Nicholas, and J.B. Owen, eds., Breeding for Disease Resistance in Farm Animals, 418pp.

Baker, R.L., Gibson, J.P., Iraqi, F.A., Menge, D.M., Mugambi, J.M., Hanotte, O., Nagda, S., Wakelin, D., and Behnke, J.M., 2003, Exploring the genetic control of resistance to gastrointestinal helminth infections in sheep and mice, *Proc. 15th Conf., Assoc. Advan. Anim. Breed. Genet.*, Melbourne, Australia, 7–11 July 2003, pp. 183–190.

Barillet, F., Arranz, J.J., and Carta, A., 2005, Mapping quantitative trait loci for milk production and genetic polymorphisms of milk proteins in dairy sheep, *Genet. Sel. Evol.* **37**: S109–S123.

Barillet, F., Carta, A., Allain, D., Amigues, Y., Bodin, L., Casu, S., Cribiu, E.P., Bed'hom, B., Boichard, D., Boscher, M.Y., Elsen, J.M., Fraghi, A., Gruner, L., Jacquiet, P., Ligios, S., Marie-Etancelin, C., Mura, L., Piredda, G., Roig, A., Rupp, R., Sanna, S.R., Scala, A., Schibler, L., Sechi, T., and Casu, S., 2003, Detection of QTL influencing present and future economic traits in dairy sheep in France and Italy, in 10èemes Rencontres autour des Recherches sur les Ruminants, Institut National de la Recherche Agronomique, Paris, France, 3–4 Décembre 2003, pp. 57–60.

Barthel, R., Feng, J.W., Piedrahita, J.A., McMurray, D.N., Templeton, J.W., and Adams, L.G., 2001, Stable transfection of the bovine NRAMP1 gene into murine RAW264.7 cells: effect on *Brucella abortus* survival, *Infect. Immun.* **69**:3110–3119.

Bayon, Y., Gutierrez-Gil, B., De La Fuente, L.F., San Primitivo Tirados, F., El Zarei, M.F., Alvarez Castelo, L., Arranz, J.J., Perez, J., and Rojo Vasquez, F.A., 2004, Búsqueda de regiones genómicas con influencia sobre la resistencia a las tricostrongilidosis en el ganado ovino de raza churra ITEA, *Prod. Anim.* **100**:197–201.

Beh, K.J., Hulme, D.J., Callaghan, M.J., Leish, Z., Lenane, I., Windon, R.G., and Maddox, J.F., 2002, A genome scan for quantitative trait loci affecting resistance to *Trichostrongylus colubriformis* in sheep, *Anim. Genet.* **33**:97–106.

Benavides, M.V., Weimer, T.A., Borba, M.F.S., Berne, M.E.A., and Sacco, A.M.S., 2002, Association between microsatellite markers of sheep chromosome 5 and faecal egg counts, *Small Ruminant Res.* **46**:97–105.

Berthier, D., Quere, R., Thevenon, S., Belemsaga, D., Piquemal, D., Marti, J., and Maillard, J.C., 2003, Serial analysis of gene expression (SAGE) in bovine trypanotolerance: preliminary results, *Genet. Sel. Evol.* **35**:S35–S47.

Brenig, B., Broad, T.E., Cockett, N.E., and Eggen, A., 2004, Achievements of research in the field of molecular genetics, WAAP Book of the Year, 2003: A Review on Developments and Research in Livestock Systems, Wageningen Universiteit (Wageningen University), Wageningen Netherlands, pp. 73–84.

Butterworth, A.E. 1984. Cell-mediated damage to helminths. *Adv. Parasit.* **23**:143–235.

Carta, A., Barillet, F., Allain, D., Amigues, B., Bibe, B., Bodin, L., Casu, S., Cribiu, E., Elsen, J.M., Fraghi, A., Gruner, L., Jacuiet, P., Ligios, S., Marie-Etancelin, C., Mura, L., Piredda, G.,

Rupp, R., Sanna, S.R., Scala, A., Scala, A., Schibler, L., and Casu, S., 2002, QTL Detection with genetic markers in a dairy sheep backcross Sarda × Lacaune Resource Population, *Proc. 7th World Congr. Genet. Appl. Livest. Prod.* Montpellier, France, August 2002, CD-Rom, Com.No. 09–07.

Cockett, N.E., Shay, T.L., and Smit, M., 2001, Analysis of the sheep genome, *Physiol. Genomics* **7**:69–78.

Coltman, D.W., Wilson, K., Pilkington, J.G., Stear, M.J., and Pemberton, J.M., 2001, A microsatellite polymorphism in the gamma interferon gene is associated with resistance to gastrointestinal nematodes in a naturally-parasitized population of Soay sheep, *Parasitology* **122**:571–582.

Crawford, A.M., and McEwan, J.C., 1998, Identification of animals resistant to nematode parasite infection New Zealand Provisional Patent 330201, New Zealand, p. 46.

Crawford, A.M., McEwan, J.C., Dodds, K.G., Wright, C.S., Bisset, S.A., Macdonald, P.A., Knowler, K.J., Greer, G.J., Green, R.S., Shaw, R.J., Paterson, K.A., Cuthbertson, R.P., Vlassoff, A., Squire, D.R., West, C.J., and Phua, S.H., 1997a, Resistance to nematode parasites in sheep: how important are the MHC genes? *Assoc. Advanc. Anim. Breeding* and *Genet. Proc. Twelfth Conf.*, Dubbo, NSW, Australia 6–10 April 1997, Part 1, pp. 58–62.

Crawford, A.M., Phua, S.H., McEwan, J.C., Dodds, K.G., Wright, C.C., Morris, C.A., Bisset, S.A., and Green, R.S., 1997b, Finding disease resistance QTL in sheep, *Anim. Biotechnol.* **8**:13–22.

Crawford, A.M., Paterson, K.A., Dodds, K.G., Tascon, C.D., Williamson, P.A., Thomson, M.R., Bisset, S.A., Beattie, A.E., Greer, G.J., Green, R.S., Wheeler, R., Shaw, R.J., Knowler, K., and McEwan, J.C., 2006, Discovery of quantitative trait loci for resistance to parasitic nematode infection in sheep: I. Analysis of outcross pedigrees, *BMC Genomics* **7**:2164–2178.

Davies, G., Stear, M.J., Benothman, M., Abuagob, O., Kerr, A., Mitchell, S., and Bishop, S.C., 2006, Quantitative trait loci associated with parasitic infection in Scottish blackface sheep, *Hered.* **96**:252–258.

Diez-Tascon, C., Keane, O.M., Wilson, T., Zadissa, A., Hyndman, D.L., Baird, D.B., McEwan, J.C., and Crawford, A.M., 2005, Microarray analysis of selection lines from outbred populations to identify genes involved with nematode parasite resistance in sheep, *Physiol. Genomics* **21**:59–69.

Diez-Tascon, C., MacDonald, P.A., Dodds, K.G., McEwan, J.C., and Crawford, A.M., 2002, A screen of chromosome 1 for QTL affecting nematode resistance in an ovine outcross population, *Proc. 7th World Congr. Genet. Appl. Livestock Prod.*, Institut National de la Recherche Agronomique (INRA), Montpellier, France, August 2002, Session 13, pp. 1–4, Communication 37.

Donaldson, L., Vuocolo, T., Gray, C., Strandberg, Y., Reverter, A., McWilliam, S., Wang, Y., Byrne, K., and Tellam, R., 2005, Construction and validation of a Bovine Innate Immune Microarray, *BMC Genomics* **6**:135.

Dukkipati, V.S.R., Blair, H.T., Johnson, P.L., Murray, A., and Garrick, D.J., 2005, A study on the association of genotypes at the interferon gamma microsatellite locus with faecal strongyle egg counts in sheep, *Proc. 16th Conf. Assoc. Advan. Anim. Breeding Genet.*, Noosa Lakes, Australia, pp. 119–122.

Elsen, J.M., Moreno, C.R., Bodin, L., François, D., Bouix, J., Barillet, F., Allain, D., Lantier, F., Lantier, I., Schibler, L., Roig, A., Brunel, J.C., and Vitezica, Z.G., 2006, Selectio for scrapie resistance in France, Is there evidence of negative effects on production and health traits?, Proc. 8th World Cong. *Genet. Appl. Livestock Prod.*, Belo Horizonte, MG, Brazil, pp. 15–15.

Fabiyi, J.P., 1987, Production Losses and Control of Helminths in Ruminants of Tropical Regions, *Int. J. Parasitol.* **17**:435–442.

Gasbarre, L.C., Sonstegard, T., VanTassell, C.P., and Araujo, R., 2004, Symposium: New approaches in the study of animal parasites, *Vet. Parasitol.* **125**:147–161.

Gibson, J.R., and Bishop, S.C., 2005, Use of molecular markers to enhance resistance of livestock to disease: a global approach, *Rev. Sci. Tech. OIE.* **24**:343–353.

GLIPHA: Global Livestock Production and Health Atlas www.fao.org/ag/aga/glipha.

Gruner, L., Aumont, G., Getachew, T., Brunel, J.C., Pery, C., Cognie, Y., and Guerin, Y., 2003, Experimental infection of Black Belly and INRA 401 straight and crossbred sheep with trichostrongyle nematode parasites, *Vet. Parasitol.* **116**:239–249.

Hill, E.W., O'Gorman, G.M., Agaba, M., Gibson, J.P., Hanotte, O., Kemp, S.J., Naessens, J., Coussens, P.M., and MacHugh, D.E., 2005, Understanding bovine trypanosomiasis and trypanotolerance: the promise of functional genomics, *Vet. Immun. Immunop.* **105**:247–258

Issaq, H.J., Veenstra, T.D., Conrads, T.P., and Felschow, D., 2002, The SELDI-TOF MS approach to proteomics: Protein profiling and biomarker identification, *Biochem. Biophys. Res. Commun.* **292**:587–592.

Janssen, M., Weimann, C., Gauly, M., and Erhardt, G., 2002, Associations between infections with *Haemonchus contortus* and genetic markers on ovine chromosome 20, *Proc. 7th World Cong. Genet. Appl. Livestock Prod.*, Montpellier, France, August 2002, Session 13, pp. 1–4, Communication 11.

Keane, O.M., Zadissa, A., Wilson, T., Hyndman, D.L., Greer, G.J., Baird, D.B., McCulloch, A.F., Crawford, A.M., and McEwan, J.C., 2006, Gene expression profiling of naive sheep genetically resistant and susceptible to gastrointestinal nematodes, *BMC Genomics* **7**:42.

Khatkar, M.S., Collins, A., Cavanagh, J.A.L., Hawken, R.J., Hobbs, M., Zenger, K.R., Barris, W., McClintock, A.E., Thomson, P.C., Nicholas, F.W., and Raadsma, H.W., 2006, A first-generation metric linkage disequilibrium map of bovine chromosome 6, *Genetics* **174**:79–85.

Khatkar, M.S., Thomson, P.C., Tammen, I., and Raadsma, H.W., 2004, Quantitative trait loci mapping in dairy cattle: review and meta-analysis, *Genet. Select. Evol.* **36**:163–190.

Lee, H.J., Tokemoto, N., Kurata, H., Kamogawa, Y., Miyatake, S., o'Garra, A., and Arai, N., 2000, GATA-3 induces Thelper cell type 2 (Th$_2$) cytokine expression and chromatin remodelling in committed Th$_i$ cells, *J. Exp. Med.* **192**:105–115.

Li, Y., Miller, J.E., and Franke, D.E., 2001, Epidemiological observation and heterosis of gastrointestinal nematode infection in Suffolk, Gulf Coast Native and crossbred lambs, *Vet. Parasitol.* **98**:273–283.

Liao, W., Collins, A., Hobbs, M., Khatkar, M.S., Luo, J., and Nicholas, F.W., 2007, A comparative location database (CompLDB): map integration within and between species, *Mamm. Genome* **18**:287–299.

Luhken, G., Stamm, V., Menge, C., and Erhardt, G., 2005, Functional analysis of a single nucleotide polymorphism in a potential binding site for GATA transcription factors in the ovine interleukin 2 gene, *Vet. Immun. Immunop.* **107**:51–56.

Maddox, J.F., Davies, K.P., Crawford, A.M., Hulme, D.J., Vaiman, D., et al., 2001, An enhanced linkage map of the sheep genome comprising more than 1000 loci, *Genome Res.* **11**:1275–1289.

Maizels, R.M., Bundy, D.A.P., Selkirk, M.E., Smith, D.F., and Anderson, R.M., 1993, Immunological modulation and evasion by helminth-parasites in human-populations, *Nature* **365**:797–805.

Marshall, K., van der Werf, J.H.J., Maddox, J.F., Graser, H.-U., Zhang, Y., Walkden-Brown, S.W., and Khan, L., 2005, A genome scan for quantitative trait loci for resistance to the gastrointestinal parasite *Haemonchus contortus* in sheep, *Proc. Assoc. Advan. Anim. Breed. Genet.*, Noosa Lakes, Queensland, Australia, pp. 115–118

Meeusen, E.N.T., and Piedrafita, D., 2003, Exploiting natural immunity to helminth parasites for the development of veterinary vaccines, *Int. J. Parasitol.* **33**:1285–1290.

Miller, J.E., Cocket, N.E., Walling, G.A., Shay, T.A., McGraw, R.A., Bishop, S.C., and Haley, C.A., 2002, Segregation of resistance to nematode infection in F2 lambs of Suffolk $\times$ Gulf Coast Native sheep and associated QTL, *Proc. Int. Soc. Anim. Genet.*, Gottingen, Germany, pp. 160.

Moreno, C.R., Gruner, L., Scala, A., Mura, L., Schibler, L., Amigues, Y., Sechi, T., Jacquiet, P., Francois, D., Sechi, S., Roig, A., Casu, S., Barillet, F., Brunel, J.C., Bouix, J., Carta, A., and Rupp, R., 2006, QTL for resistance to internal parasites in two designs based on natural and experimental conditions of infection, *Proc. 8th World Cong. Genet. Appl. Livestock Prod.*, Belo Horizonte, MG, Brazil, pp. 15–05.

Moreno, C.R., Lantier, F., Berthon, P., Gautier, A.V., Bouchardon, R., Boivin, R., Lantier, I., Brunel, J.-C., Weisbecker, J.-L., François, D., Bouix, J., and Elsen, J.-M., 2003, Genetic parameters for resistance to the salmonella abortosovis vaccinal nstrain RU6 in sheep, *Gen. Sel. Evol.* **35**:199–217.

Nicholas, F.W., 2005, Integrated and comparative maps in livestock genomics, *Austr. J. Exper. Agric.* **45**:1017–1020.

Overend, D.J., and Bowen, F.L., 1995, Resistance of *Fasciola hepatica* to Triclabendazole, *Aust. Vet. J.* **72**:275–276.

Paterson, K.A., McEwan, J.C., Dodds, K., Morris, C.A., and Crawford, A.M., 2001, Fine mapping a locus affecting host resistance to internal parasites in sheep, *Proc. Assoc. Advan. Anim. Breed. Genet.*, Queenstown, New Zealand, pp. 91.

Paterson, S., Wilson, K., and Pemberton, J.M., 1998, Major histocompatibility complex variation associated with juvenile survival and parasite resistance in a large unmanaged ungulate population (*Ovis aries* L.), *Proc. Nat. Acad. Sci. USA* **95**:3714–3719.

Pholpark, M., and Srikitjakarn, L., 1989, The control of parasitism in swamp buffalo and cattle in north-east Thailand, International Seminar on Animal Health and Production Services for Village Livestock, Khon Kaen, Thailan, pp. 244–249.

Piedrafita, D., Estuningsih, E., Pleasance, J., Prowse, R., Raadsma, H.W., Meeusen E.N.T., and Spithill, T.W., 2007, Peritoneal lavage cells of Indonesian thin-tail sheep mediate antibody-dependent superoxide radical cytotoxicity in vitro against newly excysted juvenile *Fasciola gigantica* but not juvenile *Fasciola hepatica*, *Infect. Immun.* **75**:1954–1963.

Piedrafita, D., Parsons, J.C., Sandeman, R.M., Wood, P.R., Estuningsih, S.E., Partoutomo, S., and Spithill, T.W., 2001, Anti body-dependent cell-mediated cytotoxicity to newly excysted juvenile *Fasciola hepatica* in vitro is mediated by reactive nitrogen intermediates, *Parasite Immun.* **23**:473–482.

Piedrafita, D., Raadsma, H.W., Prowse, R., and Spithill, T.W., 2004, Immunology of the host-parasite relationship in fasciolosis (*Fasciola hepatica* and *Fasciola gigantica*), *Can. J. Zoo.* **82**:233–250.

Piedrafita, D., Spithill, T.W., Dalton, J.P., Brindley, P.J., Sandeman, M.R., Wood, P.R., and Parsons, J.C., 2000, Juvenile *Fasciola hepatica* are resistant to killing in vitro by free radicals compared with larvae of *Schistosoma mansoni*, *Parasite Immun.* **22**:287–295.

Raadsma, H.W., Margawati, E.T., Piedrafita, D., Estuningsih, E., Widjayanti, S., Beriajaja, Subandriyo, Thomson, P., and Spithill, T., 2002, Towards molecular genetic characterisation of high resistance to internal parasites in Indonesian Thin Tail sheep, *Proc. 7th World Cong. Genet. Appl. Livest. Prod.*, August 2002, Institut National de la Recherche Agronomique (INRA), Montpellier, France, Session 13, pp. 1–4, Communication 18.

Raadsma, H.W., Piedrafita, D., Kingsford, N.M., Fullard, K.J., Margawati, E.T., Estuningsih, E., Widjayanti, S., Subandriyo, Clairoux, N., and Spithill, T., 2005, A functional and comparative genomics approach to characterize the high resistance of Indonesian Thin Tail (ITT) sheep to fasciolosis as a model for ruminants, In Brief and In Depth, CABI Publishing, AgBiotechNet.

Roberts, J.A., Estuningsih, E., Widjayanti, S., Wiedosari, E., Partoutomo, S., and Spithill, T.W., 1997a, Resistance of Indonesian thin tail sheep against *Fasciola gigantica* and *F. hepatica*, *Vet. Parasitol.* **68**:69–78.

Roberts, J.A., Estuningsih, E., Wiedosari, E., and Spithill, T.W., 1997b, Acquisition of resistance against *Fasciola gigantica* by Indonesian thin tail sheep, *Vet. Parasitol.* **73**:215–224.

Roberts, J.A., Widjayanti, S., Estuningsih, E., and Hetzel, D.J., 1997c, Evidence for a major gene determining the resistance of Indonesian thin tail sheep against *Fasciola gigantica*, *Vet. Parasitol.* **68**:309–314.

Roy, B., and Tandon, V., 1992, Seasonal prevalence of some zoonotic trematode infections in cattle and pigs in the north-east Montane zone in India, *Vet. Parasitol.* **41**:69–76.

Sayers, G., Good, B., Hanrahan, J.P., Ryan, M., and Sweeney, T., 2005, Intron 1 of the interferon gamma gene: its role in nematode resistance in Suffolk and Texel sheep breeds, *Res. Vet. Sci.* **79**:191–196.

Schillhorn van Veen, T.W., 1980, Fascioliasis (*Fasciola gigantica*) in West Africa: a review, *Vet. Bull.* **50**:529–533.

Schwerin, M., Czernek-Schafer, D., Goldammer, T., Kata, S.R., Womack, J.E., Pareek, R., Pareek, C., Walawski, K., and Brunner, R.M., 2003, Application of disease-associated differentially expressed genes – Mining for functional candidate genes for mastitis resistance in cattle, *Genet. Sel. Evol.* **35**:S19–S34.

Seaton, G., Haley, C.S., Knott, S.A., Kearsey, M., and Visscher, P.M., 2002, QTL Express: mapping quantitative trait loci in of simple and complex pedigrees, *Bioinform.* **18**:339–340.

Shay, T.L., Miller, J.E., McGraw, R.A., Walling, G.A., Bishop, S.C., Haley, C.A., and Cocket, N.E., 2002, Characterisation of QTL associated with resistance to nematode infection in sheep, *Int. Soc. Anim. Genet.*, Gottingen, Germany, pp. 174.

Soesetya, R.H.B., 1975, The prevalence of *Fasciola gigantica* infection in cattle in East Java, Indonesia, *Malay. Vet. J.* **6**:5–8.

Spithill, T.W., and Dalton, J.P., 1998, Progress in development of liver fluke vaccines, *Parasitol. Today* **14**:224–228.

Spithill, T.W., Piedrafita, D., and Smooker, P.M.,1997, Immunological approaches for the control of fasciolosis, *Int. J. Parasitol.* **27**:1221–1235.

Spithill, T.W., Smooker, P.M., and Copeman, D.B., 1999, Fasciola gigantica: epidemiology, control, immunology and molecular biology, in J.P. Dalton, ed., Fasciolosis, CAB International, Oxford, pp. 465–525.

Stear, M.J., Abuagob, O., Ben Othman, M., and Bishop, S.C., 2006, Major genes and resistance to nematode infection in naturally infected Scottish Blackface lambs, *Proc. 8th World Cong. Genet. Appl. Livest. Prod.*, Belo Horizonte, M.G., Brazil, pp. 15–21.

Sukhapesna, V., Tantasuvan, D., Sarataphan, N., and Imsup, K., 1994, Economic impact of fasciolosis in buffalo production, *J. Thai Vet. Med. Assoc.* **45**:45–52.

Tang, N., Tornatore, P., and Weinberger, S.R., 2004, Current developments in SELDI affinity technology, *Mass Spect. Rev.* **23**:34–44.

Wiedosari, E., and Copeman, D.B., 1990, High-resistance to experimental-infection with *Fasciola gigantica* in Javanese thin-tailed sheep, *Vet. Parasitol.* **37**:101–111.

Integrating Genomics to Understand the Marek's Disease Virus – Chicken Host–Pathogen Interaction

Hans H. Cheng

1 Introduction

Poultry is the third largest agricultural commodity (larger than any plant species) and the primary meat consumed in the US. According to the USDA Agricultural Statistics (www.nass.usda.gov), in 2004 (latest year with complete information), the US produced 45.8 billion pounds of chicken meat, 7.3 billion pounds of turkey meat, and 87.5 billion eggs for combined sales totaling $28.9 billion (up 24% from 2003!), and the industry is the largest producer and exporter of poultry meat in the world. Primarily, due to advanced breeding programs, tremendous progress in production traits has been made to meet the growing demands of consumers.

Several major issues confront the poultry industry today. Infectious diseases are certainly at or near the top of the list. Avian influenza, exotic Newcastle's disease, and Salmonella are just a few pathogens well known to the public that harm the poultry industry through loss of birds, reduced public confidence, and lost market accessibility via trade restrictions. Disease outbreaks or the potential for them to occur are enhanced by more concentrated chicken rearing and reduced genetic diversity from industry consolidation. Changes in animal husbandry (e.g., "all in, all out" rearing), new vaccines, etc. have helped to alleviate some of the problems; however, improved or alternative control measures are still needed in the near future to address current diseases and impede emerging threats.

The field of genomics offers one of the more exciting avenues for solving many of these issues. While still in its formative years, by identifying quantitative trait loci (QTL) and genes that control heritable traits of agricultural importance, it is possible to select for birds with superior agricultural traits via marker-assisted selection (MAS). Other positive attributes commonly cited for MAS include greater speed and accuracy compared to traditional breeding. Furthermore, for infectious diseases, MAS would eliminate the exposure risk to elite flocks associated with handling a hazardous pathogen. The recent release of the chicken genome sequence

H.H. Cheng
USDA, ARS, Avian Disease and Oncology Laboratory, 3606 E. Mount Hope Rd., East Lansing, MI 48823, USA

J.P. Gustafson et al. (eds.), *Genomics of Disease*,
© Springer Science+Business Media, LLC, 2008

(Hillier et al., 2004; $\sim \times 6.6$ coverage) and funded improvements to finish the assembly only increase the power of this discipline. The ultimate goal is to address the long-standing question of how genetic variation explains the observed phenotypic variation.

In this review, I will briefly describe Marek's disease (MD), the most serious chronic disease problem facing the poultry industry, and a very interesting model for cancer, vaccines, viral evolution, and host–pathogen interactions. Then I will discuss how we are using genomic and functional genomic approaches to identify genes and pathways that confer resistance to MD. Given that most labs have limited finances and resources, this integrated genomics strategy may be appealing to others. Finally, the impact of the chicken genome sequence on our approach is also described.

2 Marek's Disease

MD is a T cell lymphoma disease of domestic chickens induced by a naturally oncogenic, highly cell-associated α-herpesvirus referred to as the Marek's disease virus (MDV) (Marek, 1907; Churchill and Biggs, 1967; Nazerian and Burmester, 1968; Solomon et al., 1968). The disease is characterized by a mononuclear infiltration of the peripheral nerves, gonads, iris, various viscera, muscles, and the skin. Partial or complete paralysis is a common symptom of MD due to accumulation and proliferation of tumor cells in peripheral nerves.

During the 1960s as the industry converted to high-intensity rearing, MD generated tremendous economic losses. Since the 1970s, MD has been controlled through the use of vaccination and improved animal husbandry. However, even with vaccines, annual losses in the US. by MD due to meat condemnation and reduced egg production exceed $160 million (Purchase, 1985), which is a minimum estimate since the figure has not been revised to reflect inflation, new disease outbreaks, or MDV-induced immunosuppression. Although vaccination prevents the formation of lymphoma and other MD symptoms, it does not prevent MDV infection, replication, or horizontal spread (Purchase and Okazaki, 1971). Moreover, even though available vaccines protect chickens against the disease, MD still remains a threat due to increasingly frequent outbreaks of highly virulent strains of the MDV combined with the incomplete immunity that is elicited by vaccination (Witter et al., 1980; Schat et al., 1981; Osterrieder et al., 2006).

By inoculating susceptible chickens with oncogenic MDV, four different phases of infection have been established: (1) Early cytolytic infection of the lymphoid organs, (2) a period of latent infection, (3) late cytolytic infection, and (4) the transformation of T-lymphocytes. In genetically resistant hosts, only the early cytolytic infection followed by the establishment of lifelong latency has been observed. Transformation is known to occur only in T cells. It is also believed that T cells are susceptible to MDV infection only after activation. The mechanism that leads from latency to transformation is not well understood. Yet, current evidence suggests that latent infection is a prerequisite to transformation.

MDV, the causative pathogen, is a herpesvirus that has lymphotropic properties similar to those of γ-herpesviruses, such as Epstein–Barr virus (EBV) in humans, where the host range is restricted and latency is often evident in lymphoid tissue. Viruses of this group are capable of transforming cells in natural hosts. MDV was re-classified, however, because its molecular structure and genomic organization more closely resembles that of α-herpesviruses, such as herpes simplex virus (HSV) and Varicella-Zoster virus (VZV) (Buckmaster et al., 1988). The complete sequence of several MDV strains has been determined (Tulman et al., 2000; Lee et al., 2000; Niikura et al., 2006).

2.1 MD as a Model

In addition to its agricultural importance, MD is a fantastic model for studying vaccine efficacy and protection. In the US, ∼1 million chickens are processed per hour. The vast majority of these chickens are vaccinated *in ovo* with MD vaccines, which probably makes them the most widely administered vaccine in the world. These vaccines are very effective as the latest incidence rates for leukosis from the USDA Agricultural Statistics are <0.01%. The problem is MDV strains evolve and show regular increases in virulence (Witter, 1997). This means that the vaccine industry must vigilant and produce a new generation of MD vaccine every 10 years or so to stay ahead of the game. And it appears that the current and most protective vaccine (CVI988) has reached the maximum efficacy (Witter and Kreager, 2004), which suggests the need for either new vaccines (e.g., recombinant) or supplementation with other controls measures (e.g., genetic resistance).

Furthermore, MDV is one of very few herpesviruses that are naturally oncogenic in its host, which makes it a model for viral-induced transformation. And since the vaccines prevent tumor formation rather than viral replication or spread, it presents an opportunity to examine vaccinal immunity to cancer.

2.2 Genetic Resistance

Chickens resistant to MD are those that fail to develop characteristic symptoms upon exposure to MDV. Genetic differences in resistance to fowl paralysis, assumed to be MD, have been reported for 70 years (Asmundson and Biely, 1932). Since that time, chickens have been selected for resistance to MD and several inbred lines have been developed. In 1939, the Regional Poultry Research Laboratory (renamed ADOL in 1990) initiated the development of inbred lines to study what was then known as the avian leukosis complex. Of the fifteen lines initially developed, two proved to be the most interesting with respect to MD. When Line 6 chicks are inoculated with the JM strain of MDV at 1 day of age, less than 3% of the chicks will succumb to the disease. In contrast, similar inoculations into Line 7 chicks will result in greater than 85% mortality. These lines are maintained at ADOL and are over 99% inbred.

MD resistance is controlled by multiple genes or QTL. For example, the level of disease resistance as measured by mortality among F_1 siblings of a cross between Lines 6 and 7 is intermediate to the parents, approximately 60% (Stone, 1975). The levels of resistance observed in an F_2 population encompass a large spectrum indicating that there is more than one gene for resistance involved. Further analysis of additional crosses suggests that there are only very few loci of sizeable effect that encode resistance to MD, which makes individual gene identification difficult as most of the genes will contribute only a small portion of the measurable effect.

The best understood mechanism for the involvement of genetic resistance to MD involves the MHC or, as it is known in the chicken, the B complex. The MHC contains three tightly linked regions known as B-F, B-G, and B-L which control cell surface antigens. The B-G (class IV) locus is expressed in erythrocytes, which enables convenient typing of blood groups. By measuring the frequency of specific blood groups or using B congenic chickens (lines that have a common genetic background and differ only in the MHC), it has been observed that certain B alleles can be associated with resistance or susceptibility. In general, chickens with the B^{21} allele have been found to be more resistant than those with other B haplotypes (Bacon, 1987; Bacon and Witter, 1992). Other studies have allowed for the relative ranking of the other B alleles: moderate resistance, B^2, B^6, B^{14} and susceptibility, B^1, B^3, B^5, B^{13}, B^{15}, B^{19}, B^{27} (Longenecker and Mosmann, 1981). This relationship with MD resistance has been often cited as the classic example of the MHC haplotype effect, which is facilitated by the relative simplicity of the chicken MHC. However, the B haplotype effect is dependent on the genetic background as shown by several studies (e.g., Bacon et al., 1981; Hartman, 1989). The B-haplotype also influences vaccinal immunity as some haplotypes develop better protection with vaccines of one serotype than of a different serotype (Bacon and Witter, 1994a;b).

Besides the MHC, other genetic factors are known to exist that have a major influence on MD resistance. For example, Lines 6 and 7 chickens share the same B haplotype, B^2 (Hunt and Fulton, 1998), yet differ greatly with respect to resistance to MD. In contrast to MHC-controlled resistance, non-MHC genetic resistance may be related to the number of target cells. Spleen and thymus cells from Line 6 chickens absorb less MDV than do similar cells from Line 7 chickens (Gallatin and Longenecker, 1979; Powell et al., 1982). The size of the primary lymphoid organs (thymus and spleen) and the number of lymphocytes in Line 6 chickens are also significantly smaller than those found in Line 7 chickens (Fredericksen and Gilmour, 1981; Lee et al., 1981).

3 Integrating Genomics, Version 1.0 (Before the Genome Sequence)

Genetic resistance is another control strategy that can augment MD vaccinal protection. Ideally, selection for MD resistance would be based on simple immunological assays that are associated with MD resistance, or genetic markers for the disease resistance genes, or ones that are tightly linked to them. The latter case is known as

MAS and is expected to accelerate genetic progress by increasing the accuracy and timing of selection. Furthermore, for use in disease resistance, challenging breeding stock with hazardous pathogens could be avoided with either screen.

With this goal in mind, we have been implementing and integrating genomic approaches that identify QTL, genes, and proteins that are associated with resistance to MD. The rationale for using more than one approach is that the strengths of each system can be combined to yield results of higher confidence. Another justification is that given the large volume of data produced by genomics, each method provides an additional screen to limit the number of targets to verify and characterize in future experiments. The current methods used are briefly described.

3.1 Genome-Wide QTL Scans

Myself and others have for the past 14+ years been contributing to the development of a molecular genetic map of the chicken genome (e.g., Cheng et al., 1995; Groenen et al., 2000). With the advent of this powerful tool, it became possible to identify QTL or regions in the genome that contain one or more genes controlling complex traits, such as disease resistance. Our efforts have utilized resource populations based on experimental inbred lines (Vallejo et al., 1998; Yonash et al., 1999) and commercial strains (McElroy et al., 2005). The use of the experimental lines allows for more precisely controlled environmental conditions and trait measurements while the use of commercial strains is more agriculturally relevant and permits many more birds to be produced and evaluated. In all populations, many QTL of small-to-moderate effect were identified, which agrees well with earlier estimates of the multigenic nature of MD resistance (Stone, 1975). Many of the QTL are common across populations, which provides additional confidence on these QTL and suggests that results from experimental populations can be transferred to commercial birds.

Genome-wide scans for QTL conferring MD resistance have been with ADOL Lines 6 (MD resistant) and 7 (MD susceptible), two highly (99+%) inbred parental experimental lines. Specifically, 272 unvaccinated F_2 progeny were challenged with JM strain MDV and measured for MD as well as a variety of MD-associated traits such as viral titer, number of tumors, and length of survival (Vallejo et al., 1998; Yonash et al., 1999). Using 135+ markers (mostly microsatellites) that cover 2, 500+ cM or ~65% of the chicken genome, 14 QTL (7 significant and 7 suggestive) were discovered that explain one or more MD-associated traits. The QTL were of small-to-moderate effect as they explained 2–10% of the variance, with additive gene substitution effects from 0.01 to 1.05 phenotypic standard deviations. Collectively, the QTL explained up to 75% of the genetic variance. Interestingly, 10 of the 14 QTL displayed non-additive gene action, 3 with overdominance, and 7 were recessive. Theoretically, non-additive QTL should be among the most useful for MAS. With multiple traits being measured, the QTL could be grouped. In the first set, 3 of the QTL were associated almost exclusively with viremia levels while the remaining QTL could account for disease, survival, tumors, nerve enlargement, and other disease-associated traits. This suggests that disease resistance occurs at least at two levels: initial viral replication and cellular transformation, which occurs later. It also highlights the added value of measuring several components as it may functionally separate a complex trait as well as provide clues on positional candidate genes.

3.2 Gene Profiling

It is clear from theory and our results that it will be extremely difficult to identify positional candidate genes for MD resistance using genetic approaches only. Gene expression profiling using microarray hybridization technology is a powerful tool for gene function studies. Our hope is that DNA microarrays will identify genes and pathways involved in MD resistance, which combined with genetic mapping can reveal positional candidate genes (Liu et al., 2001a). In other words, positional candidate genes are those that have a genetic association and are identified as being relevant through gene expression analyses.

Gene profiling has been conducted to identify differentially expressed genes between Lines 6 and 7 after MDV challenge (Liu et al., 2001a), among B (MHC) congenic lines of chicken following inoculation with different MD vaccines (unpublished), and in chicken embryo fibroblasts (CEF) infected with MDV (Morgan et al., 2001). Analyses of these experiments have identified a number of genes and pathways that are consistently associated with either MD resistance or MDV infection. More importantly, the results suggest that chickens with immune systems that are more stimulated by MDV infection are more susceptible. Initially, this seems counterintuitive but upon further reflection, MDV is thought to only infect activated lymphocytes and, thus, chickens with immune systems that are more responsive may present more targets for MDV to infect and later transform.

3.3 Virus–Host Protein–Protein Interaction Screens

As the third component of our integrative approach, we have been systematically examining protein variation and interactions associated with MD resistance. Specifically, we have screened for MDV–chicken protein–protein interactions using a two-hybrid screen. Briefly describing the method, the two-hybrid system screens a cDNA (prey) library that has been fused to the activation domain (AD) of a transcriptional activator to identify proteins that interact with bait (protein of interest) that is fused with the DNA-binding domain (BD). As the AD and BD do not need to be physically connected to promote transcription, if the two fusion proteins interact, a reporter gene is expressed. Our hypothesis was that some chicken proteins that interact with MDV proteins are involved in the immune response and genetic resistance to MD. Thus, we could utilize the two-hybrid system to quickly identify interacting (and interesting) proteins, which when combined with genetic mapping would identify positional candidate genes for MD resistance.

Our initial MDV bait highlights the success we have achieved (Liu et al., 2001b). We initially chose MDV *SORF2* gene as bait since *SORF2* overexpression in the RM1 strain may account for the reduced virulence compared to its parental JM/102W strain (Jones et al., 1996). Using the yeast two-hybrid system and a splenic cDNA library, growth hormone (GH) was found to specifically interact with *SORF2* (Liu et al., 2001b). It is critical to confirm the two-hybrid results, as this method is known to have a high false-positive rate. Thus, to corroborate the detected interaction, in vitro protein binding assay using GST-fusion proteins was performed to confirm direct binding of GH to SORF2. Our results showed that, while SORF2 protein was not retained by GST protein alone, SORF2 could be retained by GST–GH fusion protein presumably due to the presence of GH. This result was in agreement with the result of the yeast two-hybrid system assay and indicated that the interaction between SORF2 and GH is a direct and specific protein–protein interaction without other intermediary factors (e.g., yeast proteins) involved.

Having confirmed the SORF2–GH interaction, we treated the GH gene (*GH1*) as a candidate gene for MD resistance. To see if *GH1* had a genetic basis, an association study was conducted in our MD resource population derived from commercial White Leghorn lines. *GH1* variation was significantly associated ($P \leq 0.01$) with a number of MD-associated traits in MHC B^2/B^{15} chicks (Liu et al., 2001b). Furthermore, to provide some functional information support, our DNA microarray results indicate that GH is differentially expressed between MD resistant (Line 6) and susceptible (Line 7) chicks following MDV challenge (Liu et al., 2001a).

Thus, the combined results of a specific MDV–chicken protein interaction, differential expression of GH between MD resistant and susceptible chickens, and association of *GH1* with MD disease-related traits and selected lines for MD resistance, all strongly suggested that *GH1* is a MD-resistance gene. This conclusion is supported by reports demonstrating that GH modulates the immune system in many species (e.g., reviews by Gala, 1991; Auernhammer and Strasburger, 1995), and *GH1* alleles change in chicken strains in response to selection for MD resistance (Kuhnlein et al., 1997). Most importantly, it exemplifies the power of combining genetic and molecular approaches to identify positional candidate genes for QTL.

While effective, the number of baits that could be screened by an investigator would be limited using the yeast two-hybrid system. Fortunately, two-hybrid systems based in *Escherichia coli* have become commercially available. With the complete sequence of the MDV genome (Lee et al., 2000; Tulman et al., 2000), it became feasible to conduct a systematic screen of the relevant MDV genes for interacting chicken partners; there are ~100 different MDV genes and proteins (Lee et al., 2000; Tulman et al., 2000; Liu et al., 2006). We screened all the MDV genes that are considered unique to serotype I (virulent) strains, and all potential MDV–host protein interactions were tested by an in vitro binding assay to confirm the initial two-hybrid results. As a result, eight new MDV–chicken protein interactions were identified (Niikura et al., 2004). More importantly, genetic mapping and association analyses of the encoding chicken genes revealed that *LY6E* [lymphocyte complex 6, locus E, aka, stem cell antigen 2 (*SCA2*) and thymic-shared antigen 1 (*TSA1*)] is another MD-resistant gene (Liu et al., 2003) and suggest that *BLB*, the gene for MHC class II β chain, is a strong positional candidate gene (Niikura et al., 2004).

4 Integrating Genomics, Version 2.0 (After the Genome Sequence)

The field of genomics is strongly influenced by technological advancements. DNA sequencing, molecular markers, and DNA microarrays are just a few examples that emphasize this point. Consequently, investigators must remain vigilant to recent developments and ready to adopt new methods.

In 2004, chicken joined the "genome sequence" club. With the draft chicken genome sequence assembly (Hillier et al., 2004) and ~2.8 SNPs (Wong et al., 2004), researchers obtained a near complete "parts list" of the chicken and a tremendous step toward understanding how genetic variation leads to phenotypic variation. The impact of these invaluable tools and resources changed or refined our strategy for identifying MD resistance genes.

4.1 Genome-Wide QTL Scans

With the genome sequence, millions of high-confidence SNPs, and cost-efficient high-throughput genotyping systems, it is possible to genotype large resource populations very quickly and economically. Recently, we were able to re-evaluate our Line 6 × 7 F_2 MD resource population. Specifically, 578 additional genetic markers were scored, which provided denser and superior coverage (Cheng et al., 2007). Most interestingly, partly due to the higher accuracy of our current genetic markers, we could conduct a search for two-locus epistatic interactions. A large number of highly significant two-way interactions were found that could account for viremia; no highly significant interactions were revealed for MD or survival. A total of 239 highly significant interactions (LRS $\geq$ 57.8; genome-wide $P \leq 0.001$) were identified (Cheng et al., 2007). The location of loci and their interacting partners appears to be distributed throughout the genome. On the other hand, loci in specific regions on chromosomes 1 and 4 are frequently involved; a single region on chromosome 1 accounted for 166 of the 239 highly significant interactions. Interestingly, most if not all of the interacting loci are not in the QTL regions previously identified.

These results imply (1) interactions between loci can make a substantial contribution to variation in complex traits, (2) because of the growing awareness of biological complexity, genome-wide QTL scans should attempt to incorporate resource populations that can utilize the genotyping and analytical capabilities available today or in the near future, and (3) the challenge remains to identify the underlying genes and genetic variation for QTL.

4.2 Gene Profiling

While DNA microarrays experiments can be very powerful and the results enlightening, our experiences also suggest that it is critical to study the response of cells in a limited or local environment. For example, MHC class I expression was increased in MDV-infected CEF (Morgan et al., 2001). However, upon closer examination of individual cells, virus-infected cells actually expressed reduced levels of MHC on their cell surfaces but neighboring uninfected cells had elevated levels of MHC class I (Hunt et al., 2001). To partially circumvent the problem of heterogeneous samples, laser-capture microdissection (LCM) and cell sorting has been used to isolate the cells of interest.

Another method that we plan to investigate further is the use of allele-specific gene expression. Specifically, we will conduct a genome-wide screen for genes that show allele-specific gene expression in response to MDV challenge using traditional DNA microarrays. For the differentially expressed genes, the 3' UTR of each gene will be sequenced to screen for SNPs, the only requirement for the assay to work. Intermatings of the parental lines will produce progeny that are heterozygous for the SNP, and RNA from several tissues and time points measured for allele-specific gene expression. Cis-acting elements will be declared for a gene when allele-specific gene expression is found. Since the gene is in the same cell and bird, monitoring the expression of each allele avoids issues like sampling, RNA quality, or environmental effects. Furthermore, because we are evaluating only allele-specific gene expression differences, the analysis of the datasets are greatly simplified and the result is a clear identification of functional elements with a genetic basis. In short, we hope to demonstrate a simple and efficient genome-wide method for identifying cis-acting elements that influence gene expression which does not require knowledge of specific regulatory variants and should be simpler and more cost effective to perform than similar eQTL experiments.

4.3 Virus–Host Protein–Protein Interaction Screens

Higher order protein–protein interactions can be revealed with gentle cell lysis and immunoprecipitation of a protein complex using antibodies directed against one member of the complex. The identity of the other interacting proteins can be quickly revealed through mass spectrometry, which is facilitated with the whole genome sequence.

To extend the MDV–chicken protein interactions, we propose to tandem affinity purification (TAP) tag most, if not all, of the MDV proteins. And rather than expressing the viral gene product alone, we plan to take advantage of our infection MDV BAC clones (Niikura et al., 2006) and the ability to make defined recombinant viruses. The use of these viruses expressing TAP-tagged MDV proteins during infection allows us to use a system closer to normal viral infection, which will hopefully give us useful information on the biology of viral infection and the role of host–virus interactions in that context.

5 Some Final Thoughts

It has been over 30 years since MDV was identified as the causative agent of MD and effective vaccines produced. In spite of our ability to detect and protect against MD, there is a surprising lack of information on what components of the chicken immune system confer disease resistance and contribute to vaccinal immunity. With the advent of the new technologies and the impending release of the whole genome sequence, we believe that our ability to tangibly improve upon MD control mechanisms is at hand. Key to making rapid gains will be the ability to integrate various methods and information. If this occurs, then we should be able to accelerate the transition from serendipity to rational, mechanism-based control of disease and food production in the poultry industry.

Acknowledgments I would like to thank current members of my laboratory (Masahiro Niikura, Taejoong Kim, Kyle MacLea, Weifeng Mao, Laurie Molitor, and Tom Goodwill) and my many collaborators especially Henry D. Hunt (ADOL); Mary Delany (UC Davis); Robin W. Morgan and Joan Burnside (University of Delaware); Janet E. Fulton (Hy-Line); Susan L. Lamont and Jack C.M. Dekkers (Iowa State University); Morris Soller (Israel); Jerry Dodgson (Michigan State University); Bill Muir (Purdue University); and Martien Groenen (The Netherlands). Financial support was provided in part from USDA awards 00-03473, 2001-52100, 2002-03407, 2003-05414, and 2004-05434.

References

Asmundson, V.S., and Biely, J., 1932, Inheritance and resistance to fowl paralysis (neuro-lymphomatosis gallinarum). I. Differences in susceptibility, *Can. J. Res.* **6**:171–176.

Auernhammer, C.J., and Strasburger, C.J., 1995, Effects of growth hormone and insulin-like growth factor I on the immune system, *Eur. J. Endocrinol.* **133**:635–645.

Bacon, L.D., 1987, Influence of the major histocompatibility complex on disease resistance and productivity, *Poult. Sci.* **66**:802–811.

Bacon, L.D., Polley, C.R., Cole, R.K., and Rose, N.R., 1981, Genetic influences on spontaneous autoimmune thyroiditis in (CSXOS) F$_2$ chickens, *Immunogen.* **12**:339–349.

Bacon, L.D., and Witter, R.L., 1992, Influence of turkey herpesvirus vaccination on the *B*-haplotype effect on Marek's disease resistance in 15.B-congenic chickens, *Avian Dis.* **36**: 378–385.

Bacon, L.D., and Witter, R.L, 1994a, Serotype specificity of *B*-haplotype influence on the relative efficacy of Marek's disease vaccines, *Avian Dis.* **38**:65–71.

Bacon, L.D., and Witter, R.L., 1994b, *B*-haplotype influence on the relative efficacy of Marek's disease vaccines in commercial chickens, *Poultry Sci.* **73**:481–487.

Buckmaster, A.E., Scott, S.D., Sanderson, M.J., Boursnell, M.E., Ross, N.L., and Binns, M.M., 1988, Gene sequence and mapping data from Marek's disease virus and herpesvirus of turkeys: implications for herpesvirus classification, *J. Gen. Virol.* **69**:2033–2042.

Cheng, H.H, Levin, I., Vallejo, R.L., Khatib, H., Dodgson, J.B., Crittenden, L.B., and Hillel, J., 1995, Development of a genetic map of the chicken with markers of high utility, *Poultry Sci.* **74**:1855–1874.

Cheng, H.H., Zhang, Y., and Muir, W.M., 2007, Evidence for widespread epistatic interactions influencing Marek's disease virus viremia levels in chicken. *Cytogen. Genome Res.*, **117**: 313–318.

Churchill, A.E., and Biggs, P.M., 1967, Agent of Marek's disease in tissue culture. *Nature* **215**:528–530.

Fredericksen, T.L., and Gilmour, D.G. 1981. Chicken lymphocyte alloantigen genes and responsiveness of whole blood cells to concanavalin A, *Fed. Proc.* **40**:977.

Gala, R.R., 1991, Prolactin and growth hormone in the regulation of the immune system, *Proc. Soc. Exp. Biol. Med.* **198**:513–527.

Gallatin, W.M., and Longenecker, B.M., 1979, Expression of genetic resistance to an oncogenic herpesvirus at the target cell level, *Nature* **280**:587–589.

Groenen, M.A.M., Cheng, H.H., Bumstead, N., Benkel, B.F., Briles, W.E., Burke, T., Burt, D.W., Crittenden, L.B., Dodgson, J., Hillel, J., Lamont, S., Ponce de Leon, A., Soller, M., Takahashi, H., and Vignal, A., 2000, A consensus linkage map of the chicken genome, *Genome Res.* **10**:137–147.

Hartmann, W., 1989, Evaluation of "major genes" affecting disease resistance in poultry in respect to their potential for commercial breeding in recent advances, in B.S. Bhogal and G. Koch. Alan R. Liss, eds, Recent Advances in Avian Immunology Research, New York, pp. 221–231.

Hillier, L. W., Miller, W., Birney, E., Warren, W., Hardison, R. C., et al. International Chicken Genome Sequencing Consortium, 2004, Sequence and comparative analysis of the chicken genome provide unique perspectives on vertebrate evolution, *Nature* **432**:695–716.

Hunt, H.D., and Fulton, J.E., 1998, Analysis of polymorphisms in the major expressed class I locus (B-FIV) of the chicken, *Immunogen.* **47**:456–476.

Hunt, H.D., Lupiani, B., Miller, M.M., Gimeno, I., Lee, L.F., and Parcells, M.S., 2001, Marek's disease virus down regulates surface expression of MHC (B complex) class I (BF) glycoproteins during active but not latent infection of chicken cells, *Virology* **282**:198–205.

Jones, D., Brunovskis, P., Witter, R., and Kung, H., 1996, Retroviral insertional activation in a herpesvirus:, transcriptional activation of Us genes by an integrated long terminal repeat in a Marek's disease virus clone, *J. Virol.* **70**:2460–2467.

Kuhnlein, U., Ni, L., Weigend, S., Gavora, J.S., Fairfull, W., and Zadworny, D., 1997, DNA polymorphisms in the chicken growth hormone gene: response to selection for disease resistance and association with egg production, *Anim. Genet.* **28**:116–123.

Lee, L.F., Powell, P.C., Rennie, M., Ross, L.J.N., and Payne, L.N., 1981, Nature of genetic resistance to Marek's disease, *J. Natl. Canc. Inst.* **66**:789–796.

Lee. L.F., Wu, P., Sui, D., Ren, D., Kamil, J., Kung, H.J., and Witter, R.L., 2000, The complete unique long sequence and the overall genomic organization of the GA strain of Marek's disease virus, *Proc. Natl. Acad. Sci. USA* **97**:6091–6096.

Liu, H.-C., Cheng, H.H., Sofer, L., and Burnside, J., 2001a, A strategy to identify positional candidate genes conferring Marek's disease resistant by integrating DNA microarrays and genetic mapping, *Anim. Genet.* **32**:351–359.

Liu, H.-C., Kung, H.-J., Fulton, J.E., Morgan, R.W., and Cheng, H.H., 2001b, Growth hormone interacts with the Marek's disease virus SORF2 protein and is associated with disease resistance in chicken, *Proc. Natl. Acad. Sci. USA* **98**:9203–9208.

Liu, H.C., Niikura, M., Fulton, J., and Cheng, H.H., 2003, Identification of chicken stem lymphocyte antigen 6 complex, locus E (*LY6E*, alias *SCA2*) as a putative Marek's disease resistance gene via a virus–host protein interaction screen, *Cytogen. Genome Res.* **102**:304–308.

Liu, H.C., Soderblom, E.J., and Goshe, M.B., 2006, A mass spectrometry-based proteomic approach to study Marek's disease virus gene expression, *J. Virol. Methods* **135**:66–75.

Longenecker, B.M., and Mosmann, T.R., 1981, Structure and properties of the major histocompatibility complex of the chicken. Speculations on the advantages and evolution of polymorphism, *Immunogen.* **13**:1–23.

Marek, J., 1907, Mutiple Nervenentzundung (Polyneuritis) bei Huhnern, Dtsch. *Tierarztl. Wschr.* **15**:417–421.

McElroy, J.P., Dekkers, J.C.M., Fulton, J.E., O'Sullivan, N.P., Soller, M., Lipkin, E., Zhang, W., Koehler, K.J., Lamont, S.J., and Cheng, H.H., 2005, Microsatellite markers associated with resistance to Marek's disease in commercial layer chickens, *Poultry Sci.* **84**:1678–1688.

Morgan, R.W., Sofer, L., Anderson, A.S., Bernberg, E.L., Cui, J., and Burnside, J., 2001, Induction of host gene expression following infection of chicken embryo fibroblasts with oncogenic Marek's disease virus, *J. Virol.* **75**:533–539.

Nazerian, K., and Burmester, B.R., 1968, Electron microscopy of a herpes virus associated with the agent of Marek's disease in cell culture, *Cancer Res.* **28**:2454–2462.

Niikura, M., Liu, H.-C., Dodgson, J.B., and Cheng, H.H., 2004, A comprehensive screen for chicken proteins that interact with proteins unique to virulent strains of Marek's disease virus, *Poultry Sci.* **83**:1117–1123.

Niikura, M., Dodgson, J, and Cheng, H.H., 2006, Direct evidence of host genome acquisition by the alphaherpesvirus Marek's disease virus, *Arch. Virol.* **151**:537–549.

Osterrieder, N., Kamil, J.P., Schumacher, D., Tischer, B.K., and Trapp, S., 2006, Marek's disease virus: from miasma to model, *Nature Rev. Micro.* **4**:283–294.

Powell, P.C., Lee, L.F., Mustill, B.M., and Rennie, M., 1982, The mechanism of genetic resistance to Marek's disease in chickens, *Int. J. Canc.* **29**:169–174.

Purchase, H.G., 1985, Clinical disease and its economic impact, in L. N. Payne, ed., Marek's Disease, Scientific Basis and Methods of Control, Martinus Nkjhoff Publishing, Boston, pp. 17–42.

Purchase, H.G., and Okazaki, W., 1971, Effect of vaccination with herpesvirus of turkeys (HVT) on horizontal spread of Marek's disease herpesvirus, *Avian Dis.* **15**:391–397.

Schat, K.A., Calnek, B.W., Fabricant, J., and Abplanalp, H., 1981, Influence of oncogenicity of Marek' disease virus on evaluation of genetic resistance, *Poult. Sci.* **60**:2559–2566.

Solomon, J.J., Witter, R.L., Nazerian, K., and Burmester, B.R., 1968, Studies on the etiology of Marek's disease. I. Propagation of the agent in cell culture, *Proc. Soc. Exp. Biol. Med.* **127**: 173–177.

Stone, H.A., 1975, Use of highly inbred chickens in research, *USDA Technical Bulletin* No. 1514.

Tulman, E.R., Afonso, C.L., Lu, Z., Zsak, L., Rock, D.L., and Kutish, G.F., 2000, The genome of a very virulent Marek's disease virus, *J. Virol.* **74**:7980–7988.

Vallejo, R.L., Bacon, L.D., Liu, H.C., Witter, R.L., Groenen, M.A., Hillel, J., and Cheng, H.H., 1998, Genetic mapping of quantitative trait loci affecting susceptibility to Marek's disease virus induced tumors in F2 intercross chickens, *Genetics* **148**:349–360.

Witter, R.L., 1997, Increased virulence of Marek's disease virus field isolates, *Avian Dis.* **41**: 149–163.

Witter, R.L., and Kreager, K.S., 2004, Serotype 1 viruses modified by backpassage or insertional mutagenesis: approaching the threshold of vaccine efficacy in Marek's disease, *Avian Dis.* **48**:768–782.

Witter, R.L., Sharma, J.M., and Fadly, A.M., 1980, Pathogenicity of variants Marek's disease isolates in vaccinated and unvaccinated chickens, *Avian Dis.* **24**:210–232.

Wong, G. K., Liu, B., Wang, J., Zhang, Y., Yang, X., et al.; International Chicken Polymorphism Map Consortium, 2004, A genetic variation map for chicken with 2.8 million single-nucleotide polymorphisms, *Nature* **432**:717–722.

Yonash, N., Bacon, L.D., Witter, R.L., and Cheng, H.H., 1999, High resolution mapping and identification of new quantitative trait loci (QTL) affecting susceptibility to Marek's disease, *Anim. Genet.* **30**:126–135.

Combining Genomic Tools to Dissect Multifactorial Virulence in *Pseudomonas aeruginosa*

Daniel G. Lee, Jonathan M. Urbach, Gang Wu, Nicole T. Liberati, Rhonda L. Feinbaum, and Frederick M. Ausubel

1 Introduction

The growing number of sequenced bacterial genomes, including those of important pathogenic isolates, has made a significant impact on the field of bacterial pathogenesis (Raskin et al., 2006). Combined with other technical advances in the laboratory, this wealth of information has made possible the development and widespread adoption of genomic tools for the study of infectious disease. This chapter focuses on the sequencing and functional analysis of PA14, a clinical isolate of the ubiquitous environmental bacterium and important opportunistic human pathogen, *Pseudomonas aeruginosa*. By comparing PA14 to the sequence of the less virulent *P. aeruginosa* isolate, PAO1, we have identified genomic sequences absent in one or the other strain in order to examine the relationship between genomic content and pathogenicity.

To assess the importance of strain-specific genes and virulence, we tested 20 diverse *P. aeruginosa* strains, including PA14 and PAO1, in a *Caenorhabdidits elegans* pathogenesis model and observed a wide range of pathogenic potential; however, genotyping these strains using a custom microarray showed that the presence of genes present in PA14 and absent in PAO1 did not correlate with the virulence of these strains. To further examine the roles in virulence of PA14 genes absent in PAO1, we utilized a full-genome non-redundant mutant library of PA14 to identify five ORFs (absent in PAO1) required for *C. elegans* killing. Surprisingly, although these five genes are present in many other *P. aeruginosa* strains, they do not correlate with virulence in *C. elegans*. Genes required for pathogenicity in one strain are neither required for nor predictive of virulence in other strains. We therefore propose that virulence in this organism is a complex process that is both multifactorial and combinatorial. Not only are multiple pathogenicity determinants acting in parallel within a given strain, but also, when comparing different strains, different combinations of pathogenicity factors may be selected to determine the ultimate virulence

D.G. Lee
Department of Molecular Biology, Massachusetts General Hospital and Department of Genetics, Harvard Medical School, Boston, MA, USA

J.P. Gustafson et al. (eds.), *Genomics of Disease*,
© Springer Science+Business Media, LLC, 2008

phenotype. Additionally, this work highlights the importance of combining multiple genomic tools to address complex biological problems. These genomic tools included traditional comparative genome alignment algorithms, a microarray-based method for rapidly genotyping multiple strains, and a non-redundant, genome-wide library of mutants to examine pathogenesis in a model high-throughput host infection system. Finally, although this work has focussed on the use of genomic tools to dissect the pathogen, model hosts that are amenable to genetic manipulation make possible the use of comparable genome-wide approaches on the host side, ultimately allowing for the combination of sophisticated tools in both organisms to elucidate the mechanisms underlying host–pathogen interactions.

2 Background

2.1 *Pseudomonas aeruginosa is an Opportunistic Human Pathogen*

Pseudomonas aeruginosa is a ubiquitous Gram negative soil bacterium that has been shown to colonize and thrive in a wide range of environments. In the clinical setting, *P. aeruginosa* is also an important opportunistic human pathogen, infecting patients that are injured, burned, immunodeficient, or immunocompromised. *P. aeruginosa* also causes persistent respiratory infections in individuals suffering from cystic fibrosis (CF) and remains the primary cause of illness and death in these patients (Doring, 1993; Wood, 1976). The first genome sequence of *P. aeruginosa* was that of strain PAO1, a clinical isolate that is currently the most widely studied strain in the laboratory. The PAO1 isolate contains a large number of genes involved in regulation, catabolism, transport, and efflux of organic compounds and several potential chemotaxis systems (Stover et al., 2000), all potentially contributing to the remarkable ability of this bacterium to adapt to diverse environmental niches.

In general, the genomes of *P. aeruginosa* isolates share a high degree of sequence similarity. In microarray-based genotyping experiments, in which genomic DNA from test strains were hybridized to an array of sequences from PAO1, between 89 and 98% of the reference sequences were detected (Ernst et al., 2003; Wolfgang et al., 2003). Furthermore, whole-genome shotgun sequencing of two CF isolates and one environmental isolate revealed that these strains share a large core of highly conserved genes with PAO1. The major differences between strains were due to the presence of strain-specific islands of genes, consisting either of genes with similar or related function but divergent DNA sequence, or genes that are entirely absent in some strains (Spencer et al., 2003). Despite this high level of overall similarity, complex phenotypes such as pathogenicity can vary greatly. The clinical isolate PA14, which is the subject of this chapter, is significantly more virulent than PAO1 (in mammalian as well as invertebrate models of infection described below; Choi et al., 2002; Rahme et al., 1995; Tan et al., 1999a).

What accounts for the significant difference in pathogenicity between these isolates? In dedicated human pathogens, the potential virulence of a strain is generally dictated by the presence or absence of pathogenicity islands (Hacker et al., 1997; Oelschlaeger and Hacker, 2004), clusters of one or more virulence-related genes often acquired by horizontal gene transfer. Screens for PA14 genes required for virulence have identified novel genes absent in PAO1 (Choi et al., 2002; He et al., 2004; Mahajan-Miklos et al., 1999; Rahme et al., 1997; Tan et al., 1999b), consistent with this model. However, other genes identified in these studies include genes common to many if not all *P. aeruginosa* strains, including global transcriptional regulators such as *gacA*, genes involved in pathogenesis-related processes such as motility, quorum sensing, and phenazine biosynthesis, and genes that encode secreted cellulytic factors and toxins such as *ExoU*, exotoxin A, phospholipase C, and elastase (Jander et al., 2000; Mahajan-Miklos et al., 1999; Miyata et al., 2003; Rahme et al., 1995; Rahme et al., 1997; Tan et al., 1999a; Tan et al., 1999b). Therefore, pathogenesis in *P. aeruginosa* appears to be multifactorial, in that multiple gene products and processes act in parallel to result in a virulent phenotype. The multiple virulence traits may be part of a common, core genome, which contributes to a base level of pathogenicity. In addition, the acquisition of strain-specific genes can modify and augment this basal activity, resulting in the enhanced virulence observed in isolates such as PA14. To test this hypothesis on a genome-wide level, we sequenced the genome of strain PA14 and compared it to that of PAO1. This traditional comparative genomic analysis was supplemented with the use of additional genomic tools developed in our lab: (1) A custom microarray for genotyping of additional strains to determine the extent to which sequences absent in either PA14 or PAO1 were conserved in other strains, and (2) a non-redundant mutant library of transposon insertions in strain PA14 to identify genes (absent in PAO1) that were required for pathogenesis. To make maximal use of these tools, we chose to study virulence in the context of a well-established model host infection system using the nematode, *C. elegans*.

2.2 The Model Host System for Studying Pathogenesis

Traditionally, the study of bacterial infectious disease has relied on one of two general (non-mutually exclusive) approaches. The first simplifies the complex host–pathogen interaction by focussing narrowly on a bacterial activity or process thought to be important for infection (such as attachment to surfaces or production of toxic compounds) that is amenable to detailed mechanistic study in the absence of a host. Such studies allow for the elucidation of mechanistic details but run the risk of over-simplifying and potentially focussing on aspects of virulence that only have a minor contribution to the overall infectious process. Alternatively, whole-animal infection models examine pathogenesis in the context of a living host with a complex set of host cell types and immune responses. However, these studies have typically used mice or other small mammals as hosts that generally impose both financial and ethical constraints on the size and scope of experiments that can feasibly be proposed. More recently, the application of genome-wide tools including signature-tagged mutagenesis (STM) and transposon site hybridization (TraSH) to the use of whole-animal infection systems allows more complex questions to be addressed than were previously possible, extracting more useful data from a smaller number of hosts (Badarinarayana et al., 2001; Hensel et al., 1995; Saenz and Dehio, 2005; Sassetti et al., 2001; Wong and Mekalanos, 2000). However, a more recent development in the study of bacterial pathogenesis, the use of non-mammalian model hosts, provides a third approach that represents a compromise between the two traditional extremes and is inherently well-suited to the use of high-throughput, genomic tools on both the host and the pathogen side of the equation.

A number of years ago, our laboratory discovered that the PA14 clinical isolate of *P. aeruginosa* could infect and kill the plant, *Arabidopsis thaliana* (Plotnikova et al., 2000; Rahme et al., 1995). Importantly, *P. aeruginosa* virulence factors known to be important in causing mammalian disease were also found to be required for infecting plants, arguing that at least some components of the pathogenesis process were conserved. Subsequently, additional model hosts were found to be susceptible to PA14 infection, including nematodes (*Caenorhabditis elegans*), insects (*Drosophila melanogaster* and the greater wax moth, *Galleria mellonella*), and amoebae (*Dictyostelium discoideum*) (Jander et al., 2000; Lau et al., 2003; Mahajan-Miklos et al., 1999; Pukatzki et al., 2002; Tan et al., 1999a; Tan et al., 1999b). Bacterial mutants with reduced virulence in these model hosts were also attenuated in mice, demonstrating that this system could be used for forward genetic screens to identify novel virulence factors important for mammalian disease. Because these model organisms are generally cheap, small, and able to reproduce quickly, genomic tools developed in the bacterium could be fully exploited using large-scale, high-throughput studies in a living host. Furthermore, the majority of these hosts are genetically tractable and a variety of sophisticated genomic tools have been developed that further enhances their utility as laboratory models. Ultimately, the use of model hosts allows for interactive genetics, combining the power of manipulating both the pathogen and the host simultaneously.

This chapter focuses on the elucidation of the genomic basis for differences in pathogenicity of *P. aeruginosa* isolates, utilizing genomic tools developed for the virulent strain, PA14, in the context of a high-throughput *C. elegans* infection system. Although our initial emphasis has been on the use of bacterial tools, this type of analysis could also be extended by the addition of reagents to dissect the *C. elegans* response to pathogen attack, such as genome-wide RNA interference (RNAi) libraries and microarray transcriptional profiling (N. Liberati, R. Feinbaum, and F. M. Ausubel, unpublished observations; Gravato-Nobre et al., 2005; O'Rourke et al., 2006; Troemel et al., 2006). In general, the multi-host infection system combined with the use of genomic tools can be applied to the study of interactions between various host organisms and microbial pathogens.

3 Genomic Sequence of *P. aeruginosa*, Strain PA14

3.1 Comparative Alignments with Strain PAO1

To determine if the acquisition of strain-specific genes contribute to the enhanced virulence of PA14, we sequenced the PA14 genome and compared it to the genome of the less virulent isolate, PAO1 (Lee et al., 2006). PA14 has a slightly larger chromosome of 6.5 MB as compared to the 6.3 MB genome of PAO1. As expected, the overall similarity at the nucleotide level is high, with approximately 91.7% of the PA14 genome present in PAO1, and 95.8% of the PAO1 genome present in PA14 (Table 1).

Various PAO1 isolates are known to differ due to a large inversion that includes over a quarter of the genome (Stover et al., 2000). The *P. aeruginosa* genome contains four dispersed copies of a large ribosomal RNA cluster; a rearrangement between the two most separated copies (which are in an inverted orientation) results

Table 1 PA14 and PAO1 Genome Comparisons

		PA14	**PAO1**
Whole Genome	**Genome Size**	**6,537,648**	**6,264,403**
	GC Content*	**66.3% (+/− 4.3%)**	**66.6% (+/− 3.9%)**
	Total # of Genes	**5973**	**5651**
	*Class 1 Genes***	*8.6%*	*9.0%*
	*Class 2 Genes***	*18.0%*	*20.0%*
	*Class 3 Genes***	*44.5%*	*26.8%*
	*Class 4 Genes***	*28.9%*	*44.2%*
Strain-Specific Regions*	**Number of Strain-Specific Regions***	**58**	**54**
	Average Length of Regions*	**9,335**	**4,847**
	Total DNA in Regions	**541,215 (8.3%)**	**261,426 (4.2%)**
	GC Content	**59.60%**	**59.80%**
	Total # of Genes	**478**	**234**
	Average Number of Genes per Region	**8.24**	**4.33**
	*Class 1 Genes***	*3.3%*	*9.8%*
	*Class 2 Genes***	*4.0%*	*3.0%*
	*Class 3 Genes***	*29.7%*	*30.8%*
	*Class 4 Genes***	*63.0%*	*56.4%*

* Standard deviation was calculated using a sliding 1 kb window.
** Class 1 to 4 refer to the confidence rating assigned to the predicted gene function. Class 1 genes are those whose function has been experimentally validated in *P. aeruginosa*. Class 2 genes are highly similar to genes whose functions have been validated in another organism. Class 3 genes have hypothesized functions based on limited similarity to other genes or structural/functional domains. Class 4 genes are ORFs of unknown function.
*** Strain-Specific Regions with at least 1 ORF.

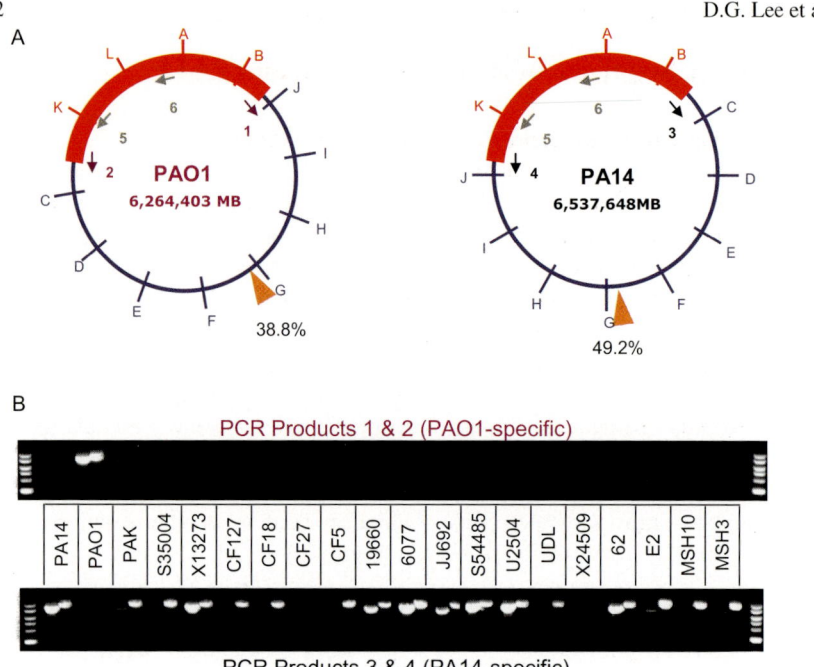

Fig. 1 Chromosomal rearrangement in PAO1 repositions the replication terminus relative to the origin. (**A**) Schematic of PAO1 and PA14 chromosomes. The region with the same orientation in both strains is shown with *a thick red line*; *a thin blue line* represents the inverted region. *Arrows* represent the positions and orientations of four ribosomal RNA clusters. PCR products spanning each rRNA cluster are indicated by numbers next to each arrow. PCR products 1 and 2 (purple numbers) are derived from sequences diagnostic for the PAO1 chromosome (with the inversion). PCR products 3 and 4 (*black numbers*) are derived from sequences diagnostic for the PA14 chromosome (without the inversion). Products 5 and 6 (*grey numbers*) are common to both PA14 and PAO1. The origin of replication is at position "A" at the top of each chromosome. The position of the presumptive terminus of replication in each strain (see text) is indicated by *an orange triangle* marked with the corresponding position along the chromosome (expressed as the percentage of the whole chromosome, starting from the origin of replication and moving in a clockwise direction). (**B**) Diagnostic long-range PCR spanning ribosomal RNA repeats demonstrates the inversion in PAO1 but not in other *P. aeruginosa* strains. Diagnostic PCR primer pairs for products 1 and 2, diagnostic of the PAO1 arrangement (*upper panels*) or for products 3 and 4, diagnostic of the PA14 arrangement (*lower panels*) were used to analyze genomic DNA isolated from PA14, PAO1, and 18 additional strains. For each strain, a pair of lanes is used for the pair of PCR products (either 1 and 2 for the *upper panels*, or 3 and 4 for the *lower panels*). Primers for PCR product 1 (*upper panel*, first lane in each pair) resulted in a faint background band present with all *P. aeruginosa* isolates tested but absent when no genomic DNA template was added to the reaction

in the inversion (Fig. 1). The genome sequence of PA14 indicated that it also contains this inversion relative to the sequenced PAO1 isolate and we used long-range PCR to confirm the presence of the inversion (Fig. 1B). Furthermore, when we surveyed 18 additional diverse *P. aeruginosa* isolates, none gave PCR products diagnostic for the PAO1 chromosomal arrangement (upper panels) and most generated one or both PCR products indicative of the PA14 arrangement (lower panels). These data suggest that the genome organization found in PA14 is more common among *P. aeruginosa* isolates than the genome organization of the sequenced PAO1 strain.

In canonical bacterial genomes, the origin and terminus of replication are at opposite ends of the circular chromosome. Furthermore, prokaryotic genomes show a bias toward containing G over C on the leading strand of DNA synthesis (Bentley and Parkhill, 2004). This bias can be measured as the GC-skew, defined as the measure of G–C/G+C for regular intervals and tends to be positive on the leading strand and negative on the lagging strand. Therefore, by beginning at the origin of replication and calculating GC-skew values on the leading strand, the putative replication terminus can be identified as the region of the genome where the GC-skew values shift from a tendency toward positive values to a tendency toward negative values, or more simply, the peak of the *cumulative* GC-skew values as one proceeds along the chromosome (Bentley and Parkhill, 2004). We therefore mapped cumulative GC-skews for both PA14 and PAO1. For PA14, the peak GC-skew was mapped opposite the replication origin (at 49.2% of the genome; Figs. 1A) as is typical for other bacterial genomes. In contrast, the position of the terminus in the sequenced PAO1 chromosome is shifted relative to the origin (at 38.8% of the genome). Combined with the PCR-based survey of additional strains, this terminus mapping analysis argues that the PA14 chromosome (with respect to this large inversion) is more representative of the canonical or ancestral *P. aeruginosa* genome than that of the sequenced PAO1 isolate. The physiological consequences of such an inversion and asymmetric replication cycle, however, are not clear.

Other than the presence of the large inversion, the genomes of PA14 and PAO1 are largely colinear. Using the MUMmer 3.0 software package (Kurtz et al., 2004), we aligned the PA14 and PAO1 genomes (Fig. 2) and observed that the two genomes were remarkably similar. The majority of the PA14 genome aligns with PAO1, in

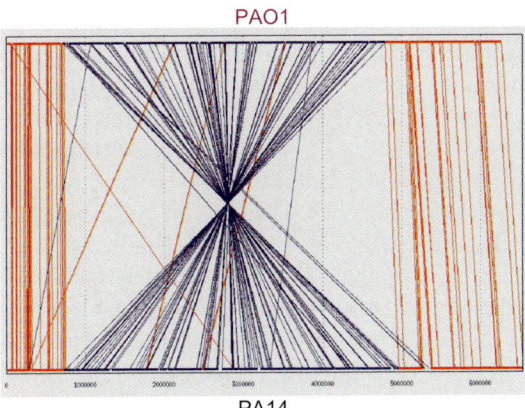

Fig. 2 Alignments of PA14 and PAO1 show that the two genomes are largely colinear. The MUMmer 3.0 software package was utilized to align PA14 and PAO1 with respect to each other based on nucleotide sequence. The PAO1 genome is represented on *top*, and the PA14 genome is on the *bottom*. *Lines* connecting the two genomes represent the boundaries of regions that align. Alignments in the forward direction are in red, and reverse alignments are in *blue*. *Gaps* in a genome represent regions of that genome that fail to align with the other genome (i.e., they are strain-specific regions)

either the same orientation (*red horizontal bar*) or in the inverted orientation (*blue horizontal bar*). Rarely, portions of PA14 fail to align with PAO1 are shown as gaps in the PA14 representation (and conversely, gaps in the PAO1 chromosome representation correspond to PAO1-regions that fail to align with PA14). These gaps were used to identify strain-specific regions (see below). In addition to the colinear aligned regions, there are rare instances of PA14 segments that appear to be translocations of PAO1 sequences to a different part of the genome.

3.2 Annotation of the PA14 Genome

We predicted ORFs in the PA14 genome (Lee et al., 2006) using a combination of automated BLAST and Glimmer2 algorithms (Delcher et al., 1999; Salzberg et al., 1998) and we annotated 5973 *PA14* genes (322 more than PAO1; Table 1). Each PA14 ORF was assigned a unique LocusName, beginning with "PA14_" followed by five numerals. ORFs were numbered sequentially with respect to their position on the chromosome, starting with PA14_00010 (*dnaA*), and increasing in increments of 10 to allow for future insertions of additional genes or functional RNAs. All annotations are available at http://ausubellab.mgh.harvard.edu/pa14 sequencing. Given the high sequence similarity between the two strains, we anticipated that the majority of PA14 genes would be true orthologs of PAO1 genes. However, the sequence of some PA14 genes appear to have diverged significantly from their counterparts in PAO1 such that they are more appropriately considered polymorphisms or variants, and other genes are completely absent in PAO1. To expedite the classification of PA14 genes as orthologs (90–100% amino acid identity), polymorphisms (50–90% identity), or genes absent in PAO1 (less than 50% identity), we devised an automated scheme to identify candidate orthologs that were then manually inspected (Fig. 3A). Briefly, PA14 genes were BLASTed, one at a time, against the collection of PAO1 genes and this process was repeated in reverse (i.e., PAO1 genes BLASTed against the collection of *PA14* genes; Fig. 3A, step 2). Best BLAST matches were determined based on BLAST scores and synteny with previous genes to maintain gene order (step 3). If a pair of genes was identified by searches starting with both the PA14 and the PAO1 ORF, a reciprocal best BLAST hit was recorded (step 6). As an additional criterion for identifying candidate orthologs (blue font), we determined if the length of the alignment represented a significant percentage of the total length of the query ORF (step 7). The best candidates for orthologous pairs were those for which (a) they were recorded as reciprocal best BLAST hits and (b) the alignment length was greater than a minimal percentage of the whole gene for both ORFs. In cases where no significant BLAST matches are found when searching ORFs in the other strain, the query sequence was BLASTed against the whole genome sequence of the other strain (steps 8 and 9; red font). A match to the raw nucleotide sequence of the other strain was examined more closely to determine if an ORF existed but was not annotated, or if a sequence polymorphism removed an ORF. Genes that resulted in no

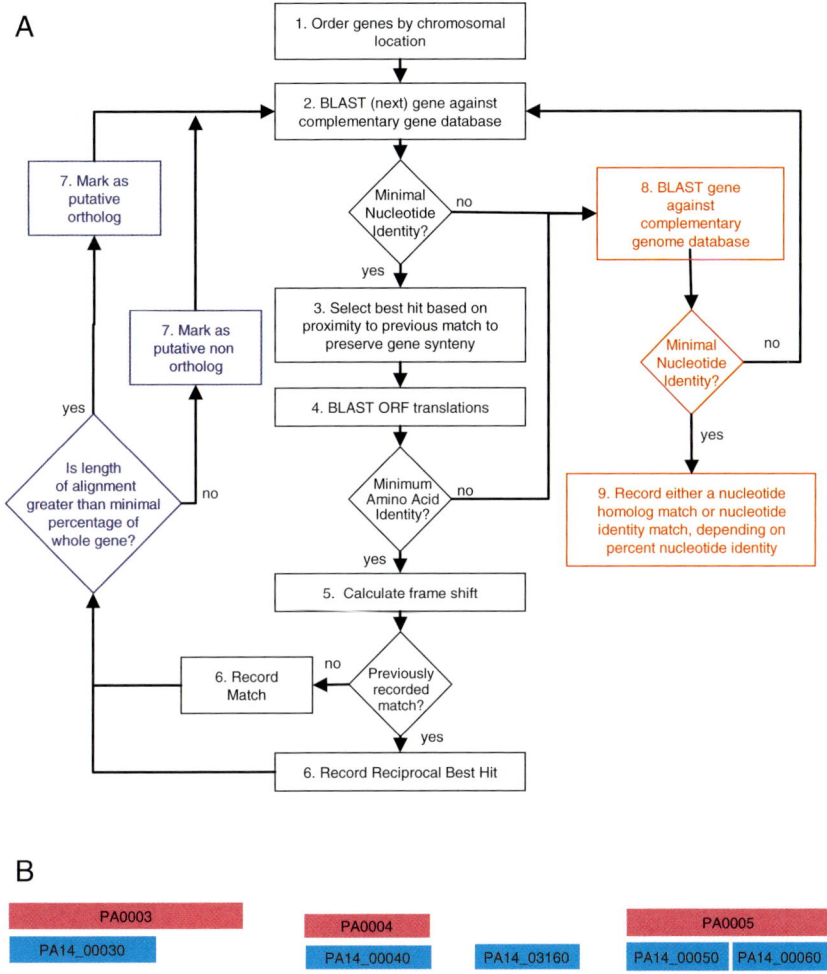

Fig. 3 Strategy for identification of putative orthologs and strain-specific genes. (**A**) Schematic for the automated identification of putative orthologs between PA14 and PAO1. Details are provided in the text. (**B**) Schematic of possible scenarios complicating ortholog identification. Putative PAO1 ORFs are shown on *top* (in *pink*) and PA14 counterparts are *below* (in *blue*). In scenario 1, a polymorphism may result in an altered gene length. In the case of the PAO1 gene (*PA0003*), the alignment length may fall below the cut off for the minimal percentage of the whole gene (step 7 in Fig. 3A) and would therefore be flagged as a putative non-ortholog. In scenario 2, a gene is duplicated in PA14. In this case, PAO1 gene *PA0004* is listed as the best BLAST match for two PA14 genes. In scenario 3, a polymorphism has resulted in a single PAO1 ORF being split into two PA14 ORFs (or the PA14 ORFs have been fused into a single ORF in PAO1). In this case, PAO1 ORF *PA0005* is listed as the best BLAST match for two PA14 genes; additionally, *PA0005* would have the putative non-ortholog flag assigned to it (step 7 in Fig. 3A) because the single best BLAST hit is only a portion of the total length of the ORF

BLAST match to both the genes and the genome of the other strain were flagged as putative strain-specific genes.

Manual inspection of the automated ortholog predictions revealed several scenarios that typically yielded ambiguous results (Fig. 3B). Scenario 1 depicts a situation in which a sequence polymorphism has truncated the PA14 gene relative to the PAO1 copy. This gene pair would be identified as reciprocal best BLAST hits, but the length of the alignment would potentially be shorter than the minimal percentage of the whole gene and would therefore be flagged as a putative NON-ortholog. In scenario 2, a PAO1 gene is duplicated in PA14. The PAO1 gene will be paired with one of the two PA14 genes as a putative orthologous set; however, the second PA14 gene would also be linked to the same PAO1 gene as its best BLAST hit (i.e., the PAO1 gene would have two entries). Manual inspection is required to ensure that the correct PA14 gene is selected as the ortholog. Scenario 3 shows a polymorphism that results in one PAO1 gene split into two PA14 ORFs (corresponding to the N- and C-termini of the PAO1 gene). Again, the PAO1 gene appears as the best BLAST hit for two different PA14 genes. By manually inspecting the ambiguous cases, we are gathering likely candidates for refinements to our search parameters that will optimize the automated analysis.

Even though the portions of PA14 absent in PAO1 are a small fraction of the total genomic content (8.3% of the genome, Table 1), we were most interested to see which genes were found in these regions and whether or not they contribute to virulence. We therefore identified 58 PA14 regions absent in PAO1 (containing 478 genes) and 54 PAO1 regions absent in PA14 (containing 234 genes). We refer to these as PA14-specific or PAO1-specific regions for the purposes of this discussion (recognizing that these genes may be present in other isolates of *P. aeruginosa* and are not strictly strain-specific). Many of these gene clusters have hallmarks of horizontally transferred DNA, including boundary sequences that include direct repeats, insertion sequences, and tRNA genes, as well as GC contents that differ significantly from the average GC content of the genome as a whole (PA14-specific clusters have an average GC content of 59.6%, more than one standard deviation below the genome average of 66.3%, Table 1). Many of the defined PA14-specific and PAO1-specific regions are small, containing only a few genes (only 22 of the 58 PA14-specific regions contain 5 or more genes, and only 18 of the 54 PAO1-specific regions contain 5 or more genes; Fig. 4). Therefore, only a subset of these putative gene clusters potentially includes functionally related genes encoding a complete pathogenicity "system" in the classical sense. Nevertheless, we included all of these strain-specific regions in our subsequent analysis to ensure that we were not excluding genes with potential roles in virulence.

These strain-specific regions contain an unusually high percentage of genes of unknown function as compared to the whole genome. Each PA14 annotation has an associated numerical value that reflects the confidence with which the gene function is described. At the highest level of confidence are genes with a class 1 designation, which are genes whose functions have been experimentally validated in *P. aeruginosa*. Class 2 genes are highly similar to genes whose functions have

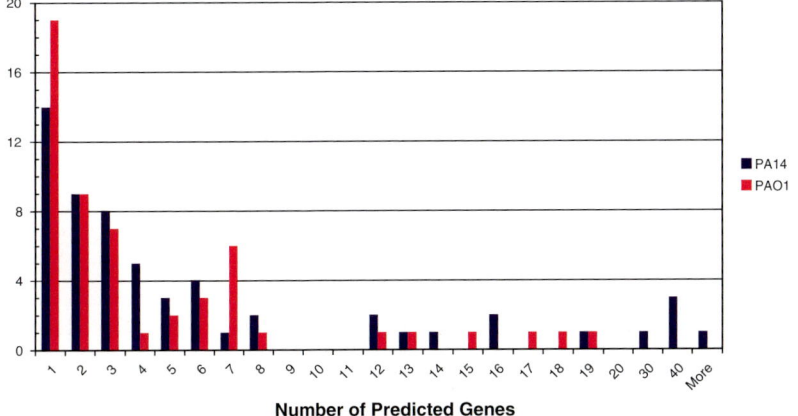

Fig. 4 Number of genes in strain-specific regions. The number (frequency) of PA14- (*blue*) and PAO1-specific regions (*pink*) containing a given number of predicted genes. On the *X*-axis, "30" refers to regions with 21 to 30 genes, "40" refers to regions with 31 to 40 genes, and "More" indicates a single PA14-specific region containing 114 predicted genes

been validated in other organisms and therefore the described gene function is quite likely to be accurate. Class 3 genes have hypothesized functions based on limited similarity to other genes or structural/functional domains, and class 4 genes are ORFs of unknown function. For the PA14-specific regions, 63% of the predicted ORFs have no known function, whereas only 28.9% of genes in the whole genome have class 4 annotations (Table 1). Therefore, gene identity alone could not suggest whether or not virulence-related genes were enriched in PA14-specific regions and we turned to a direct functional analysis of pathogenicity.

4 Relationship Between Genomic Content and Virulence

4.1 Conservation of PA14-Specific Genes and Their Potential Role in Virulence

If there were PA14 strain–specific regions functioning as canonical pathogenicity islands, they should be more prevalent among other virulent strains (and less prevalent among non-pathogenic isolates). To test this hypothesis we established an objective measure of pathogenicity to compare different *P. aeruginosa* strains and developed a high throughput method for determining the genomic content of the different strains. A model host infection system (using the nematode, *C. elegans*) was used to determine the virulence of 18 additional *P. aeruginosa* strains relative to PA14 and PAO1. These 18 strains were used in a previous study of strain diversity (Wolfgang et al., 2003) and were selected to represent a diverse

array of strain sources, including 13 clinical isolates from several types of infections (CF lung infections, urinary tract infections, ocular infections, and blood isolates), four environmental isolates, and one laboratory strain. We then determined
the genomic content and the relatedness of these 20 strains (18 plus PA14 and
PAO1) using a microarray-based genotyping assay (see below) to assess the correlation between PA14-specific genes present in other isolates and their contribution to
virulence.

In the laboratory, *C. elegans* is grown on a lawn of *E. coli*, which serves as
its food source. In the *C. elegans* model pathogenicity system (Tan et al., 1999a),
pathogens such as PA14 are used as the sole source of food in place of *E. coli*, and
the longevity of the nematodes is monitored. Remarkably, nematodes feeding on
a wide variety of human pathogens exhibit significantly shortened longevity compared to their longevity when feeding on *E. coli*. We consider shortened *C. elegans*
longevity as an indication of active killing since, in general, shortened longevity on
pathogenic bacteria is dependent on live bacteria; i.e., dead pathogens do not kill
C. elegans. The nematodes are exposed to a maximal dose of pathogen in this assay
(the bacteria are already grown to a lawn before the nematodes are added to the
assay plate). Therefore, mutations or strain differences that effect the growth rate
or overall fitness of the bacterium tend to have less of an impact on the virulence
phenotype than infection models in which an initial low titer of bacteria is injected
into the host and must survive long enough to reach a density sufficient to cause
disease and death (as, e.g., in plant, insect, and mouse models). A number of studies
have demonstrated that the relative virulence of different *P. aeruginosa* strains in
C. elegans correlates with their virulence in mice (Mahajan-Miklos et al., 1999; Tan
et al., 1999a, b).

We tested the full set of 20 *P. aeruginosa* strains using the *C. elegans* killing
model and we observed a full range of virulence phenotypes, including both the
upper and the lower limits of what this assay system can measure (Lee et al., 2006).
When we compared strains derived from a common infection type, we observed no
consistent clustering with respect to their virulence phenotype in *C. elegans*. For
example, among the CF isolates were both the most virulent and the least virulent
strains in our collection. Importantly, among the environmental isolates, two closely
related strains (MSH3 and MSH10, both obtained from the same geographical site)
were the fourth and the fifth most virulent strains tested. This latter result was interesting because it shows that strains isolated directly from the soil can have similar
infectious potential as strains isolated in a clinical setting.

To test the hypothesis that the virulence of the 20 *P. aeruginosa* strains correlates
with the presence of particular virulence islands, we determined which strain-
specific genes were present or absent in each isolate by performing a microarray-
based analysis of genomic content (a process described as genomotyping; Kim
et al., 2002). We designed a custom array of spotted (70-mer) oligonucleotides,
including 285 oligonucleotides corresponding to PA14 genes that are absent in
PAO1 and 130 oligonucleotides corresponding to PAO1 genes that are absent in
PA14 (along with additional sequences serving as positive and negative controls).
Genomic DNA was isolated from these strains, fragmented, labeled, and hybridized

to the microarray to determine if genes corresponding to each oligonucleotide were present, absent, or indeterminate (the hybridization intensity was too close to the empirically determined cut-off to make a confident present/absent call). Using this array data, we used hierarchical clustering algorithms to determine the relatedness the 20 *P. aeruginosa* strains and found no strong correlation between genomic content and neither the source of the strain nor the relative virulence of the strain (Lee et al., 2006).

4.2 Identification of PA14-specific Virulence Genes and Their Conservation in Other Strains

The experiments described above showed that there was no correlation between the PA14-specific sequences in general and virulence in the *C. elegans* killing assay. We therefore performed a functional analysis of PA14-specific ORFs, directly identifying genes required for pathogenicity and subsequently assessing their distribution among the other strains. Our laboratory has constructed a genome-wide, non-redundant transposon insertion mutant library in PA14 (http://ausubellab.mgh.harvard.edu/cgi-bin/pa14/home.cgi; Liberati et al., 2006). Using this library, we conducted a screen for mutants in PA14-specific genes that had reduced virulence in *C. elegans* (Lee et al., 2006).

We identified four mutations in genes that are present in all *P. aeruginosa* strains but whose sequences are highly divergent (i.e., different strains contain different variants or alleles of functionally similar genes). These included two mutations in genes required for the biosynthesis of the O-antigen component of the *P. aeruginosa* lipopolysaccharide, and two mutations in genes involved in type 4 fimbrial biogenesis. More importantly, we identified five genes required for PA14 virulence that are entirely absent in PAO1; four are genes of unknown function and one has hallmarks of a putative transcription factor. None of these five genes were previously identified as virulence determinants.

To determine whether or not these five PA14-specific pathogenicity genes were predictive of pathogenicity in other strains, we used the microarray genotyping data to compute Spearman's rank correlation coefficients for each PA14-specific gene, relating its presence, absence, or indeterminate status with the (rank order) virulence of the strain. Figure 5 summarizes these data. Each of the 20 strains tested are represented in columns arranged from left to right in order of decreasing pathogenicity. The PA14-specific genes tested (shown in rows) are described as present (blue), absent (yellow), or indeterminate (intermediate intensity that was too close to the cut-off between present and absent calls; red). These genes were then arranged with respect to their Spearman's rank order correlation coefficients, with those genes most tightly correlated with virulence on top and those least correlated on the bottom. This gene order therefore does not reflect a linear order on the chromosome. As shown on the left side of Fig. 5, the correlation coefficients ranged from a high of 0.63 to a low of −0.20. A perfect (positive) correlation would result from a

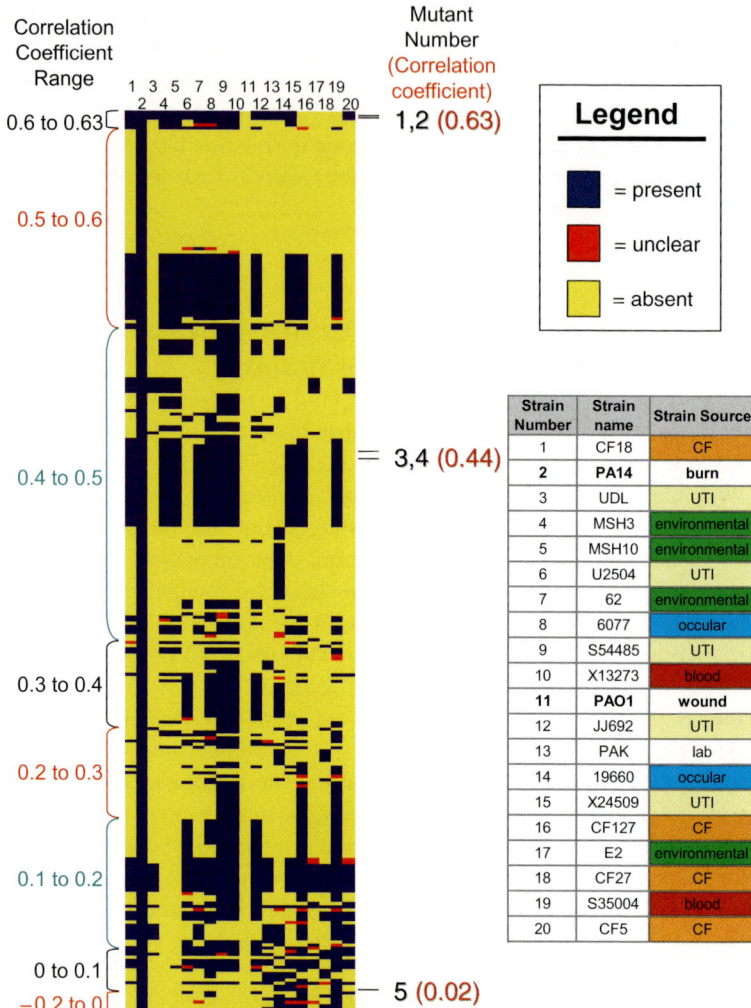

Fig. 5 Range of correlation coefficients between genomic content and virulence. A subset of the microarray-based genotyping of 20 strains is shown for PA14-specific genes. Data for each strain tested is presented in columns. Strains are arranged from left to right in order of decreasing virulence; column headers refer to the 20 strain numbers in the accompanying key listing the strain name and strain source (UTI = urinary tract infection). The strain sources are color coded to highlight that they do not cluster according to virulence. Some strains shown in adjacent columns were equally virulent (strains 4 and 5; strains 8, 9, and 10; strains 14, 15, and 16). Genes are represented as individual rows and are described as present (*blue*), absent (*yellow*), or indeterminate (*red*). Genes are arranged by Spearman's rank correlation coefficients relating the presence of PA14 genes in a strain and the strain's virulence (highest correlations at the top, lowest correlations at the bottom). *Brackets* on the left indicate groups of genes with similar correlation coefficients, with the range of coefficients shown next to the bracket. *Lines* to the right indicate genes required for maximal *C. elegans* virulence; the associated correlation coefficient for these genes is shown in *red* parentheses

row with a clustering of blue on the left (a gene present in the most pathogenic strains) and yellow clustered on the right (a gene absent in the least virulent strains).

The positions of the genes identified as novel PA14-virulence genes are indicated on the right side of Fig. 5 (with correlation coefficients shown in red parentheses). Genes 1 (unknown ORF) and 2 (putative transcription factor) are among those with the highest observed correlation coefficient (0.63). These two ORFs are found together in an 8-gene cluster known as the clone C-specific region common to clone C isolates (members of a clonal family associated with CF infections; Romling et al., 2005). This genomic region is known to be hyper-variable among clone C isolates (and absent in PAO1): two Clone C strains (C and SG17M) each contain the short clone C-specific region adjacent to large strain-specific islands PAGI-2 (strain C, 113 ORFs) or PAGI-3 (strain SG17M, 105 ORFs) (Larbig et al., 2002). The corresponding PA14 locus possesses only orthologs to the genes contained in the shorter clone C-specific region; PA14 does not have orthologs to genes described within the larger strain-specific islands. This cluster of clone C-specific genes was found to be present in 14 of the 18 tested strains, and the 4 strains that lacked these genes were among the 5 least pathogenic isolates tested. However, two exceptions to this general trend include PAO1 itself, which lacks this sequence and whose virulence is intermediate among those strains possessing this cluster, and strain CF5 which contains this cluster and is the least virulent of all isolates.

Genes 3 and 4 in Fig. 5 (both ORFs of unknown function) have intermediate correlation coefficients (0.44). They are contained within a previously identified pathogenicity island (PAPI-1, containing 112 ORFs) shown to encode genes required for pathogenicity in plants and mammals (He et al., 2004). Although the two mutations in genes that we identified in this cluster were not among those previously examined, they are adjacent to or nearby ORFs that had phenotypes in plants and mice. These two genes show a weak correlation with virulence and are present in 7 of the 9 strains more virulent than PAO1 (CF18, MSH3, MSH10, 62, 6077, S54485, and X13273, in order of most to least virulent), and absent in 6 of the 9 strains less pathogenic than PAO1 (JJ6992, PAK, 19660, CF27, E2, and CF5). However, these two genes are absent in the third and sixth most pathogenic strains (UDL and U2504), and present in three avirulent strains (X24509, CF127, and S35004).

Gene 5 (also of unknown function) is part of a 14-gene region that was previously uncharacterized; this gene shows essentially no correlation with strain virulence (correlation coefficient of 0.02). In fact, this gene tends to be more represented in avirulent strains than pathogenic isolates. Taken together, the five functionally defined PA14 genes required for *C. elegans* killing each had exceptions to the general trend (being present in some attenuated strains and/or absent in some virulent strains) and are neither required for nor necessarily predictive of another strain's ability to be pathogenic.

5 Future Directions: Testing Additional Model Hosts

We were surprised that genes identified as PA14-virulence determinants are not necessarily correlated with the virulence of other strains in which they are found, because this observation suggests that the action of some virulence determinants may depend on the total genomic content. This view contrasts with canonical pathogenicity islands that contain a set of genes acting as an autonomous set, the presence of which directly contributes to overall strain virulence. Since most studies of pathogenicity islands have been carried out in more traditional experimental systems, it is possible that our unexpected observations are a result of using a model host such as *C. elegans*. Traditional infection systems using mice or other mammals involve a single dose of bacteria, often administered at a low titer and requiring the initial population of pathogens to proliferate and avoid detection by the host's immune response long enough to reach a pathogenic or lethal dose. In contrast, the *C. elegans* infection system constantly exposes the host to a maximal does of pathogen. The former approach has conceptual similarity to an acute infection in humans, whereas the latter has aspects that more closely approximate a chronic infection.

An advantage of the model host approach to studying infectious disease is that multiple hosts can be used and compared, each with their respective strengths and weaknesses. One alternate host that we are currently examining is the greater wax moth caterpillar, *G. mellonella*. Wax moth caterpillars are commercially available, inexpensive, easy to use, and yield data rapidly, making them an ideal candidate for pilot experiments to extend our observations in *C. elegans* to other host organisms.

5.1 Wax Moth Injection Model

The wax moth caterpillar infection system involves the sub-dermal injection of a low dose of bacteria into the caterpillar and monitoring viability over several days. By injecting a dilution series of the bacteria, quantitative measurements can be made of the LD50, the initial dose required to cause 50% lethality. Wild-type PA14 is an incredibly potent killer, with an LD50 of between 1 and 2 bacteria, whereas attenuated mutants have elevated LD50s (indicating a higher initial does required for comparable disease symptoms). When comparing the virulence of various PA14 mutants in wax moth caterpillars and mice, a positive correlation is observed, making this insect an excellent model system for the identification of pathogenesis factors important in mammalian disease (Jander et al., 2000).

As a preliminary test of the viability of repeating our studies in wax moth caterpillars, we calculated the LD50s for five mutants identified as avirulent in the *C. elegans* screen (Fig. 6A and Table 2). These included mutations in genes *PA14_03370* and *PA14_27680*, both genes of unknown function in regions completely absent in PAO1 (genes 5 and 1, respectively in Fig. 5). Also shown are three mutants in

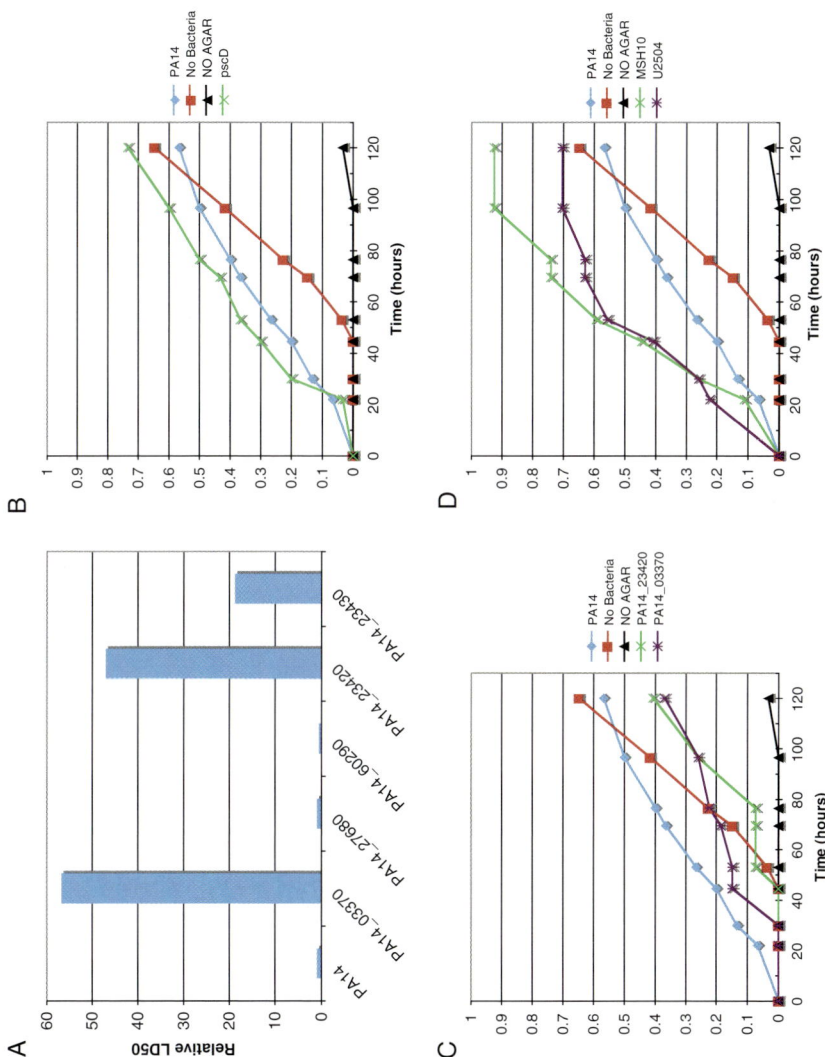

Fig. 6 (continued)

Table 2 Relative phenotypes of PA14 mutants and additional strains *C. elegans* and wax moths

Strain	Gene mutated[*]	Gene name or function	Virulence in *C. elegans*[**]	Virulence in wax moths (injection)[**]	Virulence in wax moths (feeding)[**]
PA14	none		+++	+++	+++
PA14	PA14_03370	Unknown	+	+	+
PA14	PA14_27680	Unknown	+	+++	
PA14	PA14_60290	pilW	+	+++	
PA14	PA14_23420	O-antigen biogenesis	+	+	+
PA14	PA14_23430	O-antigen biogenesis	+	+	
PA14	PA14_42340	pscD	+++	+	+++
MSH10	None		++		+++++
U2504	None		++		+++++

[*] Full annotations for listed PA14 genes are available at http://ausubellab.mgh.harvard.edu/pa14sequencing.
[**] Relative virulence in model hosts. Virulence of wild-type PA14 is assigned a value of "+++." Significantly reduced virulence in mutants is indicated as "+." Strains with slightly lower virulence relative to PA14 are indicated as "++." Enhanced virulence compared to PA14 is indicated as "+++++".

genes that are present in other strains but divergent in sequence, including a gene required for type 4 fimbrial biogenesis (*PA14_60290*) and two genes required for O-antigen biosynthesis (*PA14_23420* and *PA14_23430*). Three of these five genes required for full *C. elegans* virulence are also required in the wax moth caterpillar system, demonstrating some (but not complete) overlap between these two model host systems (Table 2).

Fig. 6 Virulence of PA14 mutants and other strains in wax moth caterpillars. (**A**) Virulence in wax moths by injection of bacteria. Five PA14 mutants identified as avirulent in *C. elegans* were examined in a wax moth caterpillar infection model by calculating a relative LD50 (the dose at which 50% lethality is observed, normalized to the dose required for wild-type PA14). An elevated LD50 indicates that a larger initial inoculum of bacteria is required to observe comparable lethality. The second and third strains tested are mutations in genes of unknown function absent PAO1 (*PA14_03370* and *PA14_27680*) and correspond to genes 5 and 1, respectively, in Fig. 5. These results and those shown in Figs 7B, C, and 7 are summarized in Table 3. (**B**) Virulence factor required for caterpillar lethality via injection is not required to kill in a feeding model. Wax moth caterpillars were exposed to LB plates containing a lawn of PA14 (*blue diamonds*) or a *pscD* mutant (*green Xs*) and the percent of all caterpillars that died (y-axis) was monitored over several days. Controls include an LB plate with no bacteria (*red squares*) or an empty Petri plate with no agar. (**C**) Two PA14 mutants avirulent in *C. elegans* are also attenuated in the wax moth feeding assay. The mutation in gene *PA14_23420* (*green Xs*) disrupts a gene required for O-antigen biosynthesis, and gene *PA14_03370* (*purple asterisks*) is an ORF of unknown function (gene 5 in Fig. 6). (**D**) Two *P. aeruginosa* strains show enhanced virulence in the wax moth feeding assay. MSH10 (*green Xs*) is an environmental isolate, and U2504 (*purple asterisks*) is a urinary tract infection strain. Both strains are virulent in the *C. elegans* infection model but neither is as pathogenic as PA14

5.2 Wax Moth Feeding Model

In addition to the established injection model of infection, we examined whether or not a chronic exposure of the wax moth caterpillars to a lawn of PA14 could result in pathogenicity. This approach is significantly faster than manually injecting each bacterial strain to be tested, and we are tentatively referring to this assay system as a wax moth "feeding" model (although we have yet to determine that bacteria are ingested by feeding rather than entering the host by some other mechanism). We placed wax moth caterpillars onto LB plates containing a lawn of PA14, LB plates with no bacteria, or an empty Petri dish with no agar (Fig. 6B). In the absence of agar, the caterpillars survived with little or no adverse effects over the course of 5 days (black triangles). When placed on an empty LB plate with no added bacteria, a lawn of an (as yet) uncharacterized microorganism grew up over the course of the experiment and the caterpillars began dying at 53 hours (red squares). We have not yet determined whether this microbe was present on the surface of the caterpillars or is initially present in the gut of the animal and seeded onto the plates by excretion. Exposure of the wax moth caterpillars to a pre-formed lawn of PA14 resulted in more rapid death, beginning at 22 hours (blue diamonds). Interestingly, we also tested a previously characterized mutant in PA14, a lesion in the *pscD* gene, which is a structural component of the type III secretion apparatus. Type III secreted bacterial products are required for mammalian pathogenesis and have also been shown to be required for virulence when bacteria are injected into *G. mellonella* caterpillars (but not required for virulence in *C. elegans*; Miyata et al., 2003). Surprisingly, the *pscD* mutant did not show a defect in the wax moth feeding system (green Xs), suggesting that the two *G. mellonella* infection systems are monitoring different sets of virulence determinants (summarized in Table 2).

We next examined two mutations in genes required for pathogenicity in *C. elegans* (Fig. 6C and Table 2). We selected mutations in genes *PA14_23420* (an O-antigen biosynthetic gene) and *PA14_03370* (an ORF of unknown function; gene #5 in Fig. 5), because they were also required for full virulence in the wax moth caterpillar injection system. Both mutants tested showed reduced pathogenicity in the feeding assay (green Xs and purple asterisks), with less overall host lethality than the control containing agar but no bacteria. In all cases in which bacteria were pre-seeded on the LB plates (either wild-type or mutant PA14), the uncharacterized microorganism was unable to grow, making it possible to observe lethality due only to the *P. aeruginosa* strain of interest. These results demonstrate that the caterpillar-feeding assay can identify mutants with reduced virulence, and that some mutations result in avirulence in all three assay systems (*C. elegans*, wax moth caterpillar injection, and wax moth caterpillar feeding; Table 2).

Finally, we examined the ability of the wax moth feeding system to detect differences in virulence among *P. aeruginosa* isolates. Strains MSH10 (an environmental isolate) and U2504 (a urinary tract infection isolate) are virulent in *C. elegans*, but neither is as pathogenic as PA14 (rank order virulence of 5 and 6, respectively, versus a rank order of 2 for PA14 within the group of 20 strains tested). Surprisingly, both strains showed elevated virulence compared to PA14 (Fig. 6D and Table 2). In the injection system, the virulence of wild-type PA14 is so potent (requiring between 1 and 2 live bacteria in the initial dose for 50% lethality) that enhanced virulence cannot be measured. However, the feeding system can potentially identify strains (or mutants) with both enhanced and decreased virulence compared to PA14.

The wax moth caterpillar pathogenicity system is a promising alternative model to extend our virulence studies that were initially carried out with *C. elegans*. Both injection and feeding assays can be used to assess the relative virulence of the 20 *P. aeruginosa* strains and to screen the PA14 mutant library. Given that the three approaches (nematode feeding, caterpillar injection, and caterpillar feeding) result in overlapping but occasionally distinct observations, it is likely that different sets of PA14 mutants will be identified as avirulent when the mutant library is tested in each model. It will be particularly interesting to see if, for a given assay system, the identified PA14-virulence factors each fail to correlate with virulence in other strains (as was the case for *C. elegans*), or whether this surprising lack of correlation is specific to the nematode system.

6 Discussion

To study the relationship between *P. aeruginosa* PA14 genomic content and virulence, we combined several genomic tools including comparative genomic analyses, functional genome-wide analyses of virulence, and a model host infection system amenable to high-throughput and low-cost screening. In the case of *C. elegans*, we have found that genes identified in PA14 as required for virulence (and absent in the less virulent PAO1) do not correlate strongly with the pathogenicity of other strains in which they are also present. In contrast, canonical pathogenicity islands tend to contain genes that contribute directly to the virulence of isolates in which they are found, such as the *P. aeruginosa* gene cluster encoding the type III secretion effector ExoU (present in PA14 but absent in PAO1), which increases the cytotoxicity of strains toward mammalian cells (Sato and Frank, 2004) and is required for pathogenicity in *G. mellonella* and the amoeba, *D. discoideum* (Miyata et al., 2003; Pukatzki et al., 2002). However, it appears that PA14 also contains virulence determinants that may not function autonomously to affect virulence. The data shown in Fig. 5 suggest that, at least in the case of *C. elegans* pathogenicity, the effects of multiple PA14 virulence factors combine in different combinations in different strains to result in an overall virulence phenotype.

What are the mechanistic underpinnings for the apparent combinatorial nature of *P. aeruginosa* virulence factors and how definitive is our analysis with respect to their conservation in other *P. aeruginosa* strains? First, our microarray-based genotyping analysis is only an initial step toward identifying strain similarities and differences at a functional level. The hybridization experiments can only identify (highly similar) sequences that are present in other isolates, but it cannot determine whether or not they encode functional proteins that are similarly regulated. Also, the apparent absence of a gene in a given strain cannot exclude the existence of a gene of divergent sequence that is nevertheless functionally analogous. Second, given the apparent multifactorial and combinatorial nature of *P. aeruginosa* virulence, elucidating the molecular basis of virulence in a particular *P. aeruginosa* strain will require an understanding of how groups of virulence factors interact with each other as well as how they interact with the core genome. With respect to the latter goal, our laboratory is currently extending the screen for PA14 mutants attenuated in *C. elegans* killing to include the entire genome (rather than focussing only on putative strain-specific genes). Third, as stated above, the examination of additional model hosts will be a key step toward determining whether the lack of observed correlations between genomic content and strain virulence is a general phenomenon or specific to nematodes. In addition to the described use of *G. mellonella* (with both an injection and a feeding model of pathogenesis), fruit flies and plants are interesting candidate hosts for future study.

In presenting this work, we have aimed to provide an example of how genomics has impacted our understanding of disease. The rapid development of genomic tools in both the pathogen and the host has greatly expanded the breadth and depth of experimental questions that can be asked in the laboratory. In particular, we believe that the model host infection system, in which multiple, genetically tractable organisms are substituted for traditional mammalian hosts, provides a powerful opportunity to utilize both pathogen and host genomic tools, either individually or combinatorially. The low cost, relatively small size, and low level of complexity of model hosts allow for high-throughput assays that are ideal for exploiting the power inherent in large reagents such as genome-wide mutant libraries. More recently, a *C. elegans–Enterococcus faecalis* infection system has been used to screen large chemical libraries for novel compounds acting as antimicrobials and anti-infectives (Moy et al., 2006); these compounds could serve as vital reagents both in clinical settings and in the lab as molecular probes to dissect the pathogenesis process. With model hosts such as *C. elegans*, genome-wide RNA interference (RNAi) libraries and transcriptional arrays can be used to identify host gene products involved in the innate immune response to pathogens (N. Liberati, R. Feinbaum and F. M. Ausubel, unpublished observations; Gravato-Nobre et al., 2005; O'Rourke et al., 2006; Troemel et al., 2006). As the use of genomic tools and model host infection systems matures, the true power of these approaches will be in the ability to combine both host and pathogen reagents to understand, at a mechanistic level, the complex interactions between the two organisms that occur during an infection.

Acknowledgments We are grateful to J. Decker, W. Brown, K. Osborn, A. Perera, R. Elliott, L. Gendal, K. Montgomery, G. Grills, and L. Li for sequencing PA14; R. Jackson for suggestions on genomic DNA preparations for Microarray analysis; N. El Massadi and J. Frietas for suggestions on sample labeling, hybridizations, and scanning for microarray experiments; and D. Park for assistance with microarray data analysis. Previously described data was obtained using funds provided by NIH grants AI064332-02 and HL66678, and DOE grant DE-FG02-ER63445.

References

Badarinarayana, V., Estep, P.W., 3rd, Shendure, J., Edwards, J., Tavazoie, S., Lam, F., and Church, G.M., 2001, Selection analyses of insertional mutants using subgenic-resolution arrays, *Nat. Biotechnol.* **19**:1060–1065.

Bentley, S.D., and Parkhill, J., 2004, Comparative genomic structure of prokaryotes, *Annu. Rev. Genet.* **38**:771–792.

Choi, J.Y., Sifri, C.D., Goumnerov, B.C., Rahme, L.G., Ausubel, F.M., and Calderwood, S.B., 2002, Identification of virulence genes in a pathogenic strain of *Pseudomonas aeruginosa* by representational difference analysis, *J. Bacteriol.* **184**:952–961.

Delcher, A.L., Harmon, D., Kasif, S., White, O., and Salzberg, S.L., 1999, Improved microbial gene identification with GLIMMER, *Nucleic Acids Res.* **27**:4636–4641.

Doring, D., 1993, Chronic *Pseudomonas aeruginosa* lung infection in cystic fibrosis patients, *Pseudomonas aeruginosa* as an opportunistic pathogen, Plenum Press, New York, pp. 245–273.

Ernst, R.K., D'Argenio, D.A., Ichikawa, J.K., Bangera, M.G., Selgrade, S., Burns, J.L., Hiatt, P., McCoy, K., Brittnacher, M., Kas, A., et al., 2003, Genome mosaicism is conserved but not unique in *Pseudomonas aeruginosa* isolates from the airways of young children with cystic fibrosis, *Environ. Microbiol.* **5**:1341–1349.

Gravato-Nobre, M.J., Nicholas, H.R., Nijland, R., O'Rourke, D., Whittington, D.E., Yook, K.J., and Hodgkin, J., 2005, Multiple genes affect sensitivity of *Caenorhabditis elegans* to the bacterial pathogen *Microbacterium nematophilum*, *Genetics* **171**:1033–1045.

Hacker, J., Blum-Oehler, G., Muhldorfer, I., and Tschape, H., 1997, Pathogenicity islands of virulent bacteria: structure, function and impact on microbial evolution, *Mol. Microbiol.* **23**: 1089–1097.

He, J., Baldini, R.L., Deziel, E., Saucier, M., Zhang, Q., Liberati, N.T., Lee, D., Urbach, J., Goodman, H.M., and Rahme, L.G., 2004, The broad host range pathogen *Pseudomonas aeruginosa* strain PA14 carries two pathogenicity islands harboring plant and animal virulence genes, *Proc. Natl. Acad. Sci. USA* **101**:2530–2535.

Hensel, M., Shea, J.E., Gleeson, C., Jones, M.D., Dalton, E., and Holden, D.W., 1995, Simultaneous identification of bacterial virulence genes by negative selection, *Science* **269**: 400–403.

Jander, G., Rahme, L.G., and Ausubel, F.M., 2000, Positive correlation between virulence of *Pseudomonas aeruginosa* mutants in mice and insects, *J. Bacteriol.* **182**:3843–3845.

Kim, C.C., Joyce, E.A., Chan, K., and Falkow, S., 2002, Improved analytical methods for microarray-based genome-composition analysis, *Genome Biol.* **3**, RESEARCH0065. 1–0065.17.

Kurtz, S., Phillippy, A., Delcher, A.L., Smoot, M., Shumway, M., Antonescu, C., and Salzberg, S.L., 2004, Versatile and open software for comparing large genomes, *Genome Biol.* **5**:R12.

Larbig, K.D., Christmann, A., Johann, A., Klockgether, J., Hartsch, T., Merkl, R., Wiehlmann, L., Fritz, H.J., and Tummler, B., 2002, Gene islands integrated into tRNA (Gly) genes confer genome diversity on a *Pseudomonas aeruginosa* clone, *J. Bacteriol.* **184**:6665–6680.

Lau, G.W., Goumnerov, B.C., Walendziewicz, C.L., Hewitson, J., Xiao, W., Mahajan-Miklos, S., Tompkins, R.G., Perkins, L.A., and Rahme, L.G., 2003, The *Drosophila melanogaster* toll pathway participates in resistance to infection by the gram-negative human pathogen *Pseudomonas aeruginosa*, *Infect. Immun.* **71**:4059–4066.

Lee, D.G., Urbach, J.M., Wu, G., Liberati, N.T., Feinbaum, R.L., Miyata, S., Diggins, L.T., He, J., Saucier, M., Deziel, E., et al., 2006, Genomic analysis reveals that *Pseudomonas aeruginosa* virulence is combinatorial, *Genome Biol.* **7**:R90.

Liberati, N.T., Urbach, J.M., Miyata, S., Lee, D.G., Drenkard, E., Wu, G., Villanueva, J., Wei, T., and Ausubel, F.M., 2006, An ordered, nonredundant library of *Pseudomonas aeruginosa* strain PA14 transposon insertion mutants, *Proc. Natl. Acad. Sci. USA* **103**:2833–2838.

Mahajan-Miklos, S., Tan, M.W., Rahme, L.G., and Ausubel, F.M., 1999, Molecular mechanisms of bacterial virulence elucidated using a *Pseudomonas aeruginosa–Caenorhabditis elegans* pathogenesis model, *Cell* **96**:47–56.

Miyata, S., Casey, M., Frank, D.W., Ausubel, F.M., and Drenkard, E., 2003, Use of the *Galleria mellonella* caterpillar as a model host to study the role of the type III secretion system in *Pseudomonas aeruginosa* pathogenesis, *Infect. Immun.* **71**:2404–2413.

Moy, T.I., Ball, A.R., Anklesaria, Z., Casadei, G., Lewis, K., and Ausubel, F.M., 2006, Identification of novel antimicrobials using a live-animal infection model, *Proc. Natl. Acad. Sci. USA* **103**:10414–10419.

O'Rourke, D., Baban, D., Demidova, M., Mott, R., and Hodgkin, J., 2006, Genomic clusters, putative pathogen recognition molecules, and antimicrobial genes are induced by infection of *C. elegans* with *M. nematophilum*, *Genome Res.* **16**:1005–1016.

Oelschlaeger, T.A., and Hacker, J., 2004, Impact of pathogenicity islands in bacterial diagnostics, *Apmis* **112**:930–936.

Plotnikova, J.M., Rahme, L.G., and Ausubel, F.M., 2000, Pathogenesis of the human opportunistic pathogen *Pseudomonas aeruginosa* PA14 in *Arabidopsis*, *Plant Physiol.* **124**: 1766–1774.

Pukatzki, S., Kessin, R.H., and Mekalanos, J.J., 2002, The human pathogen *Pseudomonas aeruginosa* utilizes conserved virulence pathways to infect the social amoeba *Dictyostelium discoideum*, *Proc. Natl. Acad. Sci. USA* **99**:3159–3164.

Rahme, L.G., Stevens, E.J., Wolfort, S.F., Shao, J., Tompkins, R.G., and Ausubel, F.M., 1995, Common virulence factors for bacterial pathogenicity in plants and animals, *Science* **268**: 1899–1902.

Rahme, L.G., Tan, M.W., Le, L., Wong, S.M., Tompkins, R.G., Calderwood, S.B., and Ausubel, F.M., 1997, Use of model plant hosts to identify *Pseudomonas aeruginosa* virulence factors, *Proc. Natl. Acad. Sci. USA* **94**:13245–13250.

Raskin, D.M., Seshadri, R., Pukatzki, S.U., and Mekalanos, J.J., 2006, Bacterial genomics and pathogen evolution, *Cell* **124**:703–714.

Romling, U., Kader, A., Sriramulu, D.D., Simm, R., and Kronvall, G., 2005, Worldwide distribution of *Pseudomonas aeruginosa* clone C strains in the aquatic environment and cystic fibrosis patients, *Environ. Microbiol.* **7**:1029–1038.

Saenz, H.L., and Dehio, C., 2005, Signature-tagged mutagenesis: technical advances in a negative selection method for virulence gene identification, *Curr. Opin. Microbiol.* **8**:612–619.

Salzberg, S.L., Delcher, A.L., Kasif, S., and White, O., 1998, Microbial gene identification using interpolated Markov models, *Nucleic Acids Res.* **26**:544–548.

Sassetti, C.M., Boyd, D.H., and Rubin, E.J., 2001, Comprehensive identification of conditionally essential genes in mycobacteria, *Proc. Natl. Acad. Sci. USA* **98**:12712–12717.

Sato, H., and Frank, D.W., 2004, ExoU is a potent intracellular phospholipase, *Mol. Microbiol.* **53**:1279–1290.

Spencer, D.H., Kas, A., Smith, E.E., Raymond, C.K., Sims, E.H., Hastings, M., Burns, J.L., Kaul, R., and Olson, M.V., 2003, Whole-genome sequence variation among multiple isolates of *Pseudomonas aeruginosa*, *J. Bacteriol.* **185**:1316–1325.

Stover, C.K., Pham, X.Q., Erwin, A.L., Mizoguchi, S.D., Warrener, P., et al., 2000, Complete genome sequence of *Pseudomonas aeruginosa* PA01, an opportunistic pathogen, *Nature* **406**:959–964.

Tan, M.W., Mahajan-Miklos, S., and Ausubel, F.M., 1999a, Killing of *Caenorhabditis elegans* by *Pseudomonas aeruginosa* used to model mammalian bacterial pathogenesis, *Proc. Natl. Acad. Sci. USA* **96**:715–720.

Tan, M.W., Rahme, L.G., Sternberg, J.A., Tompkins, R.G., and Ausubel, F.M., 1999b, *Pseudomonas aeruginosa* killing of *Caenorhabditis elegans* used to identify *P. aeruginosa* virulence factors, *Proc. Natl. Acad. Sci. USA* **96**:2408–2413.

Troemel, E.R., Chu, S.W., Renke, V., Lee, S.S., Ausubel, F.M., and Kim, D.H., 2006, p38 MAPK Regulates expression of immune response genes and contributes to longevity in *C. elegans*. PLoS, *Genetics*, 2:e183.

Wolfgang, M.C., Kulasekara, B.R., Liang, X., Boyd, D., Wu, K., Yang, Q., Miyada, C.G., and Lory, S., 2003, Conservation of genome content and virulence determinants among clinical and environmental isolates of *Pseudomonas aeruginosa*, *Proc. Natl. Acad. Sci. USA* **100**: 8484–8489.

Wong, S.M., and Mekalanos, J.J., 2000 Genetic footprinting with mariner-based transposition in *Pseudomonas aeruginosa*, *Proc. Natl. Acad. Sci. USA* **97**:10191–10196.

Wood, R.E., 1976, Pseudomonas: the compromised host, *Hosp. Pract.* **11**:91–100.

Genetic Dissection of the Interaction Between the Plant Pathogen *Xanthomonas campestris* pv. *vesicatoria* and Its Host Plants

Ulla Bonas, Doreen Gürlebeck, Daniela Büttner, Monique Egler,
Simone Hahn, Sabine Kay, Antje Krüger, Christian Lorenz,
Robert Szczesny, and Frank Thieme

Abstract The interaction between the plant pathogenic bacterium *Xanthomonas campestris* pv. *vesicatoria* (*Xcv*) and its host plants pepper and tomato depends on a type III protein secretion system (T3SS) which translocates effector proteins into the plant cell. Recent studies revealed that HpaB and HpaC are two key players in the control of protein export from *Xcv*. First identified by their avirulence activity in resistant plants, genome sequencing projects allow now the identification of more effector proteins. However, their virulence functions in the host remain elusive. The effector AvrBs3 from *Xcv* induces a hypertrophy in susceptible plants. The virulence as well as the avirulence activity of AvrBs3 depends on its eukaryotic features, i.e., nuclear localization signals and an activation domain suggesting that the effector mimics a plant transcriptional regulator. Here, we present recent progress on the identification of potential virulence targets of AvrBs3 in the plant cell.

1 Introduction

Plant pathogenic bacteria have evolved sophisticated mechanisms to suppress basal plant defense reactions and to colonize their host plants. One of the model systems to study the molecular basis of disease is the Gram-negative bacterium *Xanthomonas campestris* pathovar (pv.) *vesicatoria* (*Xcv*; also termed *X. axonopodis* pv. *vesicatoria*); (Vauterin et al., 2000), the causal agent of bacterial spot disease in pepper and tomato, which leads to great economic losses in growing regions with a warm and humid climate. In nature, *Xcv* enters the plant tissue via natural openings (stomata) or wounds and, in case of susceptible host plants, multiplies in the intercellular space. Colonization is confined to local infection sites resulting in the appearance of water-soaked lesions that later become necrotic (Stall, 1995). In case of a resistant host plant, the bacteria are recognized based on a specific plant disease resistance (*R*) gene that "corresponds" to a particular bacterial effector gene (Flor, 1971).

U. Bonas
Department of Genetics, Martin-Luther-University Halle-Wittenberg, Halle, Germany,
e-mail: ulla.bonas@genetik.uni-halle.de

J.P. Gustafson et al. (eds.), *Genomics of Disease*,
© Springer Science+Business Media, LLC, 2008

In most cases that are studied on the molecular level the outcome of resistance is a hypersensitive reaction (HR), a rapid and localized programmed cell death that restricts further spread of the pathogen (Klement, 1982).

Recently, the genome sequence of *Xcv* strain 85–10 has been elucidated. Besides the chromosome (5.17 Mb) this strain contains four plasmids 1.8 kb to 183 kb in size (Thieme et al., 2005). Essential for pathogenicity of *Xcv* is a type III protein secretion system (T3SS), which is highly conserved in many Gram-negative plant and animal pathogenic bacteria and some symbionts (Büttner and Bonas, 2006; Cornelis and Van Gijsegem, 2000; Marie et al., 2001; Tampakaki et al., 2004). Expression of the T3SS is induced in the plant and in certain minimal media. The T3SS secretes a small number of proteins into the extra cellular milieu, but most proteins (termed "effectors") are translocated into the host cell cytosol (Büttner and Bonas, 2002; Gürlebeck et al., 2006) (Fig. 1). To date, more than 20 type III effector proteins are

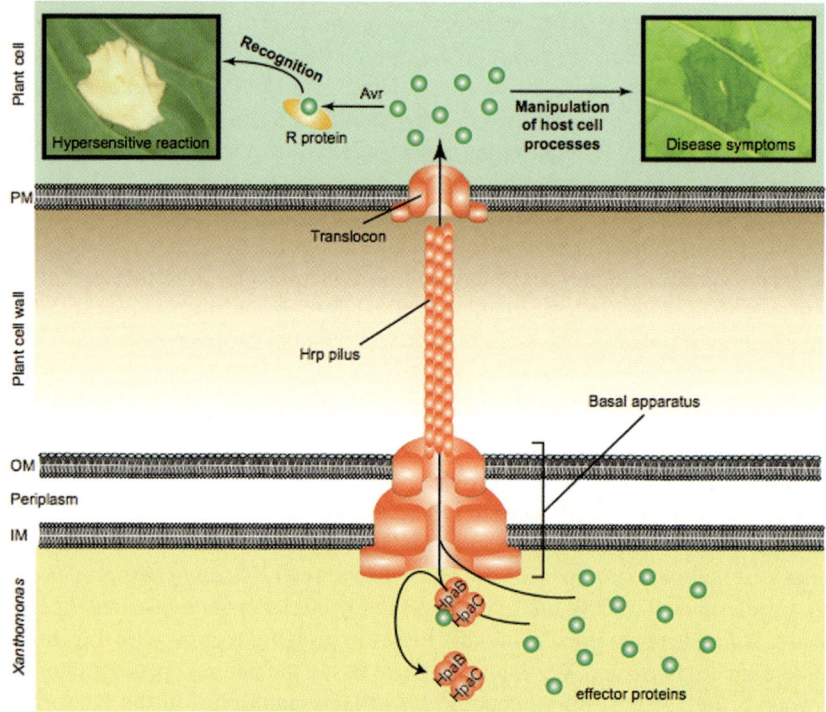

Fig. 1 Model of type III secretion of *Xcv*. The T3SS spans both bacterial membranes and is connected to an extracellular Hrp pilus and the T3S translocon, which inserts into the plant plasma membrane. The T3SS injects bacterial effector proteins into the host cell. Secretion of effectors and non-effectors is controlled by HpaB and HpaC which interact with each other and bring effectors into contact with the T3S apparatus (Büttner et al., 2006). Once in the plant cell, effector proteins modulate the host cell metabolism to the advantage of the pathogen leading to the formation of disease symptoms. Some type III effectors are recognized mediated by corresponding R proteins leading to the induction of the hypersensitive reaction. CW, cell wall; PM, plasma membrane; IM, inner membrane; OM, outer membrane

known in *Xcv* (Thieme et al., 2005). Effector proteins probably manipulate host cell processes to the need of the pathogen. Key questions studied in our laboratory are (i) regulation of type III secretion, (ii) identification of type III effector proteins and their host targets, and (iii) dissection of specific plant disease resistance to *Xcv*.

The first type III effectors were isolated some 15–20 years ago because of their role in plant resistance. These so-called "avirulence (Avr) proteins" were defined genetically as pathogen determinants that are recognized by resistant host plants expressing a corresponding resistance gene. Upon recognition, the Avr protein induces plant defense reactions, often manifested as an HR (Keen, 1990; Klement, 1982). Later, mutant studies demonstrated that the primary role of Avr proteins is in pathogen virulence, i.e., *avr* genes can contribute to bacterial multiplication in the plant and to symptom development (Gürlebeck et al., 2006). Many new effectors were identified by bioinformatic analysis of the respective bacterial genome sequence. Criteria which can be used are conserved features of the promoter regions of the corresponding genes. For example, all *P. syringae* genes for type III system components and effectors contain an *hrp* box in their promoter (Xiao and Hutcheson, 1994), whereas many *Xcv* effector genes carry a PIP-box (see below). In addition, many effector proteins can be grouped into families based on their predicted function and/or similarity to known effectors. Finally, the presence of eukaryotic features such as leucine-rich-repeats (LRR) or nuclear localization signals (NLS) might be indicative of an effector candidate.

To verify effector candidates their translocation into the plant cell has to be demonstrated. This can be monitored using a translational fusion of the N-terminal region, which presumably contains the type III secretion signal, to a suitable reporter protein. Successfully used are known effector domains, e.g., AvrBs3 that is deleted in its N-terminus but still causes the HR in a resistant plant when expression was mediated by *Agrobacterium* (Szurek et al., 2002). An alternative method is the so-called "Cya assay" using an adenylate cyclase as reporter (Casper-Lindley et al., 2002). Based on sequence conservation or predicted enzymatic functions most type III effector proteins can be grouped into families. It turned out that the bacteria often express effector proteins that use molecular mimicry to fulfill their function in the plant cell (Gürlebeck et al., 2006). Examples are XopD from *Xcv*, which is a putative cysteine protease that cleaves SUMOylated proteins (Hotson et al., 2003), and HopPtoD2, a protein phosphatase of *P. syringae* (Bretz et al., 2003; Espinosa et al., 2003). Another large effector family is the AvrBs3 family which is expressed almost exclusively in *Xanthomonas* spp.

2 Results and Discussion

2.1 The T3SS of Xcv

The genes encoding the T3SS of *Xcv* are localized in a 35.3 kb pathogenicity island on the bacterial chromosome. The core gene cluster consists of six *hrp* (hypersensitive reaction and pathogenicity) operons which encode the components of the T3SS and three Hrp-associated proteins (HpaA-C), which are non-essential virulence proteins (Gürlebeck et al., 2006). Expression of the T3SS is induced in the plant and in certain minimal media and is controlled by two regulators, HrpG and HrpX, which are encoded outside of the *hrp* region. HrpG, a member of the OmpR family of transcriptional regulators, is constitutively expressed but needs to be activated via an unknown signal from the plant. Active HrpG then induces the transcription of *hrpX*, which encodes an AraC-type regulator that in turn controls expression of most genes in the *hrp* region and genes elsewhere in the genome (for a recent comprehensive review, see Gürlebeck et al., 2006). Most HrpX-regulated promoters contain a conserved sequence element, the PIP (plant-inducible promoter; consensus TTCG-N16-TTCG) box, which is specifically bound by HrpX (Koebnik et al., 2006). More than 100 *hrpG*- and *hrpX*-induced genes have been identified by cDNA-AFLP and microarray analyses (Noël et al., 2001; Koebnik et al., 2006; B. Baumgarth, A. Becker, F. Thieme and U. Bonas, unpublished data).

On the bacterial surface the T3SS is extended by the Hrp pilus through which translocon and effector proteins are transported (Weber et al., 2005). In *Xcv* strain 85–10 more than 20 different effector proteins are known (Gürlebeck et al., 2006; Thieme et al., 2005; F. Thieme and U. Bonas, unpublished data).

The role of most effectors in virulence is not clear, because often a deletion of the respective gene does not alter bacterial virulence in the plant (Noël et al., 2003; Noël et al., 2001, 2002; F. Thieme, A. Krüger, A. Urban and U. Bonas, unpublished data).

2.2 Control of T3S by Xcv

How are bacterial proteins recognized by the T3SS? The export signal of most proteins probably resides in the N-terminal amino acid sequence of the protein. The secretion signal is not conserved although there is some amino acid bias. In contrast to proteins secreted by the type II system, the T3S signal is not cleaved (Büttner and Bonas, 2006). Recent studies revealed that HpaB and HpaC, which are not secreted, are two key players in the control of type III protein export. Interestingly, there are two classes of type III effectors based on their need for HpaB. The latter might determine the hierarchy of protein secretion; however, this intriguing process remains enigmatic. HpaB is a general T3S chaperone which interacts with HpaC and with effector proteins and hence directs them to the T3SS (Büttner et al., 2004, 2006). HpaC interacts with HpaB and appears to increase the efficiency of protein export. Once the predicted translocon complex consisting of HrpF and possibly XopA is inserted into the plant plasma membrane effector proteins are injected directly from the bacterium into the plant cell cytoplasm (Büttner et al., 2004, 2002; Gürlebeck et al., 2006; Noël et al., 2002).

2.3 The AvrBs3 Effector Protein

AvrBs3 is encoded on a self-transmissible plasmid that is present in some strains of *Xcv*. *Xcv* strains that express the 122-kDa AvrBs3 protein are specifically recognized by pepper plants carrying the *Bs3* resistance gene resulting in the HR and halt of bacterial multiplication (Bonas et al., 1989). In susceptible pepper and tomato plants the AvrBs3 protein induces hypertrophy of the mesophyll cells, i.e., cell enlargement (Marois et al., 2002). Hypertrophy is also induced in other solanaceous plants, e.g., *Nicotiana benthamiana*, when the *avrBs3* gene is delivered on a T-DNA by *Agrobacterium tumefaciens* (Marois et al., 2002). These so-called "transient expression" assays have become an important tool for the analysis of bacterial effector genes and their potential biochemical function in the plant cell.

The amino acid sequence of the AvrBs3 protein is characterized by a central region consisting of 17.5 direct 34-amino acid repeats, two NLSs and an acidic activation domain (AAD) in the C-terminal region. In addition, a leucine-zipper-like region was predicted for AvrBs3 and homologous proteins (Fig. 2; Gabriel, 1999). NLS and AAD are typical eukaryotic motifs and were both shown to be essential for the known AvrBs3 acitivities (Marois et al., 2002; Szurek et al., 2001; Van den Ackerveken et al., 1996). The direct repeats are nearly identical and are a hallmark of AvrBs3 family members, which are present in many other *Xanthomonas* strains.

Analysis of *avrBs3* repeat deletion derivatives revealed that the repeats determine the specificity of AvrBs3 activity. For instance, AvrBs3Δrep-16, in which repeats 11 to 14 are deleted, has lost the ability to be recognized by *Bs3* containing plants, but is recognized by *bs3* pepper plants (Herbers et al., 1992). Similarly, other derivatives showed also new recognition specificities. The hypertrophy symptoms, however, are only induced by the AvrBs3 wild-type protein and not by AvrBs3 deletion derivative or the close homolog AvrBs4 from *Xcv*. The AvrBs4 protein, which is recognized by tomato plants expressing the *Bs4 R* gene, is 97% amino acid sequence-identical to AvrBs3 (Bonas et al., 1993; Schornack et al., 2006).

2.4 Plant Target Proteins of AvrBs3

Due to the fact that AvrBs3 needs both NLS and AAD for its activity but lacks an obvious DNA binding domain we hypothesized that AvrBs3 might bind to DNA in a complex with proteins from the plant. To identify proteins that specifically interact

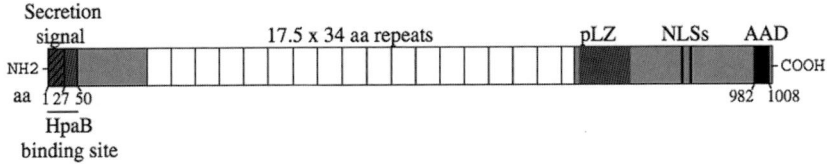

Fig. 2 Structural features of AvrBs3 from *Xcv*. The central repeat region consists of 17.5 nearly identical tandem repeats of 34 amino acids. The C-terminal region contains different eukaryotic motifs: imperfect heptad leucine zipper repeats (pLZ; *black box*; Gabriel, 1999); two functional nuclear localization signals (NLSs; *black bars*; Szurek et al., 2001; Van den Ackerveken et al., 1996) and an acidic activation domain (AAD; Szurek et al., 2001). The secretion signal and the HpaB binding site are indicated (gray hatched and gray, respectively; Szurek et al., 2002; Büttner et al., 2004)

with AvrBs3 we performed several yeast-two-hybrid screens using different AvrBs3 baits. Among the identified genes were two genes encoding importin α homologs, a methyltransferase homolog, a gene showing similarity to the thiamine biosynthesis enzyme C (ThiC), and, interestingly, one with similarity to the homeobox protein MERISTEM LAYER 1 from *Arabidopsis thaliana* (ML1). Using bimolecular fluorescence complementation we demonstrated that AvrBs3 forms complexes with the pepper importin α protein *in planta* (D. Gürlebeck and U. Bonas, unpublished data). For the ThiCH (ThiC homolog) we showed that it is indeed involved in vitamin B1 synthesis. Interestingly, recent publications suggested a role for vitamin B1 in enhancing plant defense (Ahn et al., 2005; Wang et al., 2006). Thus, ThiC is an interesting candidate virulence target of AvrBs3. AvrBs3 might block the enzymatic activity of the ThiCH and thus delay thiamine production and plant defense reactions (Fig. 3). Since the LZ-containing ML1 homolog (ML1H) was identified using the putative LZ motif of AvrBs3 as a bait (D. Gürlebeck, A. Raschke and U. Bonas, unpublished data) one can speculate that ML1H and AvrBs3 form heterodimers which interact with DNA via the LZ motifs. The latter is well known for bZIP transcription factors from plants (Jakoby et al., 2002). Thus, ML1H could be one of the "missing links" in understanding AvrBs3 function.

2.5 Plant Target Genes of AvrBs3

In vitro secretion and in vivo translocation studies demonstrated that AvrBs3 is secreted by the T3SS and localizes to the plant cell nucleus (Szurek et al., 2002). Prior to nuclear localization, AvrBs3 homodimerizes via the repeat region and interacts with importin α via the NLSs in the plant cell cytoplasm (Gürlebeck et al., 2005; Szurek et al., 2001, 2002; D. Gürlebeck und U. Bonas, unpublished data). Based on the fact that the AAD is essential for function we wondered if AvrBs3 modulates plant gene expression. Using cDNA-AFLP and suppression subtractive hybridization we isolated more than 20 *upa* (up-regulated by AvrBs3) genes expression of which is specifically upregulated in susceptible pepper plants by AvrBs3 (Marois et al., 2002; Kay et al., 2007). Among the identified *upa* genes 16 genes are also induced in the presence of cycloheximide, an inhibitor of eukaryotic protein synthesis, i.e., these genes are good candidates for being regulated directly by AvrBs3. Studies of the time course of induction demonstrated that a number of *upa* genes are induced as early as 3–4 hours post inoculation (hpi) with *Xcv*. This finding is in agreement with the observation that the pathogen requires 2–3 hours for gene transcription and translation (U. Bonas, unpublished data) to be able to induce the HR in resistant plants, i.e., to build up the T3SS and start protein translocation into the plant cell.

For technical reasons functional studies of *upa* genes are performed in *N. benthamiana*. Intriguingly, gene silencing and overexpression analyses of *upa* genes in *N. benthamiana* revealed that one of the early induced *upa* genes is crucial and sufficient for the induction of hypertrophy (Kay et al., 2007). Interestingly, this gene

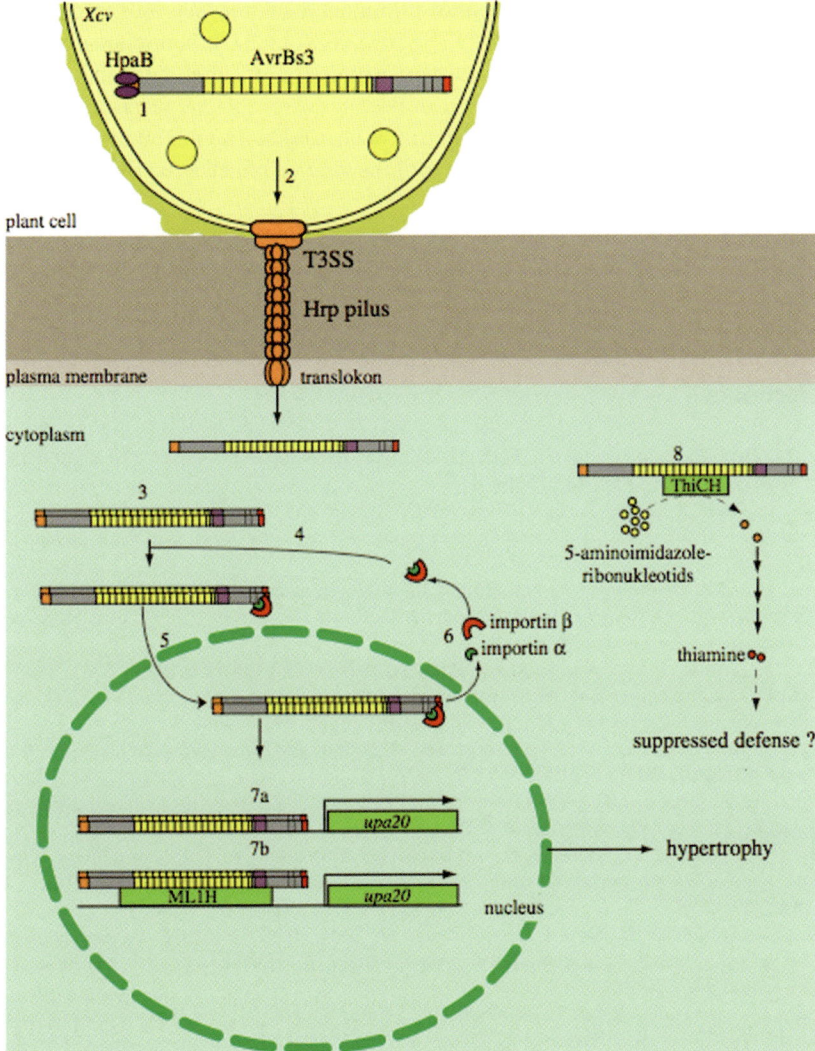

Fig. 3 Current model of action of AvrBs3. HpaB enhances the secretion and translocation of AvrBs3 by binding to the N-terminal 50 amino acids of the effector within *Xcv* (**1–2**). After injection into the plant cell, AvrBs3 dimerizes in the cytoplasm (**3**). Importin α bound to importin β binds to the NLS of AvrBs3 (**4**) and thus mediates the nuclear import of the effector (**5**). In the nucleus the importin molecules dissociate and are recycled to the cytoplasm of the plant cell (**6**) so that AvrBs3 can act as transcriptional activator (**7**). For that AvrBs3 might directly (**7a**) bind to promoters of *upa* genes or in a complex with a virulence target protein, e.g., the homeobox protein ML1 homolog (**7b**). The induction of gene expression leads then to the growth of the plant cell (hypertrophy). Prior to nuclear import, AvrBs3 molecules might transiently interact with a ThiC homolog (ThiCH; **8**) to inhibit its enzymatic activity, which could lead to a delay of plant defense reactions

encodes a protein with similarity to transcriptional regulators and thus might be the key regulator for the induction of the hypertrophy (Fig. 3). It is now interesting to investigate if this Upa protein induces the expression of other, later induced *upa* genes, e.g., genes encoding expansin α or pectate lyases that are probably involved in the development of hypertrophy (Marois et al., 2002). The model in Fig. 3 summarizes the current view on plant targets of the type III effector AvrBs3 from *Xcv*.

Acknowledgments This work has been supported by grants from the DFG (SFB 648 "Molecular mechanisms of information processing in plants") and the BMBF (GenoMik) to Ulla Bonas.

References

Ahn, I.P., Kim, S., and Lee, Y.H., 2005, Vitamin B1 functions as an activator of plant disease resistance, *Plant Physiol.* **138**:1505–1515.

Bonas, U., Stall, R.E., and Staskawicz, B., 1989, Genetic and structural characterization of the avirulence gene *avrBs3* from *Xanthomonas campestris* pv. *vesicatoria*, *Mol. Gen. Genet.* **218**: 127–136.

Bonas, U., Conrads-Strauch, J., and Balbo, I., 1993, Resistance in tomato to *Xanthomonas campestris* pv. *vesicatoria* is determined by alleles of the pepper-specific avirulence gene *avrBs3*, *Mol. Gen. Genet.* **238**:261–269.

Bretz, J.R., Mock, N.M., Charity, J.C., Zeyad, S., Baker, C.J., and Hutcheson, S.W., 2003, A translocated protein tyrosine phosphatase of *Pseudomonas syringae* pv. *tomato* DC3000 modulates plant defence response to infection, *Mol. Microbiol.* **49**:389–400.

Büttner, D., and Bonas, U., 2002, Getting across—bacterial type III effector proteins on their way to the plant cell, *EMBO J.* **21**:5313–5322.

Büttner, D., and Bonas, U., 2006, Who comes first? How plant pathogenic bacteria orchestrate type III secretion, *Curr. Opin. Microbiol.* **9**:193–200.

Büttner, D., Nennstiel, D., Klüsener, B., and Bonas, U., 2002, Functional analysis of HrpF, a putative type III translocon protein from *Xanthomonas campestris* pv. *vesicatoria*, *J. Bacteriol.* **184**:2389–2398.

Büttner, D., Gürlebeck, D., Noël, L.D., and Bonas, U., 2004, HpaB from *Xanthomonas campestris* pv. *vesicatoria* acts as an exit control protein in type III-dependent protein secretion, *Mol. Microbiol.* **54**:755–768.

Büttner, D., Lorenz, C., Weber, E., and Bonas, U., 2006, Targeting of two effector protein classes to the type III secretion system by a HpaC- and HpaB-dependent protein complex from *Xanthomonas campestris* pv. *vesicatoria*, *Mol. Microbiol.* **59**:513–527.

Casper-Lindley, C., Dahlbeck, D., Clark, E.T., and Staskawicz, B.J., 2002, Direct biochemical evidence for type III secretion-dependent translocation of the AvrBs2 effector protein into plant cells, *Proc. Natl. Acad. Sci. USA* **99**:8336–8341.

Cornelis, G.R., and Van Gijsegem, F., 2000, Assembly and function of type III secretory systems, *Annu. Rev. Microbiol.* **54**:35–774.

Espinosa, A., Guo, M., Tam, V.C., Fu, Z.Q., and Alfano, J.R., 2003, The *Pseudomonas syringae* type III-secreted protein HopPtoD2 possesses protein tyrosine phosphatase activity and suppresses programmed cell death in plants, *Mol. Microbiol.* **49**:377–387.

Flor, H.H.,1971, Current status of the gene-for-gene concept, *Annu. Rev. Phytopathol.* **9**:275–296.

Gabriel, D.W., 1999, The *Xanthomonas avr/pth* gene family, in G. Stacey and N.T. Keen, eds, Plant–Microbe Interactions, St. Paul, Minnesota, APS Press, pp. 39–55.

Gürlebeck, D., Szurek, B., and Bonas, U., 2005, Dimerization of the bacterial effector protein AvrBs3 in the plant cell cytoplasm prior to nuclear import, *Plant J.* **42**:175–187.

Gürlebeck, D., Thieme, F., and Bonas, U., 2006, Type III effector proteins from the plant pathogen *Xanthomonas* and their role in the interaction with the host plant, *J. Plant Physiol.* **163**: 233–255.

Herbers, K., Conrads-Strauch, J., and Bonas, U., 1992, Race-specificity of plant resistance to bacterial spot disease determined by repetitive motifs in a bacterial avirulence protein, *Nature* **356**:172–174.

Hotson, A., Chosed, R., Shu, H., Orth, K., and Mudgett, M.B., 2003, *Xanthomonas* type III effector XopD targets SUMO-conjugated proteins *in planta*, *Mol. Microbiol.* **50**:377–389.

Jakoby, M., Weisshaar, B., Droge-Laser, W., Vicente-Carbajosa, J., Tiedemann, J., Kroj, T., and Parcy, F., 2002, bZIP transcription factors in *Arabidopsis*, *Trends Plant Sci.* **7**:106–111.

Kay, S., Hahn, S., Marois, E., Hause, G., and Bonas, U., 2007, A bacterial effector as a plant transcription factor and induces a cell size regulator, *Science* **318**:648–351.

Keen, N.T., 1990, Gene-for-gene complementarity in plant–pathogen interactions, *Annu. Rev. Genet.* **24**:447–463.

Klement, Z., 1982, Hypersensitivity, in M.S. Mount and G.H. Lacy, eds, Phytopathogenic Prokaryotes, New York, Academic Press, pp. 149–177.

Koebnik, R., Krüger, A., Thieme, F., Urban, A., and Bonas, U., 2006, Specific binding of the *Xanthomonas campestris* pv. *vesicatoria* AraC-type transcriptional activator HrpX to plant-inducible promoter boxes, *J. Bacteriol.* **188**:7652–7660.

Marie, C., Broughton, W.J., and Deakin, W.J., 2001, *Rhizobium* type III secretion systems: legume charmers or alarmers? *Curr. Opin. Plant Biol.* **4**:336–342.

Marois, E., Van den Ackerveken, G., and Bonas, U., 2002, The *Xanthomonas* type III effector protein AvrBs3 modulates plant gene expression and induces cell hypertrophy in the susceptible host, *Mol. Plant Microbe Interact.* **15**:637–646.

Noël, L., Thieme, F., Nennstiel, D., and Bonas, U., 2001, cDNA-AFLP analysis unravels a genome-wide *hrpG*-regulon in the plant pathogen *Xanthomonas campestris* pv. *vesicatoria*, *Mol. Microbiol.* **41**:1271–1281.

Noël, L., Thieme, F., Nennstiel, D., and Bonas, U., 2002, Two novel type III-secreted proteins of *Xanthomonas campestris* pv. *vesicatoria* are encoded within the *hrp* pathogenicity island, *J. Bacteriol.* **184**:1340–1348.

Noël, L., Thieme, F., Gäbler, J., Büttner, D., and Bonas, U., 2003, XopC and XopJ, two novel type III effector proteins from *Xanthomonas campestris* pv. *vesicatoria*, *J. Bacteriol.* **185**: 7092–7102.

Schornack, S., Meyer, A., Romer, P., Jordan, T., and Lahaye, T., 2006, Gene-for-gene-mediated recognition of nuclear-targeted AvrBs3-like bacterial effector proteins, *J. Plant Physiol.* **163**:256–272.

Stall, R.E., 1995, *Xanthomonas campestris* pv. *vesicatoria*, in R.P.S. U.S. Singh, and K. Kohmoto, eds, Pathogenesis and Host-Parasite Specificity in Plant Diseases, Tarrytown, NY, Pergamon, Elsevier Science Inc., pp. 167–184.

Szurek, B., Marois, E., Bonas, U., and Van den Ackerveken, G., 2001, Eukaryotic features of the *Xanthomonas* type III effector AvrBs3: protein domains involved in transcriptional activation and the interaction with nuclear import receptors from pepper, *Plant J.* **26**:523–534.

Szurek, B., Rossier, O., Hause, G., and Bonas, U., 2002, Type III-dependent translocation of the *Xanthomonas* AvrBs3 protein into the plant cell, *Mol. Microbiol.* **46**:13–23.

Tampakaki, A.P., Fadouloglou, V.E., Gazi, A.D., Panopoulos, N.J., and Kokkinidis, M., 2004, Conserved features of type III secretion, *Cell. Microbiol.* **6**:805–816.

Thieme, F., Koebnik, R., Bekel, T., Berger, C., Boch, J., Büttner, D., Caldana, C., Gaigalat, L., Goesmann, A., Kay, S., Kirchner, O., Lanz, C., Linke, B., McHardy, A.C., Meyer, F., Mittenhuber, G., Nies, D.H., Niesbach-Klösgen, U., Patschkowski, T., Rückert, C., Rupp, O., Schneiker, S., Schuster, S.C., Vorhölter, F.J., Weber, E., Pühler, A., Bonas, U., Bartels, D., and Kaiser, O., 2005, Insights into genome plasticity and pathogenicity of the plant pathogenic bacterium *Xanthomonas campestris* pv. *vesicatoria* revealed by the complete genome sequence, *J. Bacteriol.* **187**:7254–7266.

Van den Ackerveken, G., Marois, E., and Bonas, U., 1996, Recognition of the bacterial avirulence protein AvrBs3 occurs inside the host plant cell, *Cell* **87**:1307–1316.

Vauterin, L., Rademaker, J., and Swings, J., 2000, Synopsis on the taxonomy of the genus *Xanthomonas*, *Phytopathology* **90**:677–682.

Wang, G., Ding, X., Yuan, M., Qiu, D., Li, X., Xu, C., and Wang, S., 2006, Dual function of rice OsDR8 gene in disease resistance and thiamine accumulation, *Plant Mol. Biol.* **60**:437–449.

Weber, E., Ojanen-Reuhs, T., Huguet, E., Hause, G., Romantschuk, M., Korhonen, T.K., Bonas, U., and Koebnik, R., 2005, The type III-dependent Hrp pilus is required for productive interaction of *Xanthomonas campestris* pv. *vesicatoria* with pepper host plants, *J. Bacteriol.* **187**: 2458–2468.

Xiao, Y., and Hutcheson, S.W., 1994, A single promoter sequence recognized by a newly identified alternate sigma factor directs expression of pathogenicity and host range determinants in *Pseudomonas syringae* [published erratum appears in *J. Bacteriol.* **176**:6158], *J. Bacteriol.* **176**:3089–3091.

Structure and Function of RXLR Effectors of Plant Pathogenic Oomycetes

William Morgan, Jorunn Bos, Catherine Bruce, Minkyoung Lee, Hsin-Yen Liu, Sang-Keun Oh, Jing Song, Joe Win, Carolyn Young, and Sophien Kamoun

1 Introduction

A diverse number of plant pathogens, including bacteria, fungi, oomycetes and nematodes, secrete effector proteins to different cellular compartments of their hosts to modulate plant defense circuitry and enable parasitic colonization (Birch et al., 2006; Chisholm et al., 2006; Kamoun, 2006; O'Connell and Panstruga, 2006). The current paradigm in the study of plant–microbe interactions is that unraveling the molecular function of effectors is central to a mechanistic understanding of pathogenicity. Indeed, significant progress has been made in elucidating the virulence functions of bacterial effectors and the biochemical activities that enable these proteins to perturb host defense processes and facilitate pathogenicity (Chisholm et al., 2006). In contrast, little is known about the biology of effectors of eukaryotic plant pathogens. Nonetheless, this area of research is progressing rapidly as illustrated by the recent identification of effectors from the flax rust and barley powdery mildew fungi (Catanzariti et al., 2006; Dodds et al., 2004; Ridout et al., 2006), the oomycetes *Phytophthora* and *Hyaloperonospora* (Allen et al., 2004; Armstrong et al., 2005; Rehmany et al., 2005; Shan et al., 2004), as well as root-knot nematodes (Huang et al., 2006a; Huang et al., 2006b).

Oomycetes form a distinct group of eukaryotic microorganisms that includes some of the most notorious pathogens of plants (Kamoun, 2003). Research on oomycete effectors has accelerated in recent years due in great part to the availability of resources stemmed from genomics. The emerging picture is complex and fascinating. Oomycetes are now thought to secrete hundreds of effector proteins belonging to two classes that target distinct sites in the host plant (Birch et al., 2006; Kamoun, 2006; Tyler et al., 2006). Apoplastic effectors are secreted into the plant extracellular space, whereas cytoplasmic effectors are translocated inside the plant cell, where they target different subcellular compartments (Birch et al., 2006; Kamoun, 2006). Several apoplastic effectors have been determined

S. Kamoun
Department of Plant Pathology, Ohio State University Ohio Agricultural Research and Development Center, Wooster, OH, USA
e-mail: kamoun.1@osu.edu

J.P. Gustafson et al. (eds.), *Genomics of Disease*,
© Springer Science+Business Media, LLC, 2008

to function as inhibitors of host enzymes, such as proteases and glucanases (Rose et al., 2002; Tian et al., 2005; 2004). They are thought to contribute to counter-defense by disabling host enzymes that accumulate in response to pathogen infection. In contrast, the biochemical activities of cytoplasmic effectors remain poorly understood. Oomycete cytoplasmic effectors have been discovered first through their avirulence (Avr) function, i.e. their ability to trigger hypersensitive cell death on specific host genotypes that carry particular disease resistance (R) genes (Allen et al., 2004; Armstrong et al., 2005; Rehmany et al., 2005; Shan et al., 2004). The function of these effectors in plants that do not carry the cognate R genes remains largely unknown (Kamoun, 2006).

This review summarizes recent findings on the structure and function of the RXLR class of oomycete effectors (Birch et al., 2006; Kamoun, 2006). These effectors function inside host cells and are characterized by a highly conserved region defined by the invariant sequence RXLR. This review will cover two main topics of RXLR effector research: trafficking and function.

2 The RXLR Sequence Defines a Conserved Domain of Oomycete Avr Proteins

Four oomycete *Avr* genes, $ATR1^{NdWsB}$ and *ATR13* from the downy mildew *Hyaloperonospora parasitica*, *Avr1b-1* from the soybean pathogen *Phytophthora sojae*, and *Avr3a* from *P. infestans*, have been cloned recently (Allen et al., 2004; Armstrong et al., 2005; Rehmany et al., 2005; Shan et al., 2004). The R proteins that target $ATR1^{NdWsB}$, ATR13, and $AVR3a^{KI}$ belong to the intracellular class of NBS-LRR (nucleotide binding site and leucine-rich repeat domain) proteins, suggesting that recognition of these Avr proteins occurs inside the plant cytoplasm (Allen et al., 2004; Armstrong et al., 2005; Rehmany et al., 2005). Indeed, when directly expressed *in planta* by transient transformation, $AVR3a^{KI}$, $ATR1^{NdWsB}$, and ATR13 did not require a signal peptide sequence to trigger hypersensitivity, and are therefore recognized inside the plant cytoplasm (Allen et al., 2004; Armstrong et al., 2005; Rehmany et al., 2005). How these effectors are translocated into host cells during infection is unknown but a conserved sequence centered on the RXLR motif might be implicated (Birch et al., 2006; Kamoun, 2006; Rehmany et al., 2005) (Fig. 1). All four oomycete Avr proteins carry a signal peptide followed by the RXLR region, which occurs within the N-terminal ca. 60 amino acids of these proteins (Rehmany et al., 2005). So far, definite elucidation of the function of the RXLR region has not been obtained, but the prevailing hypothesis is that RXLR functions in delivery of the effectors inside host cells (Birch et al., 2006; Kamoun, 2006; Rehmany et al., 2005). Indeed, this motif is similar in sequence and position to a host cell-targeting signal (HT/Pexel motif) that is required for translocation of proteins from malaria parasites (*Plasmodium* species) into the cytoplasm of host red blood cells (Hiller et al., 2004; Marti et al., 2004). Interestingly, the cellular biology of *Plasmodium* infection of host blood cells shares some commonalities

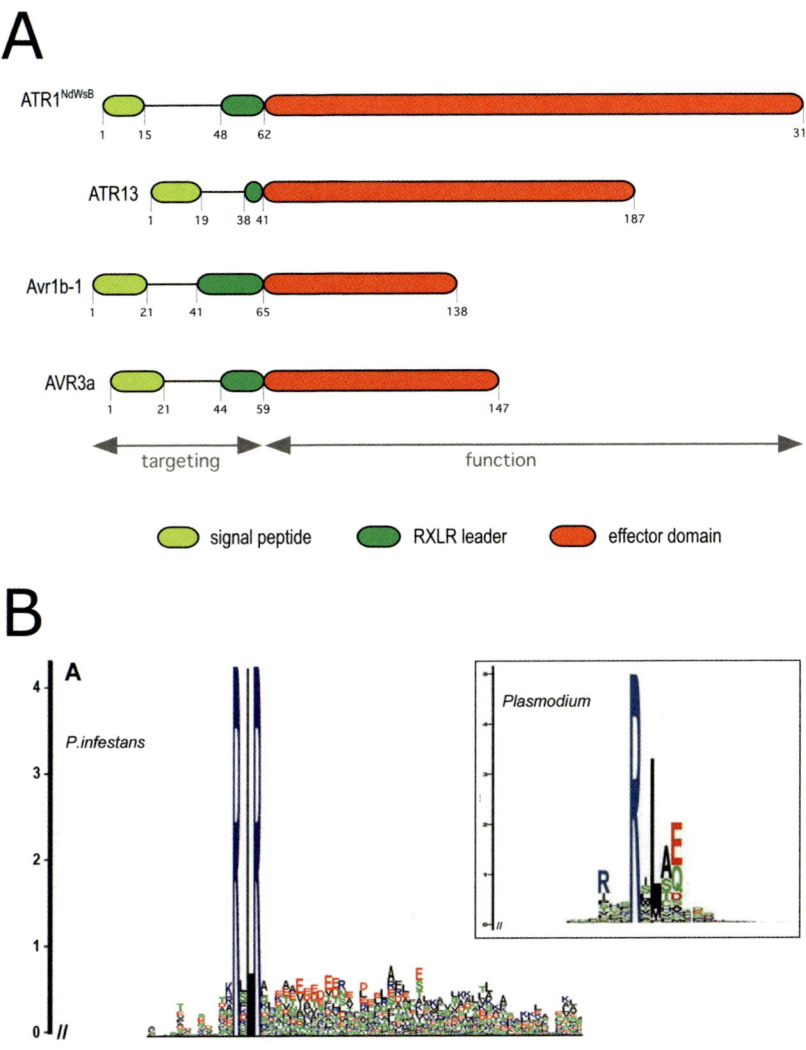

Fig. 1 (**A**)Domain organization of cytoplasmic RXLR effectors. Schematic drawings of ATR1[NdWsB] and ATR13 of *Hyaloperonsopora parasitica*, Avr1b-1 of *Phytophthora sojae*, and AVR3a of *P. infestans*. The numbers under the sequences indicate amino-acid positions. The highlighted RXLR domain includes the RXLR sequence itself and the downstream dEER sequence. The *gray arrows* distinguish the regions of the effector proteins that are involved in secretion and targeting from those involved in effector activity. (**B**) Similarity between the RXLR motif of oomycetes and the HT/Pexel motif of *Plasmodium falciparum*. Sequence logos were derived from *P. infestans* and *P. falciparum* effector proteins. Adapted from Bhattacharjee et al. (2006)

with oomycete infection structures like haustoria (Bhattacharjee et al., 2006). During the blood stages of infection, *Plasmodium* invades mature erythrocytes and develops within a parasitophorous vacuolar membrane (PVM) derived from an invagination of the erythrocyte membrane. While residing within the PVM, the parasite exports effector proteins to the erythrocyte resulting in reprogramming of the invaded blood cell. Thus, both *Plasmodium* and oomycetes need to translocate effector proteins across a host-derived membrane during colonization of host cells suggesting that the conserved sequence motif might be required to overcome this comparable hurdle (Bhattacharjee et al., 2006). Although direct demonstration that the oomycete RXLR domain is a host-targeting signal has not been reported, a number of experiments that are consistent with this hypothesis have been performed. The next three sections describe these experiments.

3 The *Phytophthora* RXLR Domain Mediates Host Targeting in *Plasmodium*

The discovery that host-targeted proteins from *Phytophthora* and *Plasmodium* share a positionally conserved sequence begged the examination of functional conservation. Bhattacharjee et al. (2006) demonstrated that the *P. infestans* AVR3a^{KI} RXLR leader sequence is sufficient to mediate the export of the green fluorescent protein (GFP) from the *Plasmodium falciparum* parasite to the host red blood cell (erythrocyte) (Fig. 2).

Mutations in the RXLR consensus abolished export. In addition, the RXLR region of PH001D5, another candidate effector identified computationally from *P. infestans* sequences, was also functional in *Plasmodium*. Interestingly, regions upstream and downstream of the RXLR motif were required for host targeting,

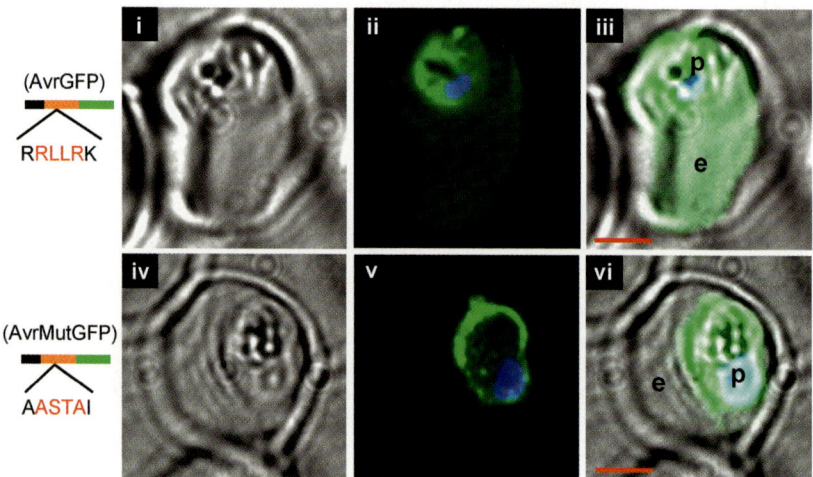

Fig. 2 The *Phytophthora infestans* AVR3a RXLR leader region mediates the export of the green fluorescent protein (GFP) from the *Plasmodium falciparum* parasite to the host erythrocyte. Erythrocytes expressing wild-type or mutated RXLR region of AVR3a (residues 21 to 69) fused to a *P. falciparum* signal peptide and GFP. Panels **ii** and **v** represent fluorescence images, **i** and **iv** bright field images, and **iii** and **vi** merged images. Parasite (p), erythrocyte (e), Hoechst stained nucleus (*blue*), scale bar represents 2 μm. Adapted from Bhattacharjee et al. (2006)

suggesting that the RXLR sequence defines a ca. 30 amino acid leader sequence. Thus the targeting domain extends beyond the RXLR sequence. Consistent with this view, sequence biases were observed in the regions flanking RXLR, particularly a high rate of E/D residues following RXLR. In summary, these findings suggest that plant and animal eukaryotic pathogens share similar secretory signals for effector delivery into host cells (Bhattacharjee et al., 2006). At this stage, it is unclear whether this functional conservation reflects conserved export machinery between these divergent eukaryotes (see below for further discussion on this topic). However, the functional analogy with bacterial type III secretion system is remarkable. In both cases, export signals are highly conserved across unrelated pathogen species but the effector secretome that is delivered to the host is highly divergent (Bhattacharjee et al., 2006).

4 The RXLR Domain Is Not Required for Effector Activities

The view that the RXLR region functions in translocation into host cells suggests that RXLR effectors are organized into two main functional domains (Kamoun, 2006) (Fig. 1). The first domain encompassing the signal peptide and RXLR leader functions in secretion and targeting, while the remaining C-terminal domain carries the effector activity. This model predicts that the RXLR region should not be required for activity when the effector is expressed inside host cells. Indeed, Bos et al. (2006) recently showed that mutation of *P. infestans* AVR3a^{KI} RXLR sequence into AXAA did not interfere with induction of R3a hypersensitivity when the protein is directly expressed in *N. benthamiana* leaves. In fact, deletion analyses of AVR3a^{KI} showed that the C-terminal 75-amino acid, which excludes the RXLR region but includes the two polymorphic amino acids K^{80} and I^{103} that are mutated in the nonfunctional allele, was sufficient for avirulence function when expressed directly inside plant cells (Bos et al., 2006). These findings are consistent with the view that the N-terminal region of AVR3a^{KI} and other RXLR effectors is involved in secretion and targeting but is not required for effector activity.

5 The C-Terminal Region of RXLR Effectors Is Typically More Polymorphic than the Signal Peptide and RXLR Domains

Higher levels of polymorphisms, particularly non-synonymous substitutions, have been detected in the C-terminal regions of RXLR effectors than in the signal peptide and RXLR leader region. For example, the C-terminal regions of *H. parasitica* ATR1 and ATR13 exhibit higher levels of non-synonymous polymorphisms than the N-terminal regions, suggesting that the effector activity is localized to the C-terminal domain (Allen et al., 2004; Rehmany et al., 2005). Also, two out of the three polymorphic residues between the two *Avr3a* alleles of *P. infestans*, amino acids 80 and 103, are located in the C-terminal effector domain (Armstrong et al., 2005). The observation that these effectors are under diversifying selection is consistent with the view that pathogen effectors with avirulence functions are caught in a coevolutionary arms race with host factors, particularly their cognate R genes (Allen et al., 2004). Signatures of selection are expected in regions involved in effector activity rather than targeting. Thus, the observation that the RXLR domain is less polymorphic than the C-terminal region of RXLR proteins is consistent with the view that it is not exposed to selection pressure by host defenses and that it functions in targeting.

6 Can RXLR Effectors Enter Host Plants in the Absence of the Pathogen?

Whether RXLR effectors require pathogen machinery or structures (e.g., haustoria) to enter plant cells is currently unclear. In our laboratory, we have failed so far to trigger R3a hypersensitivity using recombinant AVR3a^{KI} proteins (unpublished data). Shan et al. (2004) reported that infiltration of *P. sojae* RXLR effector Avr1b-1, produced in *Pichia pastoris*, into Rps1b soybean leaves resulted in cell death. However, it is unknown (although highly likely) whether Rps1b is a cytoplasmic protein and confirmation of *Avr1b-1* activity by *in planta* expression inside soybean cells has not been reported yet. Clearly, it would be highly informative to test Avr1b-1 recombinant proteins mutated in the RXLR sequence using the infiltration assay of Shan et al. (2004).

The issue of host translocation is also relevant to cytoplasmic effectors of fungal pathogens. In their work with Avr proteins of the flax rust fungus *Melampsora lini*, Catanzariti et al. (2006) concluded that the avirulence protein AvrM enters plant cells in the absence of the pathogen. AvrM is a secreted protein with a canonical signal peptide that is recognized inside flax cells in an *M* gene dependent manner. In addition, *Agrobacterium tumefaciens*-mediated expression of a full-length AvrM construct in flax plants carrying the *M* gene resulted in hypersensitive cell death. Interestingly, similar constructs carrying the ER retention sequence HDEL were unable to trigger cell death. The authors concluded from this experiment that AvrM first exits plant cells into the apoplast and then reenters through an unknown process (Catanzariti et al., 2006). We are not in absolute agreement with this interpretation and are of the opinion that this experiment is inconclusive in evaluating whether AvrM can enter plant cells in the absence of the pathogen. An equally plausible explanation is that the signal peptide carrying AvrM is translocated back from the ER into the cytosol through the well-established retrograde transport pathway (Brandizzi et al., 2003; Di Cola et al., 2001). In such case, the HDEL motif would retain the protein into the ER and prevent retrograde translocation. In summary, although the experiments of Catanzariti et al. (2006) suggest that AvrM needs to shuttle through the ER, perhaps to achieve maturation, infiltration experiments with purified AvrM proteins are necessary to obtain conclusive evidence as to the ability of this protein to enter host cells in the absence of the pathogen. Such data has been obtained conclusively with ToxA, a host-selective toxin produced by the plant pathogenic fungus *Pyrenophora tritici-repentis*. Manning and Ciuffetti (2005) demonstrated elegantly that a recombinant ToxA protein tagged with GFP internalizes inside sensitive wheat cells, where it localizes to cytoplasmic compartments and chloroplasts. This implies that effector proteins can translocate inside host cells in the absence of pathogen machinery, probably using host-derived machinery. Host-translocation of ToxA is likely to occur via receptor-mediated endocytosis, and might implicate the RGD sequence that is identical to a cell adhesion motif of mammalian extracellular matrix proteins (Manning and Ciuffetti, 2005).

In our own work with AVR3a^{KI}, we noted that high levels of expression of a full-length construct resulted in R3a-dependent hypersensitivity (Bos et al., 2006). These experiments are not particularly informative since they could be equally explained by (1) mis-targeting of the protein to the cytoplasm or (2) secretion of AVR3a^{KI} followed by re-entry of the protein inside the plant cell resulting in R3a activation. Also, in other experiments, attempts to exploit this full-length AVR3a^{KI} to evaluate the role of the RXLR motif in translocation inside host cells were inconclusive since a RXLR to AXAA mutant retained the ability to elicit R3a response (Bos et al., 2006).

7 A Model for RXLR Effector Delivery into the Host

The process through which the RXLR leader sequence might function in host targeting of effectors remains unknown. Many key questions still need to be addressed. What is the export machinery of effectors in eukaryotic pathogens? Is it derived from the pathogen or are the effectors exploiting host transport systems? Is the export machinery conserved between oomycetes and *Plasmodium*? Despite these persisting questions, some reasonable assumptions about the translocation process can be made. For instance, it seems sensible to break down the export process into two steps (Bhattacharjee et al., 2006). First, the effectors are secreted outside the pathogen cell through the general secretory pathway and endoplasmic reticulum (ER) type signal peptides. Then, the secreted effectors are transported across a host-derived membrane, most likely the haustorial membrane, via the RXLR leader. In the GFP export experiments of Bhattacharjee et al. (2006), constructs with a mutated RXLR sequence accumulated GFP outside the parasite but within the parasitophorous vacuole suggesting that the main function of the RXLR leader consists of transport across this host-derived membrane.

Here, we propose a model for effector delivery (illustrated in Fig. 3). The model is based on the fact that mechanisms of protein transport across membranes follow canonical processes involving recurrent themes (Wickner and Schekman, 2005). We propose that host translocation of the effectors via the RXLR leader involves at least a RXLR leader binding protein, one or more chaperones, and a translocon, which could be of either pathogen or plant origin. Translocation into host cells initiates with the RXLR binding protein recruiting secreted mature effectors in coordination with the chaperones. The effector cargo is then transferred to a translocon embedded in the haustorial membrane, and is then released across the membrane into the plant cytosol. The chaperones are important for maintaining the folding state of the transported effectors both prior and after transit through the translocon.

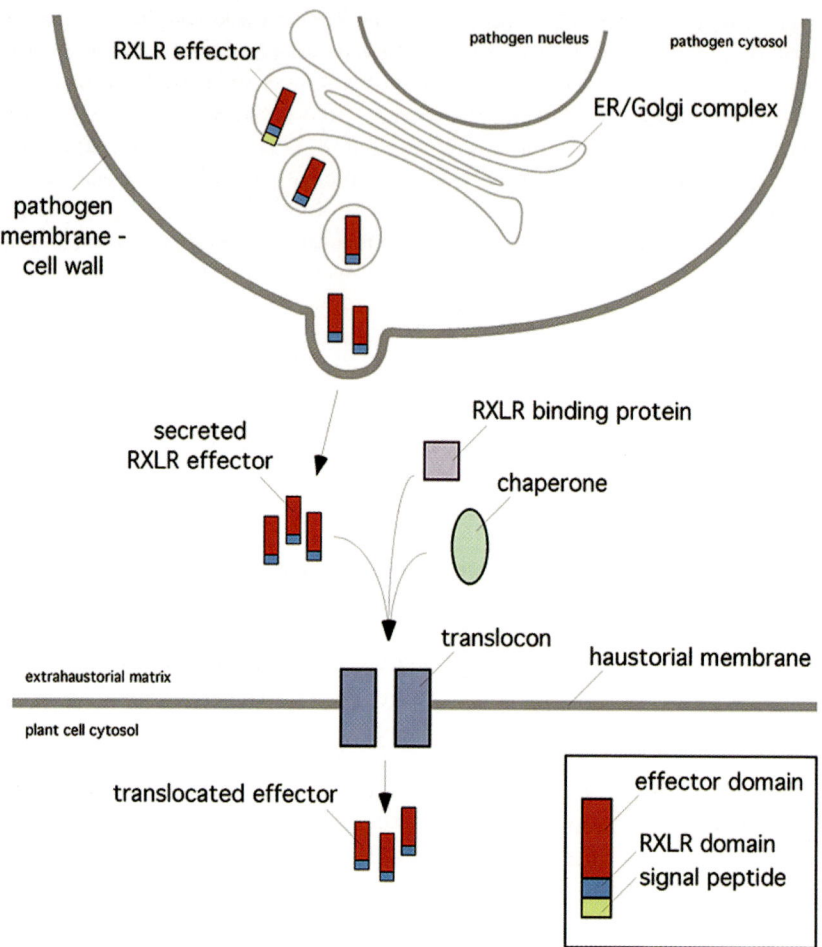

Fig. 3 A hypothetical model for RXLR effector secretion and delivery into host cells (see text for details)

At this point, this model is highly speculative. But our purpose is to outline a useful working model to serve as a hypothesis generator and help guide future research. Indeed, the model suggests immediate research avenues that would shed light on the translocation process, for instance the identification of RXLR binding proteins.

8 Virulence Functions of RXLR Effectors

The virulence function of RXLR effectors, i.e. their activity in plants that do not carry cognate *R* genes, remains in great part unknown. Bos et al. (2006), in an effort to assign virulence-related functions to AVR3a, discovered that AVR3a[KI] suppresses the hypersensitive cell death induced by the major *P. infestans* elicitin INF1 in *Nicotiana benthamiana*. The cell death suppression activity of AVR3a^{KI} exhibited some level of specificity. AVR3a^{KI} did not suppress the cell death induced by other *P. infestans* effectors, like PiNPP1 and CRN2, which elicit distinct and antagonistic cell death signaling pathways compared to INF1 (Kanneganti et al., 2006). The biological relevance of this activity of AVR3a^{KI} is unknown but could be significant considering that suppression of innate immunity is a widespread function of plant pathogen effectors, particularly the type III secretion system (TTSS) effectors of bacterial phytopathogens (Espinosa and Alfano 2004). AVR3a^{KI} could interfere with the avirulence activity of INF1 or other unidentified effectors that trigger hypersensitivity using similar pathways as INF1 (Bos et al., 2006). Future work is needed to clarify these issues and determine whether cell death suppression is a common function among RXLR effectors.

9 Outlook: Too Many Effectors, Too Little Time

Considering that five oomycete species, *H. parasitica*, *P. capsici*, *P. infestans*, *P. ramorum*, and *P. sojae*, are undergoing genome sequencing and annotation, we are moving rapidly toward genome-wide catalogues of RXLR effectors. Already it is evident that the RXLR effector secretome of plant pathogenic oomycetes is much more complex than expected, with perhaps several hundred proteins dedicated to manipulating host cells (Kamoun, 2006; Tyler et al., 2006). Recent estimates of RXLR effectors based on the draft genome sequences of *P. sojae* and *P. ramorum* range from ca. 150 to 350 per genome (Bhattacharjee et al., 2006; Tyler et al., 2006). The task of tackling the study of so many effectors is daunting. One of the challenges is to establish "effectoromics" approaches, or global studies of effector function and activity. Ultimately, comprehensive understanding of RXLR effector activities and the perturbations they cause in plants is a precondition for understanding the molecular basis of oomycete pathogenesis and disease.

Acknowledgments We thank collaborators and colleagues in the oomycete field for useful discussions and insight. This work was supported by NSF Plant Genome Research Program grant DBI-0211659. Salaries and research support were provided, in part, by State and Federal Funds appropriated to the Ohio Agricultural Research and Development Center, The Ohio State University.

References

Allen, R.L., Bittner-Eddy, P.D., Grenville-Briggs, L.J., Meitz, J.C., Rehmany, A.P., Rose, L.E., and Beynon, J.L., 2004, Host–parasite coevolutionary conflict between Arabidopsis and downy mildew, *Science* **306**:1957–1960.
Armstrong, M.R., Whisson, S.C., Pritchard, L., Bos, J.I., Venter, E., Avrova, A.O., Rehmany, A.P., Bohme, U., Brooks, K., Cherevach, I., Hamlin, N., White, B., Fraser, A., Lord, A., Quail, M.A., Churcher, C., Hall, N., Berriman, M., Huang, S., Kamoun, S., Beynon, J.L., and Birch, P.R., 2005, An ancestral oomycete locus contains late blight avirulence gene *Avr3a*, encoding

a protein that is recognized in the host cytoplasm, *Proc. Natl. Acad. Sci. USA* **102**: 7766–7771.

Bhattacharjee, S., Hiller, N.L., Liolios, K., Win, J., Kanneganti, T.D., Young, C., Kamoun, S., and Haldar, K., 2006. The malarial host-targeting signal is conserved in the Irish potato famine pathogen. *PLoS Pathog.* **2**:e50.

Birch, P.R., Rehmany, A.P., Pritchard, L., Kamoun, S., and Beynon, J.L., 2006, Trafficking arms: oomycete effectors enter host plant cells, *Trends Microbiol.* **14**:8–11.

Bos, J.I., Kanneganti, T.D., Young, C., Cakir, C., Huitema, E., Win, J., Armstrong, M.R., Birch, P.R., and Kamoun, S., 2006, The C-terminal half of *Phytophthora infestans* RXLR effector AVR3a is sufficient to trigger R3a-mediated hypersensitivity and suppress INF1-induced cell death in *Nicotiana benthamiana*, *Plant J.* **48**:165–176.

Brandizzi, F., Hanton, S., DaSilva, L.L., Boevink, P., Evans, D., Oparka, K., Denecke, J., and Hawes, C., 2003, ER quality control can lead to retrograde transport from the ER lumen to the cytosol and the nucleoplasm in plants, *Plant J.* **34**:269–281.

Catanzariti, A.M., Dodds, P.N., Lawrence, G.J., Ayliffe, M.A., and Ellis, J.G., 2006, Haustorially expressed secreted proteins from flax rust are highly enriched for avirulence elicitors, *Plant Cell* **18**:243–256.

Chisholm, S.T., Coaker, G., Day, B., and Staskawicz, B.J., 2006, Host–microbe interactions: shaping the evolution of the plant immune response, *Cell* **124**:803–814.

Di Cola, A., Frigerio, L., Lord, J.M., Ceriotti, A., and Roberts, L.M., 2001, Ricin A chain without its partner B chain is degraded after retrotranslocation from the endoplasmic reticulum to the cytosol in plant cells, *Proc. Natl. Acad. Sci. USA* **98**:14726–14731.

Dodds, P.N., Lawrence, G.J., Catanzariti, A.M., Ayliffe, M.A., and Ellis, J.G., 2004, The Melampsora lini AvrL567 avirulence genes are expressed in haustoria and their products are recognized inside plant cells, *Plant Cell* **16**:755–768.

Espinosa, A., and Alfano, J.R., 2004, Disabling surveillance: bacterial type III secretion system effectors that suppress innate immunity, *Cell Microbiol.* **6**:1027–1040.

Hiller, N.L., Bhattacharjee, S., van Ooij, C., Liolios, K., Harrison, T., Lopez-Estrano, C., and Haldar, K., 2004, A host-targeting signal in virulence proteins reveals a secretome in malarial infection, *Science* **306**:1934–1937.

Huang, G., Allen, R., Davis, E.L., Baum, T.J., and Hussey, R.S., 2006a, Engineering broad root-knot resistance in transgenic plants by RNAi silencing of a conserved and essential root-knot nematode parasitism gene, *Proc. Natl. Acad. Sci. USA*, **103**:14302–14306.

Huang, G., Dong, R., Allen, R., Davis, E.L., Baum, T.J., and Hussey, R.S., 2006b, A root-knot nematode secretory peptide functions as a ligand for a plant transcription factor, *Mol. Plant Microbe Interact.* **19**:463–470.

Kamoun, S., 2003, Molecular genetics of pathogenic oomycetes, *Eukaryotic Cell* **2**:191–199.

Kamoun, S., 2006, A Catalogue of the Effector Secretome of Plant Pathogenic Oomycetes, *Annu. Rev. Phytopathol.* **44**:41–60.

Kanneganti, T.D., Huitema, E., Cakir, C., and Kamoun, S., 2006, Synergistic interactions of the plant cell death pathways induced by *Phytophthora infestans* Nep1-like protein PiNPP1.1 and INF1 elicitin, *Mol. Plant Microbe Interact.* **19**:854–863.

Manning, V.A., and Ciuffetti, L.M., 2005, Localization of Ptr ToxA Produced by *Pyrenophora tritici-repentis* Reveals Protein Import into Wheat Mesophyll Cells, *Plant Cell* **17**: 3203–3212.

Marti, M., Good, R.T., Rug, M., Knuepfer, E., and Cowman, A.F., 2004, Targeting malaria virulence and remodeling proteins to the host erythrocyte, *Science* **306**:1930–1933.

O'Connell, R.J., and Panstruga, R., 2006, Tete a tete inside a plant cell: establishing compatibility between plants and biotrophic fungi and oomycetes, *New Phytol.* **171**:699–718.

Rehmany, A.P., Gordon, A., Rose, L.E., Allen, R.L., Armstrong, M.R., Whisson, S.C., Kamoun, S., Tyler, B.M., Birch, P.R., and Beynon, J.L., 2005, Differential recognition of highly divergent downy mildew avirulence gene alleles by *RPP1* resistance genes from two Arabidopsis lines, *Plant Cell* **17**:1839–1850.

Ridout, C.J., Skamnioti, P., Porritt, O., Sacristan, S., Jones, J.D., and Brown, J.K., 2006, Multiple avirulence paralogues in cereal powdery mildew fungi may contribute to parasite fitness and defeat of plant resistance, *Plant Cell* **18**:2402–2414.

Rose, J.K., Ham, K.S., Darvill, A.G., and Albersheim, P., 2002, Molecular cloning and character-ization of glucanase inhibitor proteins: coevolution of a counterdefense mechanism by plant pathogens, *Plant Cell* **14**:1329–1345.

Shan, W., Cao, M., Leung, D., and Tyler, B.M., 2004, The *Avr1b* locus of *Phytophthora sojae* encodes an elicitor and a regulator required for avirulence on soybean plants carrying resistance gene *Rps1b*, *Mol. Plant Microbe Interact.* **17**:394–403.

Tian, M., Benedetti, B., and Kamoun, S., 2005, A Second Kazal-like protease inhibitor from *Phytophthora infestans* inhibits and interacts with the apoplastic pathogenesis-related protease P69B of tomato, *Plant Physiol.* **138**:1785–1793.

Tian, M., Huitema, E., da Cunha, L., Torto-Alalibo, T., and Kamoun, S., 2004, A Kazal-like extra-cellular serine protease inhibitor from *Phytophthora infestans* targets the tomato pathogenesis-related protease P69B, *J. Biol. Chem.* **279**:26370–26377.

Tyler, B.M., Tripathy, S., Zhang, X., Dehal, P., Jiang, R.H., et al., 2006, Phytophthora genome sequences uncover evolutionary origins and mechanisms of pathogenesis, *Science* **313**: 1261–1266.

Wickner, W., and Schekman, R., 2005, Protein translocation across biological membranes, *Science* **310**:1452–1456.

The Biotrophic Phase of *Ustilago maydis*: Novel Determinants for Compatibility

Thomas Brefort, Kerstin Schipper, Gunther Döhlemann, and Regine Kahmann

Abstract The basidiomycete fungus *Ustilago maydis* establishes a biotrophic relationship with its host plant maize which is maintained throughout disease development. Recent insights from the genome sequence have revealed that this interaction is largely governed by a set of novel secreted proteins that are only found in *U. maydis*. Many of the respective genes are clustered and appear co-regulated during late stages of pathogenesis. Mutants in most of these gene clusters arrest development at distinct stages, suggesting that the secreted proteins fulfill discrete functions in the interaction with the host. One of the cluster mutants, however, displays increased virulence suggesting that it is not in the interest of *U. maydis* to use its full potential as a pathogen. In this chapter we will review these findings and place them in perspective for a comprehensive understanding of biotrophy.

1 Introduction

Smut diseases of grasses are caused by basidiomycetes of the order Ustilaginales. Because most smuts develop in kernel tissue they cause considerable yield losses. Smut fungi can initiate an infection only in their dikaryotic stage, which is generated by mating of compatible haploid cells. Smuts differ in host range as well as the site of symptom development, with symptoms in most cases being restricted to the inflorescences. One notable exception to this is *Ustilago maydis*, a pathogen infecting maize, which causes tumors on all aerial parts of the plant. Maize seedlings are efficiently infected and symptom development can be scored in less than a week after inoculation. This is one of the reasons why the *U. maydis*/maize system has become the model for plant pathogenic basidiomycetes. The other assets of the system are the availability of efficient techniques for gene replacement, regulatable promoters for analyzing gene function, as well as advanced cytology and live cell imaging (Basse and Steinberg, 2004). The 20.5 Mb genome of

R. Kahmann
Max Planck Institute for Terrestrial Microbiology, Department Organismic Interactions, Karl-von-Frisch-Strasse, D-35043 Marburg, Germany
e-mail: kahmann@mpi-marburg.mpg.de

J.P. Gustafson et al. (eds.), *Genomics of Disease*,
© Springer Science+Business Media, LLC, 2008

U. maydis has been sequenced and is publicly available through the Broad Institute (http://www.broad.mit.edu/annotation/fungi/ustilago_maydis/) and in manually annotated form through MIPS (http://MIPS.gsf.de/genre/proj/ustilago/) (Kämper et al., 2006).

Ustilago maydis can be propagated in the laboratory on synthetic media. Under these conditions haploid cells grow by budding. For infection-related development it is necessary that haploid cells fuse with a compatible partner (Fig. 1). Cell recognition and fusion are regulated by a pheromone/receptor system that is encoded by the *a* mating type locus. After fusion a dikaryon is generated that switches to filamentous growth if the two progenitor strains are heterozygous for genes at the *b* mating type locus (Fig. 1). The *b* mating type locus encodes two homeodomain proteins, bE and bW, that function as transcriptional regulators after dimerization. Dimerization is restricted to bE and bW proteins encoded by different alleles of the *b* locus. Thereby, it is guaranteed that active heterodimers are only formed in the dikaryon and do not arise in haploid cells (Kämper et al., 1995). The bE/bW heterodimer is the master regulator of pathogenic development and triggers a complex regulatory cascade during which a set of at least 250 genes is differentially expressed (Brachmann et al., 2001; Feldbrügge et al., 2004).

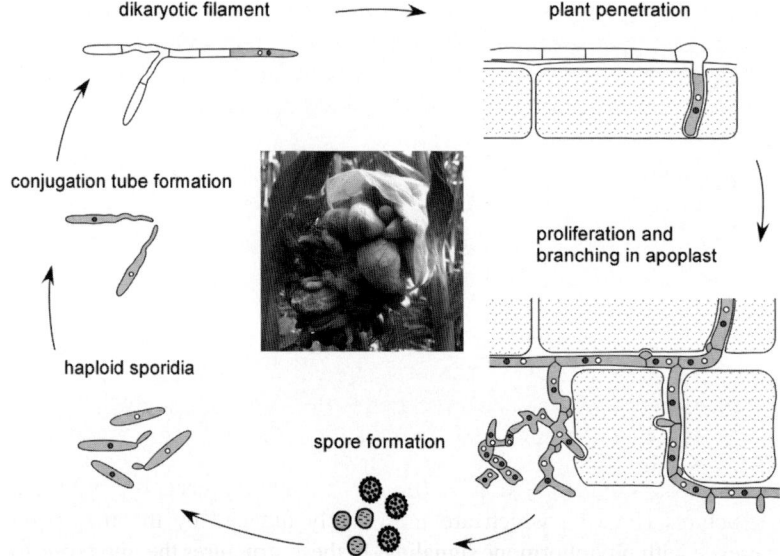

Fig. 1 The *U. maydis* infection cycle. Vegetative budding of haploid sporidia continues as long as nutrients are available. Receiving a signal from a mating partner induces formation of conjugation hyphae. They fuse at their tips and form the infectious dikaryotic filament. On the host-plant surface, dikaryotic hyphae differentiate appressoria and directly penetrate the cuticle. The biotrophic interaction is initiated by invagination of the plant plasma membrane. Invading hyphae that traverse cells are thus never in direct contact with the host cell cytoplasm. In later stages of infection, the hyphae proliferate in the apoplast, showing hyphal branching, formation of lobed structures and hyphal fragmentation. Following karyogamy teliospores mature and are released from tumor tissue (*center*). Upon germination the diploid spores undergo meiosis and produce haploid sporidia

   In nature, mating occurs on the plant surface. The resulting filamentous dikaryon forms appressoria (Fig. 1), which allow direct penetration of the plant cuticle in a process presumably aided by lytic enzymes (Schirawski et al., 2005). During this stage hyphae show tip growth but do not enlarge their cytoplasm and do not undergo nuclear division. As a result such hyphae contain cytoplasm only in the tip compartment while the older sections are highly vacuolated, and are sealed off by regularly spaced septa (Fig. 1). The dikaryon is arrested in the G2 phase of the cell cycle (Flor-Parra et al., 2006) and this block is released only after penetration (Banuett and Herskowitz, 1996). The *biz1* gene is among the genes indirectly regulated through the bE/bW heterodimer. Biz1 encodes a zinc finger transcription factor and triggers the G2 cell cycle arrest through down-regulation of the cyclin *clb1*. *biz1* mutants are unable to cause disease symptoms and this was attributed to a 10-fold reduction in appressorium formation as well as arrested growth immediately after penetration. This illustrates that the cell cycle arrest is necessary for proper infection-related development (Flor-Parra et al., 2006). Between 1 and 2 days after penetration the growth mode changes, mitotic divisions occur, and branching is observed (Fig. 1). How this is triggered is one of the unresolved mysteries. A crucial role for coordinated mitotic divisions during dikaryotic growth was recently ascribed to *clp1*, a direct target of the bE/bW heterodimer (Scherer et al., 2006), that is related to *clp1* in *coprinus cinerea* (Inada et al., 2001). *clp1* mutants have no discernible phenotype during axenic culture and mate like wild type cells. However, after plant penetration they arrest growth prior to the first mitotic division, fail to develop branch-like structures (Fig. 1), and are nonpathogenic. Conspicuously, the branch-like structures were shown to arise in the most apical cell of the dikaryotic hyphae at a point where new septa are later formed. By using strains with fluorescently labeled nuclei it was subsequently shown that one of the four mitotic nuclei becomes trapped in these structures after septum formation and moves from there to the subapical cell presumably through a septal pore (Scherer et al., 2006). This process is closely related to clamp formation in other basidiomycetes, but lacks the feature that clamps fuse with the subapical cell to donate the trapped nucleus (Brown and Casselton, 2002). This illustrates that processes regulated through the bE/bW heterodimer are at least to some extent conserved in related fungal species. In addition, these findings clearly show that the regulatory function of the bE/bW heterodimer is required during extended stages of pathogenic development.

   Around 5 days after infection by *U. maydis* the plant reacts by forming tumor-like structures (Fig. 1), which are most likely induced by the fungus through interference with phytohormone signaling. In these structures the dikaryotic hyphae show massive proliferation in the apoplast, round up, their nuclei fuse and hyphal fragmentation occurs followed by the formation of darkly pigmented spores (Fig. 1) (Snetselaar and Mims, 1994; Banuett and Herskowitz, 1996). *U. maydis* is a biotrophic pathogen, i.e. host cells stay alive throughout the course of an infection until very late when tumor tissue dries up. This leads to cell wall rupture and release of the spores (Snetselaar and Mims, 1994; Banuett and Herskowitz, 1996). Biotrophy is established after an invagination of the host plasma membrane at the site of fungal entry. This membrane surrounds the fungal hyphae and prevents direct contact with the host cytoplasm. *U. maydis* does not form the typical feeding structures (haustoria) characteristic of many other biotrophic fungi, and this has led to the hypothesis that the metabolism of its host must be altered after infection to provide the fungus with carbon and nitrogen sources at sites of fungal proliferation in the apoplast. Except accumulation of anthocyanin there are no other visible indications for an active plant defense response (Banuett and Herskowitz, 1996). For the past 15 years numerous attempts were initiated to find "true" pathogenicity genes in *U. maydis*. This has led to the identification of important signaling pathways for pathogenesis, crucial regulatory genes for disease development, and important inputs into pathogenesis via cell cycle regulators (Feldbrügge et al., 2004; Pérez-Martín et al., 2006). However, due to their regulatory roles these genes display pleiotropic effects when deleted. Therefore, it has not been possible to link discrete disease-developmental defects to the function of single gene products.

From the recently published genome sequence (Kämper et al., 2006) a number of features have been recognized which begin to provide an explanation of how the biotrophic life style is established and how disease progression is modulated. In this chapter we will review these findings, which concern the identification of a number of distinct secreted protein effectors.

2 *Ustilago maydis* Does Not Use Aggressive Infection Strategies

Hyphal penetration in the *U. maydis*/maize pathosystem is direct and rarely involves entry via natural openings like stomata or wounding sites (Snetselaar and Mims, 1994; Banuett and Herskowitz, 1996). Starting from an unmelanized appressorium which is just a swelling of the hyphal tip, stages can be visualized where the fungal cytoplasm is in part in the hyphae on the leaf surface and in part in the hyphae that has already penetrated the epidermis (Schirawski et al., 2005). Based on these findings it is assumed that penetration is largely based on the localized secretion of plant cell wall degrading enzymes. A survey of the genome sequence has revealed, however, that *U. maydis* is poorly equipped with putative plant cell wall degrading enzymes. This assertion is based on a comparison with *Magnaporthe grisea* and *Fusarium graminearum*, two ascomycete plant pathogens that adopt a necrotrophic life style at least during later infection stages. While these pathogens code for 138 and 103 putative plant cell wall degrading enzymes, respectively, *U. maydis* has only 33 such genes. In addition, many enzyme families are either absent or are represented with only one member in *U. maydis*, while respective families are amplified in *M. grisea* and *F. graminearum* (Dean et al., 2002; Kämper et al., 2006). This is likely to reflect a specific adaptation of *U. maydis* to its host, i.e. it would limit cell wall damage during penetration and thus minimize the production of cell wall fragments that could serve as elicitors for mounting a plant defense response (Bucheli et al., 1990). This limited set of enzymes would also suffice for the initial stages of growth in planta where hyphae traverse several cell layers and are surrounded by the plant plasma membrane. During massive proliferation in the apoplast at later stages of the infection the hyphae are often branched, sometimes with extensions reaching into plant cells. The plant cells appear enlarged and are often separated from their neighboring cells while their plasma membranes are still intact (Snetselaar and Mims, 1994). During these stages the fungal hyphae are imbedded in a mucilaginous material of unknown composition. We consider it likely that this material acts as a shield against antimicrobial defenses by the host. In future, it will be interesting to elucidate which of the secreted plant cell wall degrading enzymes are actually needed during penetration and which are required at the stage of tumor formation. In this regard it is of interest that it has recently been accomplished to induce appressorium formation on artificial surfaces (Mendoza-Mendoza and Kahmann, unpublished). Transcription profiles of this stage in comparison to the tumor stage is likely to provide important insights into which plant cell wall degrading enzymes play a role during these discrete stages of development.

3 *Ustilago maydis* Regulates its Interaction with the Host via a Set of Novel Secreted Protein Effectors

While the number of *U. maydis* genes encoding secreted plant cell wall degrading enzymes is small, the number of genes encoding secreted proteins of unknown function is unexpectedly high: of the 426 proteins predicted to be secreted by both SignalP and ProtComp programs, 298 are without predicted function and of these 198 are only found in *U. maydis* (Kämper et al., 2006). This prompted a closer look, which revealed that 72 genes coding for secreted proteins are arranged in 12 gene clusters (Kämper et al., 2006). Clusters are defined by at least three consecutive genes encoding secreted proteins. Within the gene clusters small gene families consisting of two to five genes were frequent. Through microarray analysis in which the expression of these genes was compared in tumor tissue and in a haploid strain it was recognized that most of the clustered genes are significantly upregulated in tumor tissue (Kämper et al., 2006). Based on these results deletion mutants were generated for all 12 gene clusters. Five of the respective mutants were significantly altered in virulence (Fig. 2). The virulence phenotypes ranged from a mere reduction of pathogenicity symptoms (clusters 6A and 10A), to dramatically reduced virulence (cluster 19A), complete failure to cause disease (cluster 5B) to increased virulence (cluster 2A) (Fig. 2). A microscopic analysis revealed that cluster 5B mutant develop like wild type strains on the plant surface and form appressoria. However, they specifically arrest development immediately after penetration. Cluster 19A mutants are indistinguishable from wild type during the early stages of infection-related development. They infect plants and proliferate, but do not induce the formation of tumors. For mutants in clusters 2A, 6A, and 10A it has not yet been determined at which stages of pathogenic development they are affected (Kämper et al., 2006). In seven cases the cluster deletions did not cause a phenotype. However, as related genes were found elsewhere in the genome for four of these cluster genes, it is likely

Clustering of genes encoding secreted proteins   Virulence of
deletion mutants

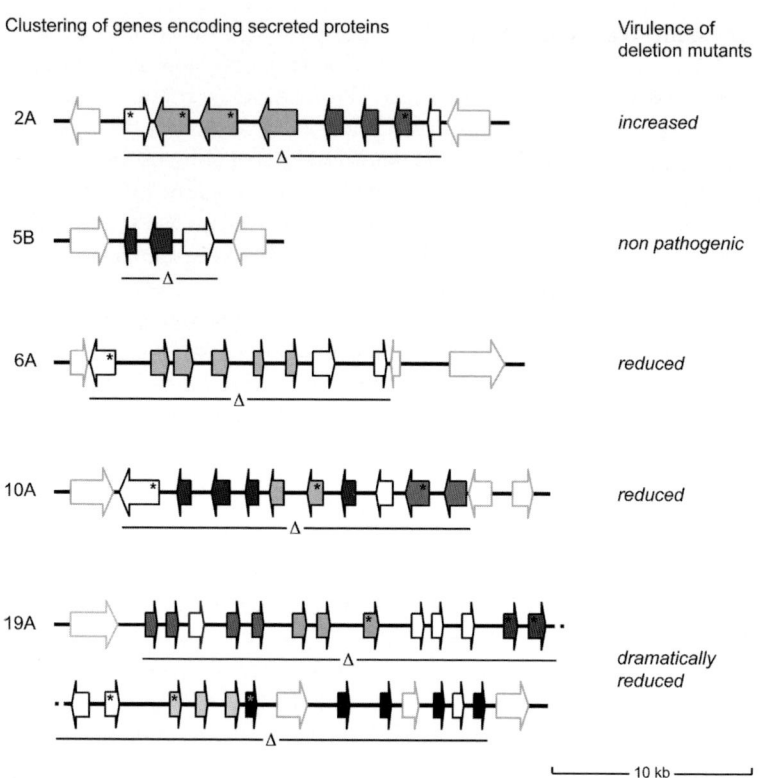

Fig. 2 Gene clusters for secreted proteins govern compatibility. *Left panel*: Schematic representation of five gene clusters for secreted proteins which play crucial roles in establishing and maintaining biotrophic development. Genes are represented by arrows indicating their direction of transcription. Predictions for secretion of the respective gene products by SIGNALP and PROTCOMP algorithms are indicated by *black line frames*. Asterisks within arrows denote a TARGETP prediction only. *Open arrows* with light gray frames indicate genes encoding products without prediction for secretion, including flanking genes. Similarities between genes within the same cluster are indicated by identical *grey filling shades*. Cluster genes deleted for assessing pathogenicity phenotypes are marked by *lines* below each scheme

that the number of clusters where the deletion causes an infection related phenotype will go up when double mutants have been generated. This set of experiments has demonstrated that the interaction of *U. maydis* with its host is largely governed by a set of secreted protein effectors of yet unknown function. The identification of these genes by virtue of their clustering and co-regulation has demonstrated the power of functional genomics. As the chosen definition of gene clusters is completely arbitrary, we are now extending our analysis to all unknown genes that are up-regulated in tumor tissue and whose products are predicted to be secreted. Preliminary analyses already indicate that many more genes with a specific function during pathogenesis can be uncovered by this approach (Brefort, Döhlemann and Kahmann, unpublished).

4 Discussion and Outlook

The identification of the *U. maydis* gene clusters encoding secreted proteins specifically required during pathogenesis raises a number of interesting questions and perspectives. In particular, it would be important to know which gene(s) from the clusters are responsible for the observed phenotype. Will it be possible to link the observed phenotype to individual genes (or gene families) or will there be cooperativity in function especially in the larger clusters? Where are the effectors localized and what is their function in the respective compartments? It is generally believed that biotrophic pathogens have to suppress plant defense responses and redirect host metabolism to the site of infection. In the *U. maydis* system there is the additional feature of tumor induction, which may require changes in phytohormone signaling in the host. Based on insights from bacterial pathogens that use type III secretion systems to translocate proteins directly into host cells (Mudgett, 2005), it is generally assumed that other types of pathogens have also developed the means to direct proteins from the pathogen to the host. Recently it was shown that a conserved pentameric sequence (Pexel motif, RXLX(E/Q)) located within 60 amino acids downstream of the signal sequence is responsible for trafficking into host cells in malaria parasites (Tonkin et al., 2006). For a group of plant pathogenic microbes, the oomycetes, which are related to brown algae and diatoms, it was shown recently that several avirulence gene products function inside host cells. These proteins, which are unrelated by primary sequence also display a conserved RxLR motif located within 30 amino acids downstream of the signal sequence (Birch et al., 2006). Hundreds of additional genes harboring this motif were identified in oomycete genome sequences, and these are therefore likely effectors for shaping the interaction with the respective plant host. Recently evidence has been provided that these host-targeting signals are equivalent in function across species (Bhattacharjee et al., 2006). Fungal plant pathogens do not code for proteins with such a motif. Nevertheless, it was demonstrated that several avirulence gene products of fungal pathogens elicit a hypersensitive response when expressed in the cytoplasm of plant cells. This is taken as evidence that these proteins are also translocated to the host and interact with their cognate cytoplasmically located receptors (Jia et al., 2000; Dodds et al., 2004; Catanzariti et al., 2006). In the bean rust fungus system a haustoria-specifically expressed protein (Rtp1) was shown to accumulate in the host cytoplasm (Kemen et al., 2005). This demonstrates that these fungal pathogens are able to translocate proteins to host cells via an as yet unknown mechanism. On these grounds we speculate that some of the secreted effectors of *U. maydis* may also be targeted to the host. As most fungal pathogens where protein translocation to the host has been shown/inferred are obligate parasites and cannot be transformed the *U. maydis* system may provide an attractive resource for gene/function analysis.

Another pertinent question is why the effectors identified in the *U. maydis* system appear to be *U. maydis* specific. Based on preliminary analyses in *Sporisorium reilianum*, a pathogen closely related to *U. maydis*, it was recently shown that at least some of the clustered genes are conserved (Schirawski, Schöning and Kahmann, unpublished). Thus, the assertion that these genes are classified as being *U. maydis* specific is likely to change with more genome sequences becoming available.

These studies begin to provide an understanding of how a biotrophic fungus modulates its interaction with its host. The goals for the future will lie in elucidating where the now identified secreted proteins function and which processes they target. By starting out with distinct pathogenesis-related phenotypes this follow-up work is expected to illuminate the molecular basis of biotrophic interactions.

References

Banuett, F., and Herskowitz, I., 1996, Discrete developmental stages during teliospore formation in the corn smut fungus, *Ustilago maydis*, *Development* **122**:2965–2976.

Basse, C., and Steinberg, G., 2004, *Ustilago maydis*, model system for analysis of the molecular basis of fungal pathogenicity, *Mol. Plant. Pathol.* **5**:83–92.

Bhattacharjee, S., Hiller, N.L., Liolios, K., Win, J., Kanneganti, T., Young, C., Kamoun, S., and Haldar, K., 2006, The malarial host-targeting signal is conserved in the Irish potato famine pathogen, *PloS Pathogens* **2**:453–465.

Birch, P.R.J., Rehmany, A.P., Pritchard, L., Kamoun, S., and Beynon, J.L., 2006, Trafficing arms: oomycete effectors enter host plant cells, *Trends Microbiol.* **14**:8–11.

Brachmann, A., Weinzierl, G., Kämper, J., and Kahmann, R., 2001, Identification of genes in the *bW/bE* regulatory cascade in *Ustilago maydis*, *Mol. Microbiol.* **42**:1047–1063.

Brown, J.B., and Casselton, L.A., 2002, Mating in mushrooms: Increasing the changes but prolonging the affair. *Trends Genet.* **17**:393–399.

Bucheli, P., Doares, S.H., Albersheim, P., Darvill, A., 1990, Host–pathogen interactions. XXXVI. Partial purification and characterization of heat–labile molecules secreted by the rice blast pathogen that solubilize plant cell wall fragments that kill plant cells, *Physiol. Mol. Plant. Pathol.* **1**:159–173.

Catanzariti, A.M., Dodds, P.N., Lawrence, G.J., 2006, Haustorially expressed secreted proteins from flax rust are highly enriched for avirulence elicitors, *Plant Cell* **18**:243–256.

Dean, R.A., Talbot, N.J., Ebbole, D.J., Farman, M.L., Mitchell, T.K. et al., 2002, The genome sequence of the rice blast fungus *Magnaporthe grisea. Nature* **434**:980–986.

Dodds, P.N., Lawrence G.J., Catanzariti, A.M., Ayliffe, M.A., and Ellis, J.G., 2004, The *Melampsora lini AvrL567* avirulence genes are expressed in haustoria and their products are recognized inside plant cells, *Plant Cell* **16**:755–768.

Feldbrügge, M., Kämper, J., Steinberg, G., and Kahmann, R., 2004, Regulation of mating and pathogenic development in *Ustilago maydis. Curr. Opin. Microbiol.* **7**:666–672.

Flor-Parra, I., Vranes, M., Kämper, J., and Pérez-Martín, J., 2006, Biz1, a zinc finger protein required for plant invasion by *Ustilago maydis*, regulates the levels of a mitotic cycle, *Plant Cell* **18**:2369–2387.

Inada, K., Morimoto, Y., Arima, T., Murata, Y., and Kamada T., 2001, The clp1 gene of the mushroom *Coprinus cinereus* is essential for A-regulated sexual development, *Genetics* **157**: 133–140.

Jia, Y., McAdams, S., Bryan, G.T., Hershey, H.P., and Valent, B., 2000, Direct interaction of resistant gene and avirulence gene products confers rice blast resistance, *EMBO J.* **19**:4004–4014.

Kämper, J., Reichmann, M., Romeis, T., Bölker, M., and Kahmann, R., 1995, Multiallelic recognition: nonself-dependent dimerization of the bE and bW homeodomain proteins in *Ustilago maydis*, *Cell* **81**:73–83.

Kämper, J., Kahmann, R., Bölker, M., Ma, L.-J., Brefort, T., et al., 2006, Living in pretend harmony: insights from the genome of the biotrophic fungal plant pathogen *Ustilago maydis*, *Nature* **444**:97–101.

Kemen, E., Kemen A.C., Rafiqui, M., Hempel, U., Mendgen, K., Hahn, M., and Voegele, R.T., 2005, Identification of proteins from rust fungi transferred from haustoria into infected plant cells, *MPMI* **18**:1130–1139.

Mudgett, M.B., 2005, New insights to the function of phytopathogenic bacterial type III effectors in plants. *Annu Rev Plant Biol.* **56**:509–531.

Pérez-Martín, J., Castillo-Lluva S., Sgarlata, C., Flor-Parra, I., Mielnichuk, N., Torreblanca, J., and Carbo, N., 2006, Pathocycles: *Ustilago maydis* as a model to study the relationships between cell cycle and virulence in pathogenic fungi, *Mol. Genet. Genomics* **276**:211–229.

Scherer, M., Heimel, K., Starke, V., and Kämper, J., 2006, The Clp1 protein is required for clamp formation and pathogenic development of *Ustilago maydis, Plant Cell* **18**:2388–2401.

Schirawski, J., Böhnert, H.U., Steinberg, G., Snetselaar, K., Adamikowa, L., and Kahmann, R., 2005, Endoplasmic reticulum glucosidase II is required for pathogenicity of *Ustilago maydis, Plant Cell* **17**:3532–3543.

Snetselaar, K.M., and Mims, C.W., 1994, Light and electron microscopy of *Ustilago maydis* hyphae in maize, *Mycol. Res.* **98**:347–355.

Tonkin, C.J., Pearce, J.A., McFadden, G.I., and Cowman, A.F., 2006, Protein targeting to destinations of the secretory pathway in the malaria parasite *Plasmodium falciparum, Curr. Opin. Microbiol.* **9**:381–387.

Virulence Evolution in Malaria

M.J. Mackinnon

Abstract One evolutionary theory of why some pathogens kill their host (i.e. are virulent) is that they need to extract host resources in order to transmit to new hosts. We have tested this theory in malaria and find it to be a likely explanation for the maintenance of this parasite's virulence in nature.

1 A Hypothesis for Pathogen Virulence

Pathogens that kill their host lose their source of ongoing survival and transmission. So why are they virulent? This question captured the attention of evolutionary biologists in the early 1980s: until then, it was generally accepted that, given enough evolutionary time, all parasites would evolve to be non-harmful to their hosts. However, this is clearly not true: many ancient host–parasite associations are still problematic for both host and pathogen. Furthermore, throughout human history, infectious diseases have remained a major cause of mortality. So what is it that maintains the pathogen's virulence in nature? If we knew the answer to this question, would this help us design vaccines and other control measures that might drive the pathogen towards lower virulence (Williams and Nesse, 1991; Dieckmann et al., 2002)?

Throughout this chapter, the term "virulence" is used generally to describe levels of morbidity or mortality: for modelling purposes, it is more strictly defined as the rate of host mortality induced by the pathogen. It is not replication rate or transmissibility or persistence – these, as it turns out, are correlates of virulence, but not virulence *per se*.

One answer to the question of why pathogens sometimes kill their hosts is that virulence is a mistake by the pathogen. For example, a mutant pathogen may have a short-term competitive advantage within a host, but is unable to transmit: alternatively, a pathogen may end up in the wrong tissue, also leading to no transmission.

M.J. Mackinnon

Department of Pathology, University of Cambridge, Tennis Court Road, Cambridge CB2 1QP and KEMRI-Wellcome Trust Research Programme, Centre for Geographic Medicine Research, Kilifi 80108, Kenya

e-mail: mjm88@cam.ac.uk

J.P. Gustafson et al. (eds.), *Genomics of Disease*,
© Springer Science+Business Media, LLC, 2008

This is called "short-sighted, or dead-end evolution" (Levin and Bull, 1994). An example of this may be the group of bacteria that cause meningitis. In this disease, the infection is asymptomatic when the bacteria remain in their normal tissue – the nasopharyngeal passages, from where they transmit. However, if the bacteria invade the cerebrospinal fluid, they cause very severe disease but do not transmit: they are therefore evolutionarily dead.

An alternative explanation for why virulence is maintained in nature is that this trait is adaptive, i.e. it has both fitness benefits, as well as fitness costs to the pathogen, and natural selection has brought about a level of virulence that carefully balances these. This is the so-called "virulence trade-off hypothesis" and is the one which is explored here in relation to malaria.

Under the trade-off hypothesis, it is assumed that there are both fitness benefits and costs associated with virulence. The benefits associated with virulence are assumed to be higher transmissibility (the rate at which the pathogen puts out transmissible forms), and/or longer duration of infection. The cost is assumed to be host

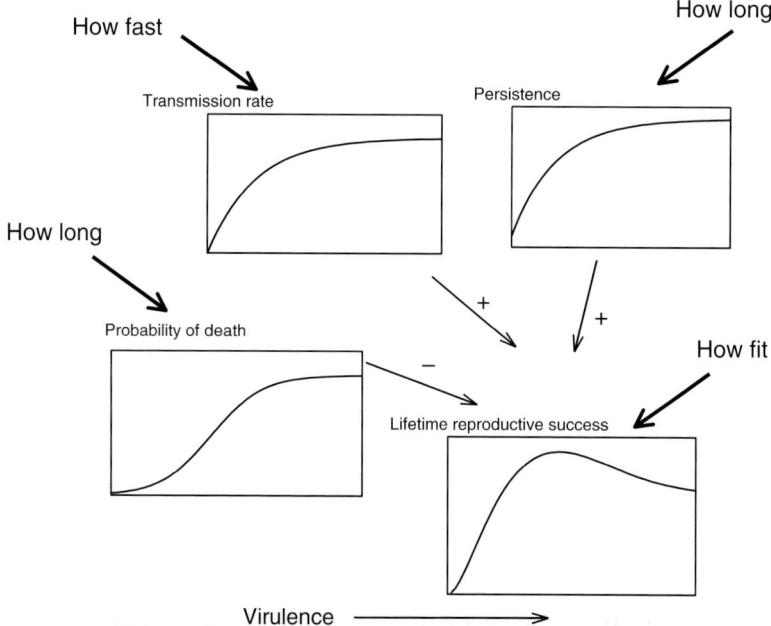

Fig. 1 The costs and benefits of virulence (or, alternatively, high levels of host exploitation) to pathogen fitness. The fitness benefits associated with virulence are higher transmissibility and persistence (duration of infection). The fitness cost of virulence is host mortality which shortens the infection. The rate of transmission and length of infection (which is determined by both persistence and the probability that the host dies) multiply together to give the total lifetime transmission of the pathogen from the host, i.e. its fitness. This reaches a maximum at an intermediate level of virulence when the negative and positive effects on virulence are combined in this way. Note that the transmissibility and duration of infection curves have to be less than linear (i.e. convex) in order to produce an intermediate optimum virulence

death because, for most pathogens, when the host dies, transmission ceases. The total transmission per lifetime of the pathogen (i.e. one infection length), which is directly related to its fitness, is the rate of transmission multiplied by the length of the infection. Thus, if death shortens the infection, the parasite's fitness is decreased. Thus, the pathogen is playing a dangerous game: it is trying to maximize transmissibility and infection length while also keeping its host alive. This, in principle, can lead to the evolution of an intermediate level of virulence, which balances the costs and benefits of virulence to the pathogen's fitness (Fig. 1).

Although mooted earlier (Topley, 1919; Levin and Pimentel, 1981), the trade-off idea only really took hold when Anderson and May (1982) applied it to the case of myxomatosis in rabbits. The myxoma virus was extremely virulent when released into rabbit populations in Australia and Britain in the early 1950s, but over the next decade, the virus evolved to an intermediate and stable level of virulence (Fenner and Ratcliffe, 1965). Anderson and May (1982) and Massad (1987) showed that this was because, although the more virulent strains of the virus had longer infections and higher transmissibility, their risk of killing the host was high enough to outweigh these transmission benefits. Therefore the virus evolved to a lower level of virulence where the transmission costs of mortality balanced the transmission benefits of longer infections and higher transmissibility.

2 Malaria

The myxomatosis example is somewhat artificial because the released virus was not found naturally in the rabbit populations and so had not co-evolved with its host. Its relevance to human disease is therefore questionable (Ebert and Bull, 2003). Unfortunately, there are few data in any host–pathogen system, medically relevant or not, that can be used to support or refute the trade-off hypothesis. Examples of pathogens where positive relationships between virulence, transmissibility and persistence (hereon termed V–T–P), as assumed by the hypothesis, have been found include trypanosomes in mice (Diffley et al., 1987; Turner et al., 1995; Masumu et al., 2006a; Masumu et al., 2006b), microsporidia in *Daphnia* (Ebert, 1994; Ebert and Mangin, 1997), bacteria in *Daphnia* (Jensen et al., 2006), *Sarcosystis* (an apicomplexan protozoa) in rats (JŠkel et al., 2001), schistosomes in mice (Davies et al., 2001), bacteria in mice (Greenwood et al., 1936), and phage in bacteria (Messenger et al., 1999). Examples that directly illustrate the cost of host death to lifetime transmission are even more difficult to find (Jensen et al., 2006). This lack of empirical data, particularly for medically relevant pathogens, spurred us to formally test the hypothesis for a major pathogen of humans – the malaria parasite.

The causative agent of the most pathogenic form of human malaria, *Plasmodium falciparum*, is a typical microparasite, i.e. it replicates rapidly within its host, as do many viral, bacterial and other protozoal pathogens. Unlike viruses and bacteria, however, the malaria parasite has a separate transmission stage form in its life cycle called gametocyte. Gametocytes, the sexual form of the parasite, are differentiated from the asexual, replicating blood-stage forms. Gametocytes do not themselves replicate, and thus are seen at low densities in the bloodstream. But, as they are the only forms that survive in the mosquito vector, their mass production is critical to the pathogen's transmission.

2.1 Mouse Malaria

We began by testing the assumptions of the trade-off hypothesis using the mouse malaria model, *Plasmodium chabaudi*, as an experimental model. Using parasite clones obtained from their natural host in the wild (Beale et al., 1978) to infect groups of inbred mice in the laboratory, we measured the transmissibility, virulence and persistence of these clones and the genetic (i.e. across-clone) relationships among these traits (Mackinnon and Read, 1999a, 2003; Ferguson et al., 2003). Our experiments showed that clones that had higher levels of virulence had higher transmissibility (both production of gametocytes and infectivity to mosquitoes) and cleared their infections more slowly in the absence of host death (Fig. 2) (Ferguson and Read, 2002; Mackinnon and Read, 1999a, b, 2003). Virulent parasites were also those that generated highest parasite densities and thus exploited more of the host's resources (red cells). Thus the V–T–P relationships assumed by the trade-off hypothesis were supported in this experimental model. Furthermore, if the host died, the total amount of gametocytes produced during the infection (i.e. the potential transmission) was less than if the host lived (Fig. 2e), thus demonstrating a clear cost of host mortality to the parasite's fitness (Mackinnon et al., 2002). We found the V–T–P relationships to be qualitatively robust to host genotype (Mackinnon et al., 2002), host sex (Mackinnon and Read, 2004a), serial passage (Mackinnon and Read, 1999b, 2004a; Mackinnon et al., 2002), mosquito passage (Mackinnon and Read, 2004a; Mackinnon et al., 2005) and, importantly, to levels of host immunity (Mackinnon and Read, 2003, 2004a). Immunity acted to reduce all three traits – transmissibility, virulence and persistence – in line with that expected from the relationships among these traits in naive hosts. In other words, the parasite lines that replicated well, transmitted faster, lasted longer and caused more morbidity when infecting naive hosts also did so in semi-immune hosts, but just at a lower level.

2.2 Human Malaria

We next turned our attention to human malaria, asking the question as to whether the relationships we observed in our mouse model in the laboratory also existed in human malaria in its natural environment. Using data from a large longitudinal cross-sectional population survey in Nigeria (the Garki project), it was observed that virulence, persistence and transmissibility were positively correlated when assessed across the whole population (Fig. 3). In these data, however, the relationships are undoubtedly driven by age-acquired immunity, i.e. by levels of host defence rather

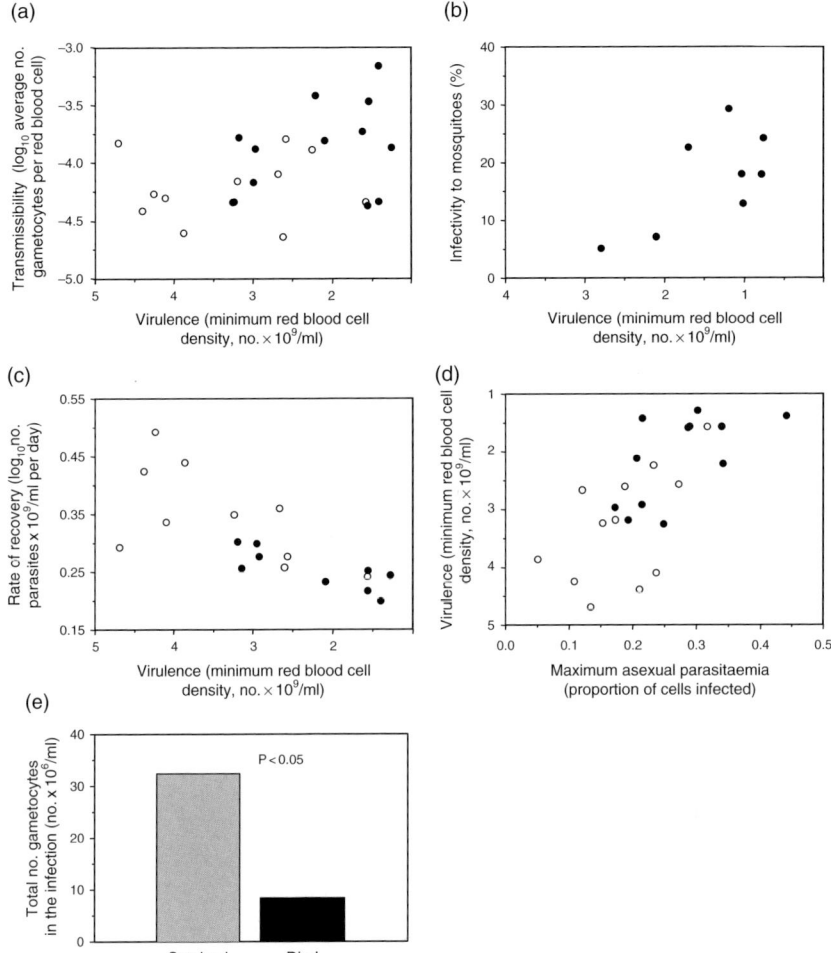

Fig. 2 Virulence–transmissibility–persistence relationships in the mouse model of malaria, *P. chabaudi*. Each point is the average value of groups of mice ($n = 5$–10) infected with one parasite clone: thus a line fitted to the points would represent the parasite genetic relationship among the traits. *Filled symbols* indicate naive mice and *open symbols* indicate mice made semi-immune by previous infection and drug clearance. Virulence was measured as the mouse's minimum red blood cell density reached during the infection. It was shown to positively relate to transmissibility as measured by (**a**) daily average gametocyte density and (**b**) infectivity to mosquitoes on 2–4 days during the peak of gametocyte production. (**c**) Recovery rate, which is inversely related to infection length, was measured as the rate at which the infection declined after peak parasitemia and was negatively related to virulence, as expected. (**d**) Virulence was positively related to maximum parasitemia, an indicator of host exploitation. Clones broadly retained their rankings for all traits when infecting naïve vs. immunised mice (Pearson correlations across treatments of 0.59–0.66 for the four traits used here.) In the experiments described in panels (**a**–**d**), very few mice died and so a virulence cost to transmission was not observable. However, in another experiment using less resistant host genotypes, mortality was high (23%) and was shown to severely reduce the total number of gametocytes produced by the infection, shown in (**e**). Data were reproduced from Mackinnon et al. (2002) and Mackinnon and Read (2003)

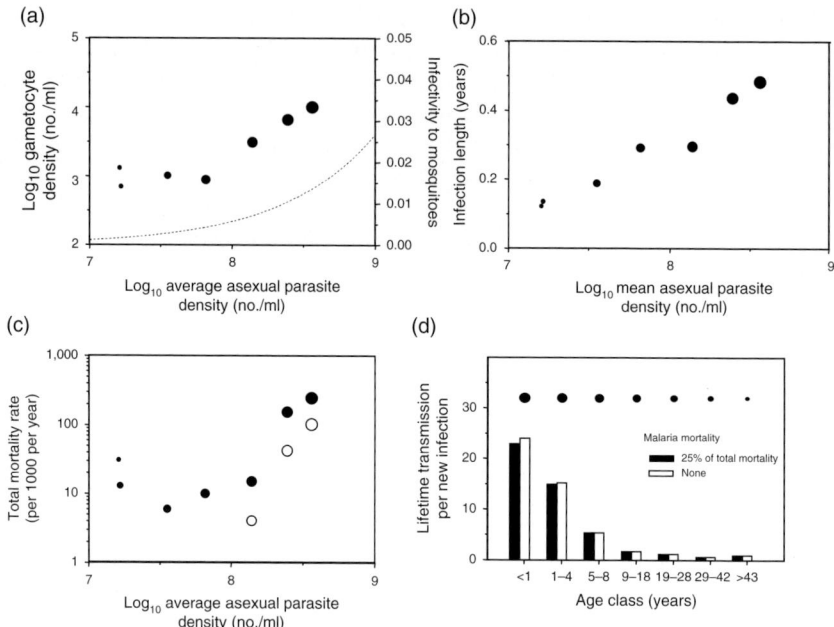

Fig. 3 Relationships between level of host exploitation, transmissibility, persistence and virulence (mortality) in human malaria in the field. Data were taken from a longitudinal study in Nigeria in the 1970s (the Garki Project; Molineaux and Gramiccia, 1980). Host exploitation is on the *x*-axis and is represented by average asexual parasite density. Each symbol represents an average for a group of people from the same age-group measured in eight surveys and decreases in size with age. (**a**) Transmissibility is represented by gametocyte density, and then converted to infectivity (*dotted line*, right axis) using logistic regression fitted to data from experimental mosquito infectivity data from the literature (data not shown). (**b**) Persistence is represented by infection length estimated from consecutive 10-week surveys (Bekessy et al., 1976) and adjusted for superinfections to lengths of individual infections using the method described by Dietz et al. (1980). (**c**) All-cause mortality rates are shown as closed symbols: these are due to other factors as well as malaria, and include all the malaria infections that occur in a host per year, i.e. including superinfections. The open symbols represent the values of mortality assuming that 25% of all deaths under the age of 9 years, and 1% in older categories, are due to malaria: these values are representative of studies where these proportions have been estimated, albeit with considerable uncertainty (Snow et al., 1999). (**d**) The expected amount of total transmission from hosts in different age-classes in the Garki project based on the data shown in Figs 3a–c: this gives some indication of the parasite's relative fitness in different host types. Total transmission (*black bars*) was calculated as infectivity multiplied by vectorial capacity (the expected number of infectious bites to new hosts that result from a mosquito feeding on one host, infected or not, assumed to have a value of 8 here) multiplied by the expected infection length. Infection length here is the reciprocal of the sum of all-cause mortality rate (Fig. 3c) and recovery rate (inverse of individual infection length, Fig. 3b). The length of an individual parasite infection is used in this calculation because we are interested in the relative transmission, or fitness, of an individual parasite strain rather than the total transmission from a group of co-infecting strains occupying the same host. By contrast, mortality rate in this calculation is all-cause mortality because individual parasite genotypes suffer from the mortality caused by co-infecting parasites as well as that caused by themselves. To illustrate the cost of host death on the parasite's total lifetime productivity, we have performed the same calculation as above but assuming that malaria causes no mortality, i.e. the mortality rate is all-cause mortality reduced by 25% in children under 9 years (*white bars*). The different-sized symbols above the bars correspond to those for the different age-classes in Figs 3a–c

than by parasite variation in virulence. They do not reflect the parasite V–T–P genetic correlations observed in mice in a uniform host type, and therefore do not directly support the assumptions of the trade-off model. Nevertheless, the human data are consistent with the general pattern of higher parasite densities giving rise to more transmission forms, longer infections and being associated with higher mortality, with all of these being reduced by immunity in an apparently straight-forward way. Thus these phenotypic relationships from the field data, which cover the entire spectrum of immunity (and host and parasite genetics), may well be simple extensions of the intrinsic biological links among V–T–P that are rooted in the parasite's ability to replicate. In other words, a pathogen's ability to repli-cate in the face of host defence mechanisms is correlated to its ability to trans-mit to new hosts, and hence its fitness. Thus host immunity favours more virulent parasites.

The other requirement for the trade-off hypothesis to be accepted as applicable to human malaria is a direct demonstration of the cost to total transmission of host death. To investigate this, we calculated the expected lifetime transmission from each host assuming that death did or did not occur (Fig. 3d). The penalties to total transmission were around 5% in young children. Whether this cost is sufficient to put the brake on virulence evolution, as supposed by the trade-off hypothesis, is not yet known. This requires an analysis of both the benefits, as well as the costs, of higher virulence, and this at the parasite genetic level: we are not yet able to do this. At the moment however, it seems plausible, at least, that death in children imposes some selection against higher virulence.

The analysis of total lifetime transmission (Fig. 3d) also reveals that adults trans-mit around 20 times less than children, but suffer little death. The latter means that more virulent parasites, which can "get away with it" in these protected hosts, may be maintained in the parasite population. Thus, while these more virulent parasites may help the parasite transmit from the semi-immune hosts, they would do so at the expense of the children. Furthermore, if the shape of the trade-off is as we assume (Fig. 1), parasites in such hosts have higher selection coefficients per unit increase in virulence and would therefore be more strongly selected, thus putting further pressure on increased virulence. On the other hand, adult hosts contribute much less to the total transmission population, and so this source of selection may not have a significant impact. The question is, then, what will happen to virulence if we effectively turned all children into adults by immunizing with a (as yet hypothetical) malaria vaccine?

2.3 Consequences of Malaria Vaccination

The data so far indicate that in malaria the common factor underlying the V–T–P relationships is asexual parasite density. We reasoned, therefore, that any control measures that brought about a reduction in asexual parasite densities, such as an asexual stage, anti-replication vaccine, would bring about evolution for increased intrinsic virulence because of its consequences to transmissibility, persistence and host mortality (Gandon et al., 2001). If this evolution were to happen, the unprotected (e.g. unvaccinated) people in the population would then be exposed to a more virulent parasite and their risk of death would be higher than before. On the other hand, it might be argued, the vaccine, if used widely, would protect many more people from disease so that the population-wide reduction in mortality may well outweigh the increased mortality among the unvaccinated few. Thus, as is often the case in vaccination programmes, even without evolution, the benefits to the majority of the population have to be weighed against the risk to a few individuals. Therefore, to determine the benefit to the whole population of vaccination allowing for an evolutionary response in the parasite's virulence, we modelled the case of virulence evolution under an imperfect asexual stage malaria vaccine incorporating the feedbacks between the epidemiology (force of infection) and the parasite evolution (Gandon et al., 2001). Parameter values were chosen to mimic an area of high year round malaria transmission such as central Kenya or Tanzania. As predicted, the parasite evolved higher virulence under vaccination so that the case fatality rate amongst unvaccinated naive hosts was higher than if evolution had not occurred. Moreover, the total mortality across the whole population was also higher, especially when the vaccine was given to a moderate fraction of the population. Thus, all the expected benefits of vaccination were eroded by parasite evolution.

We also modelled other types of vaccines or devices that block infection (e.g. liver stage vaccines), or stop it transmitting from humans to mosquitoes (e.g. bednets), but that do not act directly on the asexual stage parasites and hence the V–T–P relationships. These control measures were predicted to have no impact or, as explained below, could even lead to a decrease in virulence (Gandon et al., 2001). Thus, priority should be given to control measures that prevent pathogens from entering or leaving the host rather than measures that directly target the infection itself. In malaria, the former would include bednets, vector control and liver-stage vaccines (unless they also reduce multiplication rates): the latter includes asexual-stage vaccines, anti-toxin vaccines and drug use. Thus, in the face of a large potential for pathogen evolution, the hygiene principle would appear to be an excellent one to follow (where possible) for evolutionary reasons as well as for its more obvious epidemiological applications.

In nature, malaria infections are frequently genetically diverse, raising the question of the relative fitness of virulent and avirulent strains when present in the same host. Theoreticians have envisaged a range of possible outcomes, and hence effects on virulence evolution (reviewed in Read and Taylor 2001). In the *P. chabaudi* model, more virulent clones have a competitive advantage: they are able to competitively suppress or even exclude less virulent clones during the acute phase of mixed-clone infections (de Roode et al., 2005a). The competitive outcome is reflected in relative gametocyte production (Wargo et al., in preparation) and hence transmission to mosquitoes (de Roode et al., 2005b). The competitive outcome is qualitatively robust to host genotype (de Roode et al., 2004) and host immune status (Raberg et al., 2006).

The theoretical conclusion that using devices that prevent infection or block transmission will lead to reduced virulence results from the lower force of infection caused by the control programme. This leads to lower levels of co-infection or super-infection and hence a lower level of competition among parasites occupying the same host. The reduction in virulence due to reduced competition can be understood by taking the view of a resident parasite that is either ousted by (super-infection) or forced to share with (co-infection) another parasite type. In the case of super-infection, the resident parasite has its infection shortened by the invading parasite: the parasite then compensates by evolving a higher level of host exploitation, or intrinsic virulence, up to the point when the benefits are balanced by the costs. In the case of co-infection, the risk of mortality for the resident parasite is increased by the presence of another parasite, thus increasing the costs, or the infection of a less virulent genotype is curtailed by that of a more virulent genotype. This is especially so where virulence is associated with competitive ability. Whether the latter is true is still being worked out, but the data from the *P. chabaudi*-mouse model so far suggest that it is (Read and Taylor, 2001; De Roode et al., 2004; De Roode et al., 2005a,b; Raberg et al., 2006). Thus so far, we believe that reducing the force of infection through vaccination will relieve the amount of within-host competition on the parasite and hence selection pressure on high virulence.

As we cannot test the vaccination hypothesis in the field until it is too late, we would ideally test it in the laboratory by experimental evolution. However, it is extremely difficult in the laboratory to repeatedly transmit multiple parasite lines of malaria through mosquitoes for the entire period of their transmissibility: it is also unethical to allow host mortality to act as the selecting agent. Therefore, we designed an experiment to test one component of the vaccine hypothesis, namely that immunity selects for higher rates of host exploitation (regardless of the trans-mission consequences). This was done by transferring a standard number of parasites between mice every 7 days for 20 passages, thus bypassing the mosquito stage of the life cycle. At the end of this selection phase, the replicated lines were passaged once through mosquitoes and tested for virulence and other characteristics in either naïve or semi-immune mice. We found that parasite lines that had evolved in immunised hosts were more virulent than parasite lines that had evolved in naïve hosts (Fig. 4). This was true whether the lines were infecting naïve or immunised mice, and whether or not the lines had been passaged through mosquitoes prior to testing them (Mackinnon and Read, 2004b). The immune-selected lines also had higher intrinsic replication rates than the naïve-selected lines (Fig. 4). Thus immunity had more efficiently selected the parasites with greater ability to exploit the host and thus cause more virulence. This result is consistent with our theory that blood-stage vaccines would select for more virulent parasites, but the mechanism of selection in this experiment must have been different to that we proposed for human malaria in the field, since the mortality costs of virulence were unaltered in our experiments. The results from this experiment nevertheless demonstrate rapid response to selection imposed by immunity, and support the underlying basis of the trade-off model, that replication rate is the key to maximizing parasite fitness in the face of immunity, with virulence being an unfortunate side-effect, especially for non-immune hosts. It will be impossible to do a direct test of the vaccination

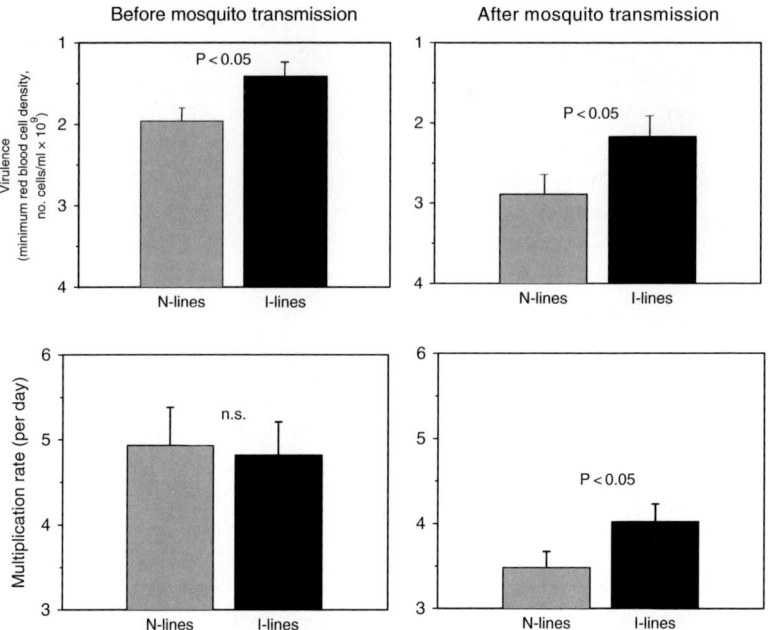

Fig. 4 Results of serial passage of *P. chabaudi* through either immunised (I-lines) vs. naïve (N-lines) mice. Starting with an avirulent clone, five independent parasite lines were blood-passaged in mice every 7 days for 20 passages (see Mackinnon and Read, 2004a, for further details of the experimental design). At the end, the lines were passaged through mosquitoes once, and then compared for their virulence (measured by the minimum blood cell density reached during the infection) and the rate of population growth during the first 6 days using real-time PCR (see Mackinnon et al., 2005 for details of methods). Heights of bars represent means ($\pm$ 1.s.e.) of groups of mice infected with the five selection lines per selection treatment. Selection in immunised mice generated more virulent parasites (upper panels) than in naïve mice indicating that immunity selects more intensively for virulent forms during within-host selection. These differences were present in the lines both before (graphs on *left*) and after (*right*) mosquito transmission indicating that this evolution was genetically stable. These parasites also multiplied faster (*lower panels*) during the early, non-growth-limiting stage of the infection before the onset of disease and a strong immune response. This difference was not apparent in the lines before mosquito transmission, probably due to the maximum growth potential (around 5-fold per day, Mackinnon et al., 2005) having been reached. Differences between pre-and post-mosquito transmission were significant for both traits (virulence, $P < 0.05$; multiplication rate, $P < 0.001$). Data in the *top* two figures are reproduced from Mackinnon and Read (2004a)

hypothesis in human malaria before the uncontrolled experiment is done on a vast scale. But, there are three approaches that could be taken in the meantime. First, one could examine the outcome of the natural experiment that has been done. According to the theory, parasites from highly immune populations (naturally acquired) should be intrinsically more virulent than parasites from less immune populations. Thus a comparison of mortality rates and disease severity in naïve cases could be made from high transmission intensity areas (where population immunity is high) from low transmission areas. There is some limited evidence to support this argument

(Murphy and Breman, 2001; Idro et al., 2006), but see Reyburn et al. (2005), though more data are required and it will be a substantial challenge to disentangle the effects of immunity and infection rate, which both covary with transmission rate. Another approach might be to compare mortality rates of malaria-naïve travellers who have acquired their infection from high vs. low transmission areas, or to compare in vitro the virulence-related molecular phenotypes of parasites from areas with contrasting transmission intensities.

3 Vaccine-Driven Virulence Evolution in Other Diseases

Are there any examples where vaccines have driven the pathogen towards higher virulence? Vaccines against smallpox, measles and polio have been remarkably successful. Among these, only polio virulence has been shown to increase in response to vaccination pressure (Kew et al., 2002): in this case it was the attenuated vaccine strain that reverted to virulence and transmissibility rather than the wild-type strain changing in response to vaccination. In the case of measles and smallpox, the vaccine induces near-sterilizing immunity, and so it is therefore not surprising that evolution seems not to have occurred.

On the other hand, for diseases where vaccines are more imperfect (i.e. less effective), changes in pathogen virulence following vaccination have been documented, but not always in the direction we predict here, e.g. diphtheria, pneumococcal disease and whooping cough (pertussis). Such cases highlight an important distinction between the type of virulence evolution we have been discussing here for malaria ("generalised virulence" evolution) and that which has been discussed far more widely in the literature in relation to vaccine-driven evolution ("antigenic escape" evolution) (McLean, 1995; Lipsitch and Moxon, 1997; Gupta et al., 1997). In the case of diphtheria, virulence has been observed to decrease after vaccination (Pappenheimer, 1984). However, in this case, the vaccine was directed *only* at the virulence-causing (tox+) strain, thus causing replacement by the non-toxin producing, avirulent strain. A similar pattern has been observed for *Streptococcus pneumoniae*, for which the vaccine targets only the most virulent serotypes (Kaplan et al., 2004). With this sort of antigenic escape evolution (also know as "serotype replacement"), the successful genotypes following vaccination are those that do not carry the particular antigenic type in the vaccine. However, a priori, they may or not be more intrinsically virulent than those that they replace: this will depend on which strains the vaccine targets. In this chapter on malaria we have been discussing the possibility of "generalised virulence" evolution in which the successful types are those that are able to "grow over" vaccine-induced immunity by virtue of their higher intrinsic replication rate: these types are therefore also more virulent in unvaccinated hosts. In doing so, this does not discount the importance of the antigenic escape form of evolution in malaria: indeed it has already been observed following a malaria vaccine trial (Genton et al., 2002), and its potential is implied from field studies of naturally acquired immunity in which selection away from the immunising types is observed (Bull et al., 1999, 2000; Nielsen et al., 2002, 2004).

In pertussis, there appears to have been changes in the pathogen's virulence genes in response to vaccination (Mooi et al., 2001), but it is not clear whether these changes constitute the "antigenic escape" or the "generalized virulence" form of evolution. On the other hand, there is a very clear example of vaccine-driven evolution in the Marek's disease virus of chickens where vaccines have been in widespread use for 40 years (Witter, 1998): while the precise reasons for this evolution are not yet known, it has all the hallmarks of being driven by the "generalized virulence" form of evolution that we have been discussing here (Read et al., 2004). A further example of generalized virulence evolution is the increased virulence of myxomatosis virus in response to the evolution of higher levels of resistance in rabbits in Australia (Fenner and Fantini, 1999).

4 Conclusions

In the case of malaria, for which a vaccine is not yet available but is being intensively pursued, the case has been presented for why generalised virulence may evolve in response to vaccines, what might be done to avoid it and what needs to be done to further determine the likelihood of it happening. The key points are virulence, transmissibility and persistence appear to be intrinsically linked through the parasite's biology/life history, and these links are maintained over different levels of host immunity. Immunity has a very strong impact on parasite fitness through its effect on reducing all three V–T–P traits in accordance with their biological links. More data are required in human malaria to determine whether these links are encoded by the parasite's genes and to determine how much virulence variation exists in the field.

Immunity is expected to select for higher virulence because, under the cover of immunity, the most virulent parasites can "get away with it". In naïve hosts, they cannot. Thus highly immune host populations should select for more virulent pathogens. In malaria, children suffer as a result of selection by immune adults.

Devices that reduce asexual density but do not block infection, such as anti-asexual stage vaccines and drugs, are expected to lead to evolution of higher virulence, while hygiene (i.e. preventing infection and transmission) is expected to lead to evolution of lower virulence.

Ineffective control measures are expected to minimize undesirable virulence evolution. Therefore, age-targeted interventions that minimize selection while protecting the most vulnerable may be desirable.

Acknowledgments This chapter is published with the permission of the director of KEMRI. The work was supported by the Leverhulme Trust and the University of Edinburgh.

References

Anderson, R.M., May, R.M., 1982, Co-evolution of hosts and parasites, Parasitology **85**:411–426.

Beale, G.H., Carter, R., Walliker, D., 1978, Genetics, in R. Killick-Kendrick, W. Peters, eds., Rodent Malaria, Academic Press, London, pp. 213–245.

Bekessy, A., Molineaux, L., and Storey, J., 1976, Estimation of incidence and recovery rates of *Plasmodium falciparum* parasitaemia from longitudinal data, *Bull. World Health Organ.* **54**:685–693.

Bull, P.C., Kortok, M., Kai, O., Ndungu, F., Ross, A., Lowe, B.S., Newbold, C.I., and Marsh, K., 2000, *Plasmodium falciparum*-infected erythrocytes: agglutination by diverse Kenyan plasma is associated with severe disease and young host age, *J. Infect. Dis.* **182**:252–259.

Bull, P.C., Lowe, B.S., Kortok, M., and Marsh, K., 1999, Antibody recognition of *Plasmodium falciparum* erythrocyte surface antigens in Kenya: evidence for rare and prevalent variants, *Infect. Immun.* **67**:733–739.

Davies, C.M., Webster, J.P., and Woolhouse, M.E.J., 2001, Trade-offs in the evolution of virulence in an indirectly transmitted macroparasite, *Proc. R. Soc. Lond. B.* **268**:251–257.

De Roode, J.C., Culleton, R., Bell, A.S., and Read, A.F., 2004, Competitive release of drug resistance following drug treatment of mixed *Plasmodium chabaudi* infections, *Malar. J.* **3**:33.

De Roode, J.C., Helsinki, M.E.H., Anwar, M.A., and Read, A.F., 2005a, Dynamics of multiple infection and within-host competition in genetically diverse malaria infections, *Am. Nat.* **166**:531–542.

De Roode, J.C., Pansini, R., Cheesman, S.J., Helsinki, M.E.H., Huijben, S., Wargo, A.R., Bell, A.S., Chan, B.H.K., Walliker, D., and Read, A.F., 2005b, Virulent clones are competitively superior in genetically diverse malaria infections, *Proc. Natl. Acad. Sci. USA* **102**:7624–7628.

Dieckmann, U., Metz, J.A.J., Sabelis, M.W., and Sigmund, K., 2002, Virulence Management: the Adaptive Dynamics of Pathogen-Host Interactions, Cambridge University Press, Cambridge.

Dietz, K., Molineaux, L., and Thomas, A., 1980, The mathematical model of transmission, in L. Molineaux, G. Gramiccia, eds., The Garki Project. Research on the Epidemiology and Control of Malaria in the Sudan Savanna of West Africa, World Health Organization, Geneva, pp. 262–289.

Diffley, P., Scott, J.O., Mama, K., and Tsen, T.N.R., 1987, The rate of proliferation among African trypanosomes is a stable trait that is directly related to virulence, *Am. J. Trop. Med. Hyg.* **36**: 533–540.

Ebert, D., 1994, Virulence and local adaptation of a horizontally transmitted parasite. *Science* **265**:1084–1086.

Ebert, D., and Bull, J.J., 2003, Challenging the trade-off model for the evolution of virulence: is virulence management feasible? *Trends Microbiol.* **11**:15–20.

Ebert, D., and Mangin, K.L., 1997, The influence of host demography on the evolution of virulence of a microsporidian gut parasite, *Evolution* **51**:1828–1837.

Fenner, F., and Fantini, B., 1999, Biological Control of Vertebrate Pests, CABI Publishing, Wallingford, UK.

Fenner, F., and Ratcliffe, R.N., 1965, Myxomatosis, Cambridge University Press, London.

Ferguson, H.M., Mackinnon, M.J., Chan, B.H.K., and Read, A.F., 2003, Mosquito mortality and the evolution of malaria virulence, *Evolution* **57**:2792–2804.

Ferguson, H.M., and Read, A.F., 2002, Genetic and environmental determinants of malaria parasite virulence in mosquitoes, *Proc. R. Soc. Lond. B.* **269**:1217–1224.

Gandon, S., Mackinnon, M.J., Nee, S., and Read, A.F., 2001, Imperfect vaccines and the evolution of parasite virulence, *Nature* **414**:751–755.

Genton, B., Betuela, I., Felger, I., Al-Yaman, F., Anders, R.F., Saul, A., Rare, L., Baisor, M., Lorry, K., Brown, G.V., Pye, D., Irving, D.O., Smith, T.A., Beck, H.-P., and Alpers, M.P., 2002, A recombinant blood-stage malaria vaccine reduces *Plasmodium falciparum* density and exerts selective pressure on parasite populations in a phase 1–2b trial in Papua New Guinea, *J. Infect. Dis.* **185**:820–827.

Greenwood, M., Bradford Hill, A., Topley, W. W. C., and Wilson, J., 1936, Experimental Epidemiology. Medical Research Council. 209. London, HMSO. Special Report Series.

Gupta, S., Ferguson, N.M., and Anderson, R.M., 1997, Vaccination and the population structure of antigenically diverse pathogens that exchange genetic material, *Proc. R. Soc. Lond. B.* **264**:1435–1443.

Idro, R., Aloyo, J., Mayende, L., Bitarakwate, E., John, C.C., and Kivumbi, G.W., 2006, Severe malaria in children in areas with low, moderate and high transmission intensity in Uganda, *Trop. Med. Int. Health* **11**:115–124.

JŠkel, T., Khoprasert, Y., and Mackenstedt, U., 2001, Immunoglobulin subclass responses of wild brown rats to *Sarcocystis singaporensis, Int. J. Parasitol.* **31**:273–283.

Jensen, K.H., Little, T.J., Skorping, A., and Ebert, D., 2006, Empirical support for optimal virulence in a castrating parasite. *PLoS Biol.* **4**:e197

Kaplan, S.K., Mason, E.O., Wald, E.R., Schutze, G.E., Bradley, J.S., Tan, T.Q., Hoffman, J.A., Givner, L.B., Yogev, R., and Barson, W.J., 2004, Decrease in invasisve pneumococcal infections in children among 8 children's hospitals in the United States after the introduction of the 7-valent pneumococcal conjugate vaccine, *Paediatrics* **113**:443–449.

Kew, O.M., Sutter, R.W., De Gourville, E.M., Dowdle, W.R., and Pallansch, M.A., 2002, Vaccine-derived polioviruses and the endgame strategy for global polio eradication, *Annu. Rev. Microbiol.* **59**:587–635.

Levin, B.R., and Bull, J.J., 1994, Short-sighted evolution and the virulence of pathogenic microorganisms, *Trends Microbiol.* **2**:76–81.

Levin, S.A., and Pimentel, D., 1981, Selection of intermediate rates of increase in parasite-host systems, *Am. Nat.* **117**:308–315.

Lipsitch, M., and Moxon, E.R., 1997, Virulence and transmissibility of pathogens: what is the relationship? *Trends Microbiol.* **5**:31–36.

Mackinnon, M.J., Bell, A.S., and Read, A.F., 2005, The effects of mosquito transmission and population bottlenecking on virulence, multiplication rate and rosetting in rodent malaria, *Int. J. Parasitol.* **35**:145–153.

Mackinnon, M.J., Gaffney, D.J., and Read, A.F., 2002, Virulence in malaria parasites: host genotype by parasite genotype interactions, *Infect. Genet. Evol.* **1**:287–296.

Mackinnon, M.J., and Read, A.F., 1999a, Genetic relationships between parasite virulence and transmission in the rodent malaria *Plasmodium chabaudi, Evolution* **53**:689–703.

Mackinnon, M.J., and Read, A.F., 1999b, Selection for high and low virulence in the malaria parasite *Plasmodium chabaudi. Proc. R. Soc. Lond. B.* **266**:741–748.

Mackinnon, M.J., and Read, A.F., 2003, Effects of immunity on relationships between growth rate, virulence and transmission in semi-immune hosts, *Parasitology* **126**:103–112.

Mackinnon, M.J., and Read, A.F., 2004a, Immunity promotes virulence in a malaria model, *PLoS Biol.* **2**:1286–1292.

Mackinnon, M.J., and Read, A.F., 2004b, Virulence in malaria: an evolutionary viewpoint. *Phil. Trans. R. Soc. Lond. B.* **359**:965–986.

Massad, E., 1987, Transmission Rates and the evolution of pathogenicity, *Evolution* **41**:1127–1130.

Masumu, J., Marcotty, T., Geysen, D., Geerts, S., Vercruysse, J., Dorny, P., and Van den Bossche, P., 2006a, Comparison of the virulence of *Trypanosoma congolense* strains isolated from cattle in a trypanosomiasis endemic area of eastern Zambia, *Int. J. Parasitol.* **36**:497–501.

Masumu, J., Marcotty, T., Ndeledje, N., Kubi, C., Geerts, S., Veycruysse, J., Dorny, P., and Van den Bossche, P., 2006b, Comparison of transmissibility of *Trypanosoma congolense* strains, isolated in trypanosomiasis endemic area of eastern Zambia, by *Glossina morsitans morsitans, Parasitology* **133**:331–334.

McLean, A.R., 1995, Vaccination, evolution and changes in the efficacy of vaccines: a theoretical framework, *Proc. R. Soc. Lond. B.* **261**:389–393.

Messenger, S.L., Molineux, I.J., and Bull, J.J., 1999, Virulence evolution in a virus obeys a trade-off, *Proc. R. Soc. Lond. B.* **266**:397–404.

Molineaux, L., and Gramiccia, G., 1980, The Garki Project: Research on the Epidemiology and Control of Malaria in the Sudan Savanna of West Africa, World Health Organization, Geneva.

Mooi, F.R., Van Loo, I.H.M., and King, A.J., 2001, Adaptation of *Bordetella pertussis* to vaccination: a cause for its reemergence, *Emerg. Infect. Dis.* **7**:S526–S528.

Murphy, S.C., and Breman, J.G., 2001, Gaps in the childhood malaria burden in Africa: cerebral malaria, neurological sequelae, anemia, respiratory distress, hypoglycaemia, and complications of pregnancy, *Am. J. Trop. Med. Hyg.* **64**:57–67.

Nielsen, M.A., Staalsoe, T., Kurtzhals, J.A.L., Goka, B.Q., Dodoo, D., Alifrangis, M., Theander, T.G., Akanmori, B.D., and Hviid, L., 2002, *Plasmodium falciparum* variant surface antigen expression varies between isolates causing severe and nonsevere malaria and is modified by acquired immunity, *J. Immunol.* **168**:3444–3450.

Nielsen, M.A., Vestergaard, L.S., Lusingu, J.P., Kurtzhals, J.A.L., Giha, H.A., Grevstad, B., Goka, B.Q., Lemnge, M.M., Akanmori, B.D., Theander, T.G., and Hviid, L., 2004, Geographical and temporal conservation of antibody recognition of *Plasmodium falciparum* to variant surface antigens, *Infect. Immun.* **72**:3531–3535.

Pappenheimer, A.M., 1984, Diphtheria,in R. Germanier, ed., Bacterial Vaccines, Academic Press, US, pp. 1–36.

Raberg, L., De Roode, J.C., Bell, A.S., Stamou, P., Gray, D., and Read, A.F., 2006, The role of immune-mediated apparent competition in genetically diverse malaria infections, *Am. Nat.* **168**:41–53.

Read, A.F., Gandon, S., Nee, S., and Mackinnon, M.J., 2004, The evolution of parasite virulence in response to animal and public health interventions,in D. Dronamraj, ed., Evolutionary Aspects of Infectious Diseases, Cambridge University Press, Cambridge.

Read, A.F., and Taylor, L.H., 2001, The ecology of genetically diverse infections, *Science* **292**:1099–1102.

Reyburn, H., Mbatia, R., Drakeley, C.J., Bruce, J., Carneiro, I., Olomi, R., Cox, J., Nkya, W.M.M.M., Lemnge, M.M., Greenwood, B.M., and Riley, E.M., 2005, Association of transmission intensity and age with clinical manifestations and case fatality of severe *Plasmodium falciparum* malaria, *JAMA* **293**:1461–1470.

Snow, R.W., Craig, M., Deichmann, U., and Marsh, K., 1999, Estimating mortality, morbidity, and disability due to malaria among Africa's non-pregnant population, *B. World Health Organ.* **77**:624–640.

Topley, W.W.C., 1919, The spread of bacterial infection, *Lancet* **194**:1–5.

Turner, C.M.R., Aslam, N., and Dye, C., 1995, Replication, differentiation, growth and the virulence of *Trypanosoma brucei* infections, *Parasitology* **111**:289–300.

Williams, G.C., and Nesse, R.M., 1991, The dawn of Darwinian medicine, *Q. Rev. Biol.* **66**:1–22.

Witter, R.L., 1998, The changing landscape of Marek's disease, *Avian Pathol.* **27**:S47–S63.

The Ins and Outs of Host Recognition
of *Magnaporthe oryzae*

Sally A. Leong

Flor first proposed the gene-for-gene hypothesis to describe the relationship of races of the flax rust and cultivars of its flax host (Flor, 1955). In its simplest form, this hypothesis states that for every resistance gene in the host plant, there exists a complementary avirulence (cultivar specificity) gene in the pathogen that allows the host to recognize the pathogen and resist development of the diseased state. Since its first proposal, the gene-for-gene hypothesis has been found to be applicable to many host-pathogen interactions including that of the rice blast fungus *Magnaporthe oryzae* and its host *Oryza sativa* (e.g., Ellingboe, 1992; Leung et al., 1988; Silue et al., 1992a,b; Smith and Leong, 1994; Valent et al., 1991) as well as subspecific groups of *M. oryzae* and their respective hosts (Kato, 1983; Valent et al., 1986; Valent and Chumley, 1994). This fundamental relationship is of great practical interest as *M. oryzae* is rapidly able to overcome new disease resistances in rice soon after their deployment (Bonman et al., 1992). Moreover, *M. oryzae* and the closely related *Magnaporthe grisea* together exist as complex taxa with numerous subspecific groups that are sometimes interfertile but differ in their host range (Couch and Kohn, 2002; Kato et al., 2000; Valent et al., 1991). How these different subspecific groups interrelate evolutionarily is of great concern as some of these alternate hosts are frequently found growing in close proximity to or in rotation with rice, and *M. oryzae* isolates infecting these alternate hosts can sometimes also infect rice (Kato, 1983; Kato et al., 2000; Mackill and Bonman, 1986; Y. Jia, personal communication). In Japan, barley and rice are alternatively cultivated in the same field, and rice-infecting strains of *M. oryzae* also infect barley. Blast disease of barley is apparently minimized when *M. oryzae* infection is limited by cooler temperatures (Yukio Tosa, personal communication). With the advance of global warming, these environmental limitations of blast prevalence may soon disappear, as is already being seen with the emergence of blast in California. Pathogens and pests that were a problem only in more southern regions of the US are now becoming serious in northern US regions like Wisconsin (Judy Reith-Rozelle, personal

S.A. Leong
USDA, ARS CCRU, Department of Plant Pathology, University of Wisconsin, 1630 Linden Dr., Madison, WI 53706, 608-262-6309
e-mail: saleong@wisc.edu

J.P. Gustafson et al. (eds.), *Genomics of Disease*,
© Springer Science+Business Media, LLC, 2008

communication). This observation can be likely attributed to the warmer winters that have occurred in recent years.

The molecular bases of host/cultivar specificity and pathogenic variability in *M. orzyae* is only beginning to be understood with the cloning of the avirulence genes for cultivar specificity: *AVR2-YAMO/* (*AVR- Pi-ta*) (Orbach et al., 2000; Jia et al., 2000; Valent and Chumley, 1994), *AVRI-CO39* (Farman and Leong, 1998), and *AVR-IRAT7* (Böhnert et al., 2004), and an avirulence gene for host specificity *PWL2* (Sweigard et al., 1995) from rice pathogenic isolates of *M. orzyae*. These genes function as classical avirulence genes (Flor, 1955) by preventing infection of a specific cultivar or host. *AVR-IRAT7* encodes a polyketide synthetase (Böhnert et al., 2004) and other genes may be involved in cultivar specificity. By contrast, *AVR2-YAMO/(AVR-Pi-ta)* encodes a 223 amino acid protein with homology to a zinc metalloprotease (Jia et al., 2000) while *PWL2* encodes a 145 amino acid polypeptide that is glycine-rich (Sweigard et al., 1995); both genes are predicted to have secretion signals. Homologs of both genes appear to be widely distributed in rice as well as grass-infecting isolates of *M. grisea* (e.g., Kang et al., 1995; Valent and Chumley, 1994) and confirm the prediction obtained through genetic analysis that *M. grisea* isolates infecting monocots other than rice contain cultivar specificity genes for rice (Yaegashi and Asaga, 1981; Valent and Chumley, 1994; Valent et al., 1991) as well as host specificity genes that preclude infection of other hosts (Valent et al., 1986; Yaegashi, 1978). Rice pathogenic isolates of *M. orzyae* contain cultivar specificity genes for rice (see above) and other grass hosts (Kato et al., 2000) as well as host specificity genes that prevent infection of other monocot hosts (Valent and Chumley, 1994).

Both *AVR2-YAMO/(AVR-Pi-ta)* and *PWL2* are unusually unstable (Valent and Chumley, 1994). Molecular characterization of a number of mutant alleles has revealed point, deletion, or insertion mutations (Valent and Chumley, 1994). In one case, an inverted repeat transposon was found in the promoter of *AVR2-YAMO/(AVR-Pi-ta)*. These studies illustrate the potential array of mutational events that can lead to increased virulence and host range in *M. orzyae*. How frequently these events contribute to pathogenic variability in *M. orzyae* existing in a natural agroecosystem is beginning to be understood for the *AVR2-YAMO/(AVR- Pi-ta)* gene (Y. Jia, personal communication). The genome of *M. orzyae* contains a number of transposable elements (e.g., Nitta et al., 1997; Dobinson et al., 1993 Farman et al., 1996; Kachroo et al., 1994) and genomic alterations associated with transposon activity have been documented. Further examination of how these elements affect genome evolution and expression in the context of pathogenic variability deserves close attention.

The cloning and molecular characterization of numerous disease resistance genes from different plant species is beginning to provide clues on how plants perceive invading pathogens and how these genes may be evolving (e.g., Meyers et al., 1999; Nimchuk et al., 2001). The many common structural features of these genes, nucleotide binding sites (NBS) and leucine-rich repeats (LRR), suggest that common mechanisms of perception and transmission of the pathogen signals exist in these diverse plant species (McDowell and Dangl, 2000). The finding that functional homologs of disease resistance genes exist in different plants species (Dangl et al., 1992; Innes et al., 1993) and that the same disease resistance locus can show specificity for different *AVR* genes as well as pathogens further supports this conclusion (e.g., Bisgrove et al., 1994; Grant et al., 1995; Rossi et al., 1998; van der Vossen et al., 2000). Finally, expression of resistance genes from one species of plant can prevent infection of pathogens carrying the corresponding avirulence gene when introduced as a transgene in another species of plant (e.g., Tai et al., 1999; Whitham et al., 1996).

The structure of the rice *Xa21* (Song et al., 1995) and *Xa-1* (Yoshimura et al., 1998) loci and *Pi-b* (Wang et al., 1999) and *Pi-ta* (Jia et al., 2000) loci conferring resistance to bacterial blight or rice blast diseases, respectively, indicates that resistance genes from rice share these common NBS and LRR features. The recent genome sequence of rice variety Nipponbare has provided for the first time a comprehensive view of the number and distribution of the NBS/LRR of genes in rice (International Rice Genome Sequencing Consortium, 2005; Consortia for the Sequencing of Rice Chromosomes 11 and 12, 2005; Fritz-Laylin et al., 2005). The LRR domain was found 1589 times and was one of the 50 most frequently observed classes of domains in the rice genome (International Rice Genome Sequencing Consortium, 2005). Furthermore, the NBS–LRR class of genes was found to occur predominantly in clusters, which has allowed continuing evolution of the genes through recombination. The function of these candidate resistance genes remains unknown; however, not all LRR-type genes are expected to function exclusively or solely in disease resistance as genes controlling plant development are also known to have LRR (e.g., *Arabidopsis ERECTA*, Godiard et al., 2003; Fritz-Laylin et al., 2005; Arabidopsis CLAVATA, Rojo et al., 2002).

Recent comparative sequencing analysis of alleles of *AVR- Pi-ta* in *M. oryzae* isolates and that of *Pi-ta* in Oryza species has provided evidence to support a model of "trench warfare" for this host–parasite interaction. Most sequence diversity is found in the coding region of *AVR- Pi-ta* leading to functional alleles while the *Pi-ta* protein is extremely conserved with a single functional nucleotide polymorphism in a conserved protease motif that distinguishes resistant from susceptible alleles and the majority of sequence diversity localizing to the intron region (Jia, 2007; Jia et al., 2003).

The molecular genetic characterization of additional disease resistance genes for rice blast is needed to provide a better understanding of how rice recognizes *M. oryzae* and how changes in the corresponding *M. oryzae AVR* genes defeat this resistance, as well as to clarify the relationship of host and cultivar specificity. Moreover, the availability of cloned disease resistance genes to blast may facilitate the genetic identification and cloning of other genes conferring resistance to this disease in other cereals as well as provide insight on how rice resists infection by other host-specific groups of *M. oryzae*. The genomes of members of the Poaceae family, which includes the many hosts of *M. oryzae*, are highly syntenic and can be considered as essentially one genome (Bennetzen and Freeling, 1993; Devos, 2005). Disease resistance gene homologs have been mapped to syntenic locations across grass genomes (Leister et al., 1998). It will be interesting to learn whether homologs of disease resistance genes to blast are found in the same genomic location of other monocot hosts of *M. orzyae* and how these genes functionally relate. This information is also essential to the design of resistance to rice blast disease that is based on expression of *AVR* genes in host plants (Hammond-Kosack et al., 1994; Keller et al., 1999) and may lead to the identification of new disease resistance loci in these allied monocot hosts. As noted above, homologs of *AVR2-YAMO/(AVR-Pi-ta)* and *PWL2* are broadly distributed in subspecific groups of *M. orzyae* (Couch et al., 2005; Valent and Chumley, 1994). In some cases these have been shown to be functional and to exhibit the same host or cultivar specificity as *AVR2-YAMO/(AVR-Pi-ta)* or *PWL2* (Kang et al., 1995). Whether other alleles confer new host/cultivar specificities will be important to learn. By analogy, homologs of the corresponding disease resistance genes may specify novel resistances that are recognized by these cultivar and host specificity gene alleles. The availability of cloned cultivar and host specificity genes from *M. orzyae* and the corresponding disease resistance genes provides an experimental avenue to test this hypothesis.

Over the last several years, my laboratory has been engaged in the development of a genetic map and molecular karyotype for *M. orzyae* (Farman and Leong, 1995; Nitta et al., 1997; Skinner et al., 1993), the identification and genetic mapping of a cultivar specificity gene *AVR1-CO39* to rice cultivar CO39 (Smith and Leong, 1994), and the map-based cloning of *AVR1-CO39* and the phylogenetic distribution and structure of its alleles (Farman et al., 2002; Tosa et al., 2005). Our published and unpublished work in these areas are presented in the following sections.

1 Sequence Analysis of the AVR1-CO39 Locus

The minimum region of the *AVR1-CO39*-containing cosmid able to confer avirulence when transformed into virulent strain Guy11 was 1.05 kb (Farman and Leong, 1998). DNA sequence analysis of this fragment for open reading frames (ORF) of at least 25 amino acids, beginning with a methionine and having a fungal consensus sequence for the start of translation surrounding the ATG (Ballance, 1991) revealed eight ORFs (M. Farman, N. Punekar, D. Lazaro, and S. A. Leong, unpublished findings). No evidence for paired intron splice sites flanking lariat motifs could be found in the sequence. *Avr9* from *Cladosporium* is only 63 amino acids (Van den Ackerveken et al., 1992) and *AVR2-YAMO/ (AVR- Pi-ta)* and *PWL2* of *M. orzyae* encode small polypeptides specified by ORFs with multiple exons. Thus, it is not unreasonable for any of these ORFs to encode all or part of *AVR1-CO39*.

Orf3 encodes the largest predicted polypeptide in this region having 89 amino acids and has been the focus of our attention. Orf3 has a codon use which is similar to other *M. orzyae* genes, contains a putative signal sequence that is predicted with a probability of 0.999 (Fig. 1) by Signal P analysis (Nielsen and Krogh 1998; Nielsen et al., 1997), has a good match of sequence surrounding the initiator ATG with other fungal genes, and is flanked at the 5' end by a carbon catabolite regulatory sequence (GTCCATTTA) and at the 3' end by a post-transcriptional regulatory sequence

```
CAACGTACTAGAAATGACTAATAAGTACCCAGTCAAGTCAACTTGCTGTAGTATTATATTTAACGAAGCGTCCATTTACTGCCAGG
GCAAGTTTATCAATGGGACCAGTGTTCTCCCTCCTCTGGACAACTCAGTTCTTTGCAAACCCTAGACAGTCTACCTCTCTGCCACC
ATTTTTACTTTTCAAAAATTTACTCCTTGCCGCTACTGAAACTTCTACAATTGAAAGAGCCCACAATGAAAGTCCAAGCTACATTC
GCCACCCTTATCGCCCTTGCGGCTTACTTTCCAGCAGCCAATGCTTGGAAAGATTGCATCATCCAACGTTATAAAGACGGCGATGT
CAACAACATATATACTGCCAATAGGAACGAAGAGATAACTATTGAGGAATATAAAGTCTTCGTTAATGAGGCCTGCCATCCCTACC
CAGTTATACTTCCCGACAGATCGGTCCTTTCTGGCGATTTTACATCAGCTTACGCTGACGACGATGAGTCTTGTTGATCAATAAGA
GTCCAGGTTGAAAAATTCGCCACCATGGTAATAGAGGGTTATTTATCTCGGAATAGCAGCCGTGTGTGCAATTATCACGGCTGTTC
CTCTGCGATAGGGATATTAGAAGCAGGACAAATTTACGGCAATAGCAACCAATTGTCCTTGTCTATGGATTCGCCCGTCGAATGGA
GGCGACGGCGGATCC
```

Fig. 1 DNA sequence of the Orf3-containing region of *AVR1-CO39*

(TTATTTAT) for inflammatory response in mammalian cells (Caput et al., 1986) (Fig. 1). Analysis of mutant forms of orf3 carrying mutations in the ATG start codon or a frame-shift mutation indicate that this orf or its overlapping sequences are critical to the function of *AVR1-CO39* in infection assays. Analysis of orf3 using BlastP revealed a potential, but limited relationship with several mammalian vasopressin receptors as well as shrimp sarcoplasmic calcium-binding protein and a mast cell protease. Closer examination of these identities suggest that orf3 may contain an EF-hand-type calcium-binding site (Bateman et al., 2004); however, 2 out of 13 amino acid residues vary from the prototype consensus sequence (Fig. 2).

Analysis of the full length *AVR1-CO39* transcript by 5' and 3' RACE revealed that Orf3 is transcribed only *in planta* (data not shown, Lazaro and Leong, unpublished findings) and includes 65 bp that extends beyond the 1.05 kb fragment to the 3' end of orf3. No evidence for transcription of the other orfs was found under the conditions used.

Fusion of the C terminal–coding region of orf3 to a *GFP* reporter gene and introduction into an expression vector with the CAMV 35S promoter enabled expression of the fungal gene in rice leaf cells bombarded with these constructs (Fig. 3; Lazaro and Leong, unpublished findings). As a control for efficiency of transformation, the GUS gene was co-bombarded. *GFP* alone was bombarded as a negative control. Resistant rice lines showed a marked reduction of the ratio of *GFP*-expressing cells to GUS-expressing cells when compared to susceptible rice 48 hours postbombardment using the mature form of orf3 (Fig. 3). The full-length form of orf3 was poorly detected in both resistant and susceptible germplasm and may relate the prediction that the secretion signal would direct the protein to the vacuole (data not shown). This finding is similar to that obtained with the *AVR- Pi-ta* gene where full length *AVR-Pi-ta* gene was toxic to yeast, bacteria, and host, experimental results were difficult to understand (Bryan et al., 2000; Jia et al., 2000).

```
X Y Z  -Y-X  -Z
TAN RNE EITIEEY        orf3

DFNKDGEVTVDEF          Sarcoplasmic Calcium Binding
Protein
```

Fig. 2 DNA sequence of Orf3 putative EF hand. Consensus sequence is shown in dark grey, Orf3 sequence in light grey, and sarcoplastic calcium-binding protein in black. Atypical amino acids differing from consensus are shaded in gray

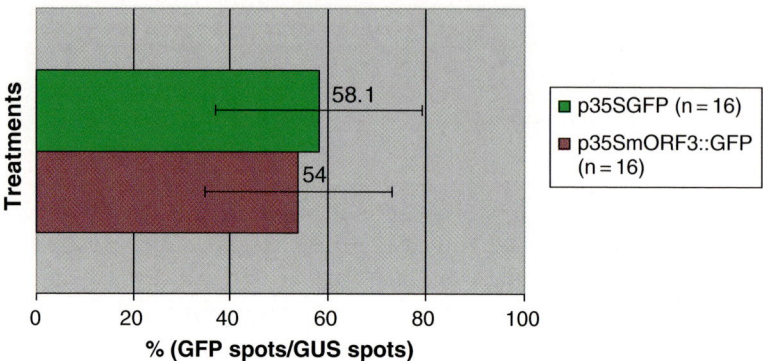

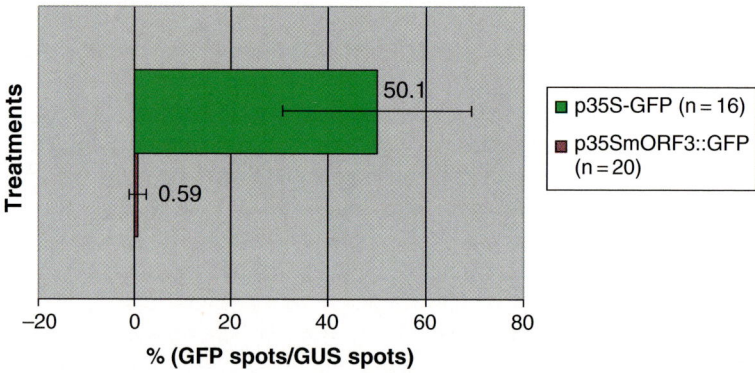

Fig. 3 (**A**) The susceptible rice variety *M202* was bombarded with gold micro-carrier treatments designed to deliver p35SmORF3::*GFP* or p35S-*GFP* on separate fractions of gold particles mixed with a second fraction of pAHC25-coated gold particles (Lazaro, D, 2003). No significant difference in the percentage of GUS-normalized *GFP* expressing cells was observed between p35S-*GFP* and p35SmORF3::*GFP* in rice variety *M202*. (**B**) When the same gold micro-carrier treatments were used on the resistant variety *DREW*, a significant reduction in percent activity was observed for p35SmORF3::*GFP* treatments than for the p35S-*GFP* treatment. Similar tests with rice varieties 51583 (S) and CO39 (R) demonstrated that the mORF::*GFP* fusion is recognized by a resistant rice variety known to recognize *AVR1*-CO39 expressed from *M. orzyae*

Based on these analyses orf3 encodes the active function in *AVR1-CO39*. All AVR genes characterized to date from *M. orzyae* give rise to a putative-secreted product (Böhnert et al., 2004; Sweigard et al., 1995; Jia et al., 2000). Finally, functional alleles of *AVR1-CO39* having the same specificity of *AVR1-CO39* but isolated from *M. orzyae* infecting other grasses have the identical orf3-translated sequence (see below). By contrast, non-functional alleles of *AVR1-CO39* from rice isolates lack an orf3 sequence or contain a mutated form of orf3 (see below).

2 Distribution of AVR1-CO39-Like Sequences in Grass-Infecting Isolates of M. orzyae

A hybridization survey of genomic DNA from 85 different phylogenetically characterized *M. orzyae* isolates with the 1.05 kb probe containing *AVR1-CO39* and by PCR using flanking and internal primers demonstrated that few rice-infecting isolates contained homologous sequences while all other cereal-infecting isolates showed strong hybridization signals and PCR products (Tosa et al., 2005) (Figs. 4 and 5). This survey also identified homologs in isolates infecting turf grasses. Similar findings were obtained by Peyyala and Farman (2006). These isolates are causing

```
MKVQATFATLIALAAYFPAANARKDCVIQRYKDGDVDNIYTANRNEMITIEEYKVFVNEACHPYPVILPDKSVLSGDFTSAYADDDESC
MKVQATFATLIALAAYFPAANAWKDCIIQRYKDGDVNNIYTANRNEEITIEEYKVFVNEACHPYPVILPDRSVLSGDFTSAYADDDESC
MKVQATFATLIALAAYFPAANAWKDCIIQRYKDGDVNNIYTANRNEEITIEEYKVFVNEACHPYPVILPDKSVLSGDFTSAYADDDESC
MKVQATFATLIALAAYFPAANAWKDCIIPRYKDGDVNNMYTANRNEEITIEEYKVFVNEACHPYPVILPDRSVLSGDFTSAYADDDESC
MKVQATFATLIALAAYFPAANAWKDCIIQRYKDGDVNNIYTANRNEEITIEEYKVFVNEACHPYPVILPDRSVLSGDFTSAYADDDESC
MKVQATFATLIALAAYFPAANAWKDCIIQRYKDGDVNNIYTANRNEEITIEEYKVFVNEACHPYPVILPDRSVLSGDFTSAYADDDESC
MKVQATFATLIALAAYFPAANAWKDCIIQRYKDGDVNNIYTANRNEEITIEEYKVFVNEACHPYPVILPDRSVLSGDFTSAYADDDESC
```

Fig. 4 Sequence alignment of predicted translation products of orf3 from rice grass-infecting strains of *M. orzyae*. Listed in order: *AVR1-CO39* alleles from strains infecting *Avena sativa*, *Lolium perenne*, *Eleusine corana*, *Setaria Italica*, *Eleusine coracana*, *Panicum milliaceum*, and *Triticum aestivum*. *Italic characters* show putative signal sequence. *Grey characters* show potential EF hand Ca-binding motif. Other colored characters show varied amino acids relative to the original reference weeping love grass allele *AVR1-CO39*

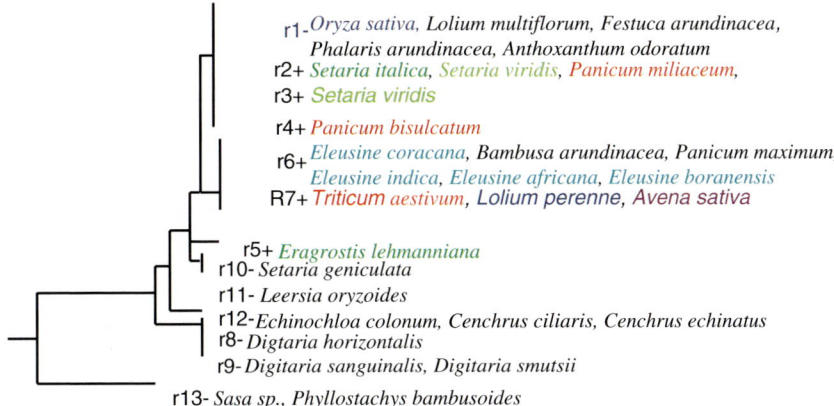

Fig. 5 Phylogeny of host-specific forms of *M. orzyae* (Kato et al., 2000) based on 85 isolates with analysis for *AVR1-CO39* by Southern hybridization and PCR tests (Tosa et al., 2005). Ribosomal groups (r) of *M. orzyae* and presence (+) or absence (−) of a *AVR1-CO39* allele are shown. Colored hosts show host-specific forms investigated belonging to *Pyricularia oryzae*

widespread death of turf in golf course fairways in Japan and in the central and northeastern United States.

3 Structure of AVR1-CO39 in Non-rice-Infecting Isolates of M. orzyae

A 4 kb region containing the *AVR1-CO39* locus was cloned and sequenced from non-rice-infecting isolates of *M. orzyae* by our collaboration with Yukio Tosa and his coworkers (Figs 4 and 5, Tosa et al., 2005). The DNA sequences of orf3 from these isolates were over 95% identical to the original allele cloned from a rice-infecting laboratory strain 2539 (Leung et al., 1988). Orf3 polypeptides of these alleles were identical or contained one or more amino acid changes relative to the *AVR1-CO39* orf3 (Fig. 4). A more limited analysis revealed similar results by Payyela and Farman.

At least two of these are known to map in the same location on chromosome 1 of *M. oryzae* as *AVR1-CO39* from strain 2539 (Y. Tosa, personal communication). The laboratory of Yukio Tosa has constructed an RFLP map of non-rice-infecting isolates of *M. oryzae* using an F1 population derived from a cross between GFSI1-7-2 (Setaria isolate) and Br48 (Triticum isolate) (Y. Tosa, personal communication). Chromosome number was assigned using the chromosome-specific probes of a map for rice-infecting *M. oryzae* (Nitta et al., 1997). *AVR1-CO39* was located on a region of chromosome 1, which was syntenic to that in the rice-infecting isolate's map. To determine which chromosome *AVR1-CO39* is associated with in the other subgroups of *M. oryzae*, chromosomes of isolates from wheat, foxtail millet, finger millet, common millet, and perennial ryegrass were separated by CHEF gel electrophoresis, transferred to a membrane and hybridized with *AVR1- CO39*. In most of these isolates, *AVR1-CO39* was located on chromosome 1 (Y. Tosa, personal communication).

Functional analysis of these alleles by transformation of the virulent rice isolate PO-12-7301-2 showed that these alleles were fully functional having the same apparent specificity as *AVR1-CO39* (Fig. 6) (Tosa et al., 2005). Similar results were

Fig. 6 Pathogenicity of pathogenic rice isolate (PO-12-7301-2)(*left*) and its transformant (PAS1-3-1)(*right*) with the *AVR1-CO39* homolog derived from a *Setaria* isolate on rice cultivars, CO39 (R) and Yashiromochi (S), 7 days after inoculation

obtained by Payyela and Farman. However, the function of these alleles in non-rice isolates is unknown. They may function in host resistance to rice but few rice lines carry the corresponding *PiCO39 (t)* (Tosa et al., 2005; S. Leong, unpublished findings). Recent work has suggested evidence for race-specific resistance in barley to the blast fungus indicating that *AVR1-CO39* might function in race-specific resistance in other cereals (Chen et al., 2003). It will be interesting to learn if homologs of *Pi-CO39 (t)* in barley exist and if they are involved in blast resistance.

4 Structure of AVR1-CO39 Locus in Rice Isolates of M. oryzae

Further analysis of the structure of the *AVR1-CO39* locus from 45 rice-infecting isolates from several geographical regions indicated that this region has undergone a number of changes involving insertion of transposable elements, single base pair changes, and deletion events (Farman et al., 2002) giving rise to null alleles. At least three classes of alleles were discovered through PCR and DNA sequence analyses (Fig. 7). In all cases, orf3 was either deleted completely or had undergone mutation leading to an extension of the polypeptide as a result of a nonsense mutation. These data suggest that this locus has been under strong selection for loss of function and that these changes may have occurred early in or prior to the evolution of the rice lineage of *M. oryzae* since other lineages of cereal-infecting isolates all contain a functional copy of the gene.

5 Genetic and Physical Mapping of the Pi-CO39 (t) Locus

Crosses were made between the rice lines CO39 (resistant) and 51583 (susceptible) to map the *Pi-CO39 (t)* gene. The resistance phenotype in resistant offspring was reaction type 1, while that of susceptible seedlings was reaction type 4 (range of

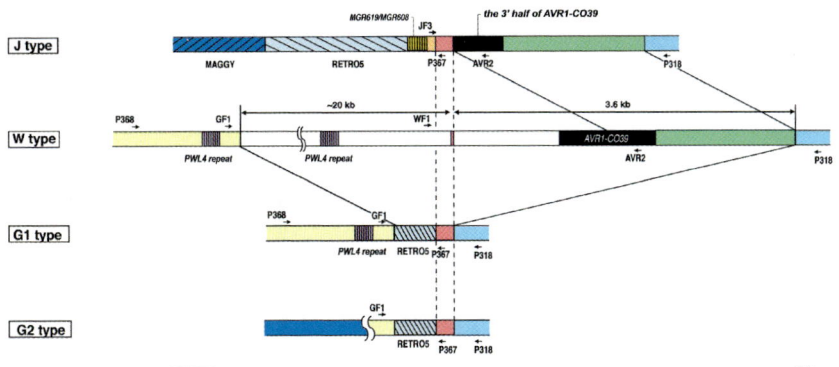

Fig. 7 Structure of *avr1-CO39* loci from rice-infecting isolates of *M. orzyae* (J, G1, G2 types) compared with a weeping love grass allele (W) (Farman et al., 2002). The 1.05 kb *AVR1-CO39* gene region is shown in *black*. The G1-type is found in Guy11

typing scale 0–4). Both of the reciprocal F_1s tested (i.e., CO39 X 51583 or 51583 X CO39) were resistant to *M. orzyae* strain 6082 that contains *AVR1-CO39*, indicating that resistance is controlled by a dominant locus in the nuclear genome of CO39. Resistance to *M. orzyae* strain 6082 in 604 F_2 progenies derived from three different F1 plants segregated in a 3:1 ratio, characteristic of single dominant locus. Resistance to a Guy 11 (*AVR1-CO39*) transformant in one F_2 population consisting of 235 F_2 progenies also segregated as a single dominant locus. Inheritance of resistance was also confirmed in F_3 families derived from the F_2 populations used for mapping the resistance locus. Testing of 78 F_3 families with the Guy11 (*AVRI-CO39*) transformant and 59 F_3 families with the *M. orzyae* progeny 6082 also showed that resistance is controlled by a single dominant locus. The segregation ratio of F3 families was all resistant: segregating for resistance: all susceptible; and fit a 1:2:1 ratio. The disease resistance locus in CO39 was designated as *Pi-CO39(t)* (Chauhan et al., 2002).

6 Comparative DNA Sequence Analysis of Resistant and Susceptible Cultivars at the Pi-CO39 (t) Locus

A large insert DNA (100 kb average insert size) library of CO39 was constructed in a binary plant transformation cosmid vector pCLD04541 and screened with co-segregating marker RGA38 as well as BAC end sequence probes from the Nipponbare BAC libraries of Clemson University Institute for Genomics. Single BAC clones hybridizing to co-segregating markers were identified in the Nipponbare (susceptible) BAC library.

The DNA sequence of 6 BAC clones (phrap score >30) at the *Pi-CO39 (t)* locus was determined in collaboration with Frederick Blattner of the University of Wisconsin, Madison. Comparative sequence analysis of blast resistant (CO39 indica) and susceptible (Nipponbare japonica) rice cultivars at genomic regions co-segregating with *Pi-CO39(t)* showed that these two haplotypes have diverged with respect to the relative type, number of paralogs, and the orientation and location of resistance gene homologs within each cluster (Fig. 8). Several disease resistance (NBS–LRR)-like genes were identified with similarity to the other rice resistance-associated genes *RPR1, Xa1, Pi-ta,* and *Pib. RPR1* (rice probenazole-responsive) is associated with probenazole-induced resistance to blast in rice (Sakamoto et al., 1999); *Xa1* confers resistance to race 1 of bacterial blight (*Xanthomonas oryzae* pv. *oryzae*) (Yoshimura et al., 1998); *Pi-ta* is a major rice blast resistance gene related to *RPM1* of *Arabidopsis* (Boyes et al., 1998; Bryan et al., 2000); *Pib* is another major rice blast resistance gene (Wang et al., 1999). All four genes belong to a non-TIR NBS–LRR subfamily of plant disease resistance genes (S.A. Leong, R. Chauhan, T. Durfee, F. Blattner, and J. Holt, unpublished data; Consortia for the Sequencing of Rice Chromosomes 11 and 12, 2005; and Meyers et al., 1999). Polymorphic regions were found in LRR units of *Pi-CO39*, whether one of these LRR units is the determinant of resistance in CO39 is under investigation.

Currently, the gene identified as *COR8* is the most promising candidate in this region based on its presence by PCR and Southern blot analysis in all resistant germplasm tested and preliminary complementation studies by transient expression. *COR8* is predicted to be the product of a recombination event involving three LRR-type genes or pseudogenes in the susceptible haplotype (Fig. 8).

Other candidate genes exist at the *Pi-CO39 (t)* locus. *COR6* is a unique NBS–LRR type gene present only in the CO39 haplotype. Receptor kinases are present in the locus and can also function as resistance genes (Becraft, 1998; Brueggeman et al., 2002; Nimchuk et al., 2001). Finally, a new non-LRR-type, class of resistance gene was recently described from *Arabidopsis* that shares common signaling pathways with NBS–LRR type genes (Xiao et al., 2005), and the pepper *Bs3* resistance gene was reported to be a flavin monooxygenase (*Genetic Dissection of the Interaction Between the Plant Pathogen* Xanthomonas campestris pv. vesicatoria *and its Host Plants*) requiring caution in assignment of gene function at the *Pi-CO39 (t)* locus based on classic structural features of plant disease resistance genes.

The LRR of four members of the *RPR1* family in this region including *COR8* were compared and revealed that most variation occurs outside of the inner core of the LRR, which is attributed to specificity in other LRR-containing proteins. In addition, the *COR8* gene has two extra LRR that are separated by a putative transmembrane domain (Fig. 9).

Cysteine proteases have been clearly associated with disease resistance in plants (e.g., Coaker et al., 2005; Holt et al., 2005; Kruger et al., 2002; Solomon et al., 1999), while serpins (serine/cysteine protease inhibitors) have been associated with innate immunity in insects (Huntington 2006). In Drosophila, the serine protease Persephone cleaves a Toll ligand named Spatzel, thus activating binding to Toll, which in turn signals the production of antifungal peptides. The serpin Necrotic inhibits Persephone (Fig. 10). Protease mutants have increased fungal infections while serpin mutants are melanized at wound sites and die early in life. These observations are of interest in the context of finding that the rice pathogen's *AVR1-CO39* has evolved by insertion/deletion and nucleotide changes

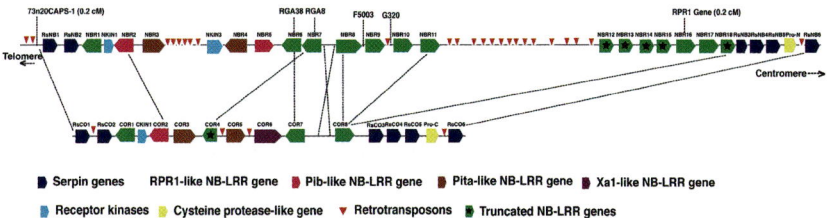

Fig. 8 Comparative analysis of CO39 (Indica) (*bottom*) and Nipponbare (Japonica) (*top*) genomes at regions co-segregating with *Pi-CO39(t)*. Shown are R_SNB (Nipponbare serpin-like genes), R_SCO (CO39 serpin- like genes), NBR (Nipponbare NBS–LRR disease resistance-like genes), COR (CO39- NBS–LRR disease resistance-like genes), NKin (Nipponbare kinase-like genes), CKin (CO39 kinase-like genes), ProN (Nipponbare protease genes), ProC (CO39 protease genes). Putative orthologous gene relationships are shown with *dotted lines*

```
SILSESKYL   TVLELQD   SDITEVP
ACIGKLFNL   RYIGLRR   TRLCSLP
ESIEKLSNL   QTLDIKQ   TKIE
KLPRGITKI   KKLRHLL   ADRYEDEKQSVFRYF
IGMQAPKDL   SKLEELQ   TLETVEASKDLA
EQLKELMQI   RSIWIDN   ISSADCGNIF
ATLSTMPLL   SSLLLSA   RDENEPLCFE
ALQPMSKEL   HRLIIRG   QWAKGTLDYP
IFRSHTTHL   KYLALSW   CNLGEDPLG
MLASHLSNL   TYLRLNN   VHSSKTLV
LDAEAFPHL   KTLVLMH   MPDVNQIN
ITDGALPCI   EGLYIVS   LWKLDKVP
QGIESLASL   KKLWLKD   LHKDFKTQWKGDGMHQKMLHVAELK

IISGALPVI   EGLYIVA   LSGLESVP
PGIETLRTL   KKLWLVG   LHWDFEAHWIESEMDQKMADCVGD
```

Fig. 9 Consensus LRR domain for two putative orthologous gene pairs (*COR7/NBR6*; *COR8/NBR11*) of the *RPRI* family. All family members have 13 LRR with the exception of *COR8*, which has two additional LRR (*underlined*) separated by a putative transmembrane domain. *Black*, all members; other colors have three or less members with the same amino acid with reference to *COR8*

to become non-functional (Fig. 7). Thus no diversifying selection would have been applied to the *PiCO39 (t)* resistance locus in rice in recent times (e.g., Lavashina et al., 1999; Tang et al., 2006). Like some disease resistance genes in plants, Toll is a membrane-associated NBS–LRR type protein. This suggests the possibility that rice serpins may play a role in regulating processing of the fungal AVR proteins and/or plant host components required for resistance (Fig. 11).

Future work will focus on validation of the role of *COR8* in transgenic rice plants and the identification of all relevant functional component(s) of the *Pi-CO39(t)* locus of rice through transient expression with *AVR1-CO39* in F-BMV

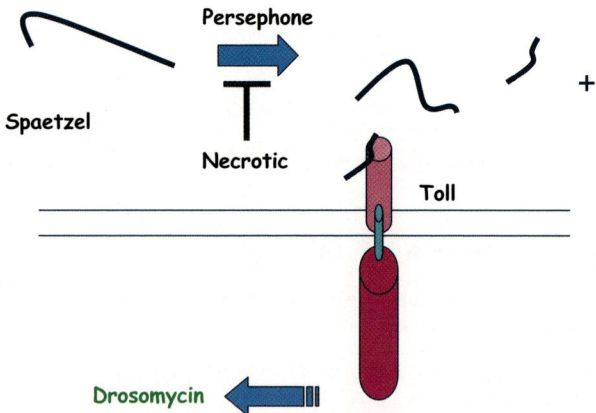

Fig. 10 Innate immunity for fungal infection in Drosophila involves the serine protease Persephone, which cleaves and activates the Toll ligand Spatzel. Necrotic is a serpin, serine protease inhibitor, which specifically modulates the activity of Persephone

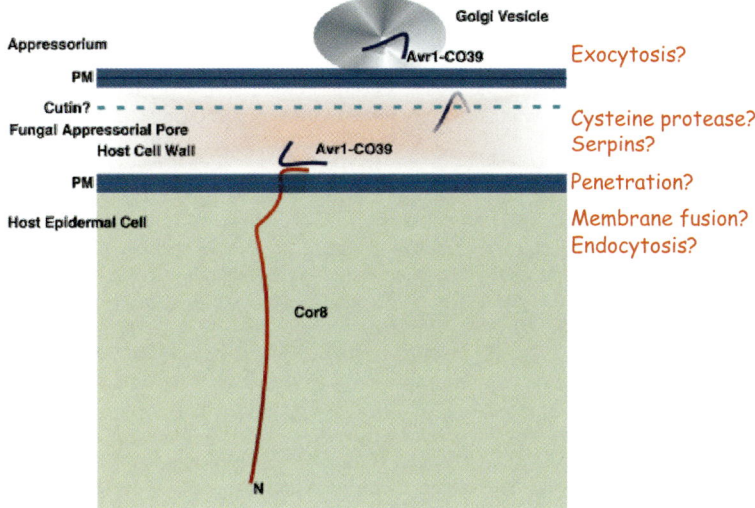

Fig. 11 Hypothetical model of the putative functional components in signaling of *AVR1-CO39* at the appressorial pore of *M. oryzae* on a resistant rice plant

(Ishikawa et al., 1997; Rouf Mian et al., 2005) or plasmid expression vectors (Leister et al., 1996), gene silencing (Miki and Shimamoto, 2004; Miki et al., 2005), and cytological examination (e.g., Koga, 1994). The results of this work will be applied to improvement of rice and other cereals in an effort to reduce reliance on chemical control (Bonman, 1998; Koizumi, 1998) for blast disease control and to improve the durability of genetic resistance to this widespread and often economically devastating disease (Baker et al., 1997).

Acknowledgments The author wishes to thank all former lab members and cooperators, in particular Mark Farman and Yukio Tosa, for their many contributions to this work, and Yulin Jia and Bob Fjellstrom for their critical reading of the manuscript. This work was supported by USDA CRIS project 3655-22000-015-00D, the Graduate School of the University of Wisconsin, grants from the Rockefeller Foundation to SAL, and a fellowship from the Department of Biotechnology of India to Rajinder S. Chauhan.

References

Baker, B., Zambryski, P., Staskawicz, B., and Dinesh-Kumar, S.P., 1997, Signaling in plant-microbe interactions, *Science* **276**:26–733.

Ballance, J., 1991, Transformation systems for filamentous fungi and an overview of fungal gene structure, in S. A. Leong and R. Berka, eds., Molecular Industrial Mycology: Systems and Applications for Filamentous Fungi, Marcel Dekker, New York, pp. 1–29.

Bateman, A., Coin, L., Durbin, R., Finn, R.D., Hollich, V., Griffiths-Jones, S., Khanna, A., Marshall, M., Moxon, S., Sonnhammer, E.L., Studholme, D.J., Yeats, C., and Eddy, S.R., 2004, The Pfam protein families database, *Nucleic Acids Res.* **32**(Database issue):D138–D141.

Becraft, P.W., 1998, Receptor kinases in plant development, *Trends Plant Sci.* **3**:384–388.

Bennetzen, J.L., and Freeling, M., 1993, Grasses as a single genetic system: genome composition, collinearity and compatibility, *Trends Genet* **9**:259–260.

Bisgrove, S.R., Simonich, M.T., Smith, N.M., Sattler, A., and Innes, R.W., 1994, A disease resistance gene in *Arabadopsis* with specificity for two different pathogen avirulence genes, *Plant Cell* **6**:927–933.

Böhnert, H.U., Fudal, I. Dioh, W., Tharreau, D., Notteghem, J.L., and Lebrun, M.H., 2004, A putative polyketide synthase/peptide synthetase from *Magnaporthe grisea* signals pathogen attack to resistant rice, *Plant Cell* **16**:2499–2513.

Bonman, M., 1998, Rice disease management: industry approaches and perspectives. Abstract 3.6.7S of the International Congress on Plant Pathology, Edinburgh.

Bonman, M., Khush, G.S., and Nelson, R.J., 1992, Breeding rice for resistance to pests, *Annu. Rev. Phytopathol.* **30**:507–528.

Boyes, D.C., Nam, J., and Dangl, J.L., 1998, The *Arabidopsis thaliana RPM1* disease resistance gene product is a peripheral plasma membrane protein that is degraded coincident with the hypersensitive response, *Proc. Natl. Acad. Sci. USA* **95**:15849–15854.

Bryan, G.T., Wu, K.S., Farrall, L., Jia, Y.L., Hershey, H.P., McAdams, S.A., Faulk, K.N., Donaldson, G.K., Tarchini, R., and Valent, B., 2000, A single amino acid difference distinguishes resistant and susceptible alleles of the rice blast resistance gene *Pi-ta*, *Plant Cell* **12**:2033–2045.

Brueggeman, R., Rostoks, N., Kudrna, D., Kilian, A., Han, F., Chen, J., Druka, A., Steffenson, B., and Kleinhofs, A., 2002, The barley stem rust-resistance gene *Rpg1* is a novel disease-resistance gene with homology to receptor kinases, *Proc. Natl. Acad. Sci. USA* **299**:9328–9333.

Caput, D., Beutler, B., Hartog, K., Thayer, R., Brown-Shimer, S., and Cerami, A., 1986, Identification of a common nucleotide sequence in the 3′-untranslated region of mRNA molecules specifying inflammatory mediators, *Proc. Natl. Acad. Sci. USA* **83**:1670–1674.

Chauhan, R.S., Farman, M.L., Zhang, H.-B., and Leong, S.A., 2002, Genetic and physical mapping of a rice blast resistance locus, *Pi-CO39(t)*, Corresponding to *AVR1-CO39* of *Magnaporthe grisea*, *Mol. Genet. Genomics* **267**:603–612.

Chen, H., Wang, S., Xing, Y., Xu, C., Hayes, P.M., and Zhang, Q., 2003, Comparative analyses of genomic location of specificities of loci for quantitative resistance to *Pyricularia grisea* in rice and barley, *Proc. Natl. Acad. Sci. USA* **100**:2544–2549.

Coaker, G., Falick, A., and Staskawicz, B., 2005, Activation of a phytopathogenic bacterial effector protein by a eukaryotic cyclophilin, *Science* **308**:548–550.

Couch, B.C., Fudal, I., Lebrun, M-H., Tharreau, D., Valent, B., van Kim, P., Notteghem, J.-L., and Kohn, L. M., 2005., Origins of host-specific populations of the blast pathogen, Magnaporthe oryzae, in crop domestication with subsequent expansion of pandemic clones on rice and weeds of rice, *Genetics* **170**:613–630.

Couch, B.C., and Kohn, L.M., 2002, A multilocus gene genealogy concordant with host preference indicates segregation of a new species, *Magnaporthe oryzae*, from *M. grisea, Mycologia*, **94**:683–693.

Dangl, J.L., Ritter, C., Gibbon, M.J., Mur, L.A.J., Wood, J.R., Goss, S., Mansfield, J., Taylor, J.D., and Vivian, A., 1992, Functional homologs of the *Arabadopsis RPM1* disease resistance gene in bean and pea, *Plant Cell* **4**:1359–1369.

Devos, K.M., 2005, Updating the 'crop circle,' *Curr. Opin. Plant Biol.* **8**:155–162.

Dobinson, K., Harris, R., and Hamer, J.E., 1993, Grasshopper, a long terminal repeat (LTR) retroelement in the phytopathogenic fungus *Magnaporthe grisea, Mol. Plant Microbe Interact.* **6**:114–126.

Ellingboe, A., 1992, Segregation of avirulence/virulence on three rice cultivars in 16 crosses of *Magnaporthe grisea, Phytopathology* **82**:597–601.

Farman, M.L., Eto, Y., Nakao, Y., Tosa, Y., Nakayashiki, H., Mayama, S., and Leong, S.A., 2002, Analysis of the structure of the Avr1-CO39 avirulence locus in virulent rice-infecting isolates of *Magnaporthe grisea, Mol. Plant Microbe Interact.* **15**:6–16.

Farman, M.L., and Leong, S.A., 1995, Physical and genetic mapping of telomeres of *Magnaporthe grisea*, *Genetics* **140**:479–492.

Farman, M.L., and Leong, S.A., 1998, Chromosome walking to the *AVR1-CO39* avirulence gene of *Magnaporthe grisea*: discrepancy between the physical and genetic maps, *Genetics* **150**: 1049–1058.

Farman, M.L., Tosa, Y., Nitta, N., and Leong, S.A., 1996, MAGGY, a retrotransposon in the genome of the rice blast fungus *Magnaporthe grisea*, *Mol. Gen. Genet.* **251**:665–674.

Flor, H.H., 1955. Host-parasite interaction in flax rust: Its genetics and other implications, *Phytopathology* **45**:680–685.

Fritz-Laylin, L.K., Krishnamurthy, N., Tor, M., Sjolander, K.V., and Jones, J.D.G., 2005, Phylogenomic analysis of the receptor-like proteins of rice and Arabidopsis, *Plant Physiol.* **138**: 611–623.

Godiard, L., Sauviac, L., Torri, K.U., Grenon, O., Mangin, B., Griimsley, N.H., and Marco, Y., 2003, ERECTA, an LRR receptor-like kinase protein controlling development pleiotropically affects resistance to bacterial wilt, *Plant J.* **36**:353–365.

Grant, M.R., Godiard, L., Straube, E., Ashfield, T., Lewald, J., Sattler, A., Innes, R.W., and Dangl, J.L., 1995, Structure of the Arabidopsis *RPM1* gene enabling dual specificity disease resistance, *Science* **269**:843–846.

Hammond-Kosack, K.E., Staskawicz, B.J., Jones, J.D.G., and Baulcombe, D.C., 1994, Functional expression of a fungal avirulence gene from a modified potato virus X genome, *Mol. Plant Microbe Interact.* **8**:181–185.

Holt, B.F., Belkahdir, Y., and Dangl, J.L., 2005, Antagonistic control of disease resistance protein stability in the plant immune system, *Science* **309**:929–932.

Huntington, J.A., 2006, Shape-shifting serpins – advantages of a mobile mechanism, *Trends Biochem. Sci.* **31**:427–435.

Innes, R.W., Bisgrove, S.R., Smith, N.M., Bent, A.F., Staskawicz, B.J., and Liu,Y-C., 1993, Identification of a disease resistance locus in *Arabidopsis* that is functionally homologous to the RPG1 locus of soybean, *Plant J* **4**:813–820.

International Rice Genome Sequencing Project, 2005, The map-based sequence of the rice genome, *Nature* **436**:793–800.

Ishikawa, M., Janda, M., Krol, M.A., and Ahlquist, P., 1997, In vivo DNA expression of functional brome mosaic virus RNA replicons in *Saccharomyces cerevisiae*, *J. Virology.* **71**:7781–7790.

Jia, Y., 2007, Plants and pathogens engage in trench warfare-knowledge learned from natural variation of rice blast resistance gene *Pi-ta*. Abstract PAGXV, San Diego, CA, January 2007.

Jia, Y., Bryan, G.T., Farrall, L., and Valent, B., 2003, Natural variation at the Pi-ta rice blast resistance locus, *Phytopathology* **93**:1452–1459.

Jia, Y., McAdams, S.A., Bryan, G.T., Hershey, H.P., and Valent, B., 2000, Direct interaction of resistance gene and avirulence gene products confers rice blast resistance, *EMBO J.* **19**: 4004–4014.

Kachroo, P., Leong, S.A., and Chattoo, B.B., 1994, Pot2, an inverted repeat transposon from the rice blast fungus *Magnaporthe grisea*, *Mol. Gen. Genet.* **245**:339–348.

Kang, S., Sweigard, J., and Valent, B., 1995, The PWL host-species specificity gene family in the blast fungus *Magnaporthe grisea*, *Mol. Plant Microbe Interact.* **8**:939–948.

Kato, H., 1983, Responses of Italian millet, oat, timothy, Italian ryegrass and perennial ryegrass to Pyricularia species isolated form cereals and grasses, *Proc. Kanto-Tosan Plant Protect. Soc.* **30**:22–23.

Kato, H., Yamamoto, M., Yamaguchi-ozaki, T., Kadouchi, H., Iwamoto, Y., Nakayashiki, H., Tosa, Y., Mayama, S., and Mori, N., 2000, Pathogenicity, mating ability and DNA restriction fragment length polymorphisms of *Pyricularia* populations isolated from Gramineae, Bambusideae and Zingiberaceae plants, *J. Gen. Plant Pathol.* **66**:30–47.

Keller, H., Pamboukdjian, N., Ponchet, M., Poupet, A., Delon, R., Verrier, J.L., Roby, D., and Ricci, P., 1999, Pathogen-induced elicitin production in transgenic tobacco generates a hypersensitive response and nonspecific disease resistance, *Plant Cell* **11**:223–235.

Koga, H., 1994, Hypersensitive death, autofluorescence, and ultrastructural changes in cells of leaf sheaths of susceptible and resistant near isogenic lines of rice (*Pi-z^t*) in relation to penetration and growth of *Pyricularia grisea, Can. J. Bot.* **72**:1463–1477.

Koizumi, S. 1998, New fungicide use on rice in Japan. Abstract 5.6.3S of the International Congress on Plant Pathology, Edinburgh.

Kruger, J., Thomas, C.M., Golstein, C., Dixon, M.S., Smoker, M., Tang, S., Mulder, L., and Jones, J.D., 2002, A tomato cysteine protease required for *Cf-2*-dependent disease resistance and suppression of auto-necrosis, *Science* **296**:744–747.

Lavashina, E.A., Langley, E., Green, C., Gubb, D., Ashburner, M., and Hoffmann, J.A., 1999, Constitutive activation of Toll-mediated antifungal defense in serpin-deficient Drosophila, *Science* **285**:1917–1919.

Lazarro, D., 2003, Characterization of the *AVR1-CO39* Locus of *Magnaporthe grisea*. Master's Thesis, University of Wisconsin, Madison.

Leister, D., Kurth, J., Laurie, D.A., Yano, M., Sasaki, T., Devos, K., Graner, A., and Schulze-Lefert, P., 1998, Rapid reorganization of resistance gene homologues in cereal genomes, *Proc. Natl. Acad. Sci. USA* **95**:370–375.

Leister, R.T., Ausubel, F.M., and Katagiri, F., 1996, Molecular recognition of pathogen attack occurs inside of plant cells in plant disease resistance specified by the *Arabidopsis* genes *RPS2* and *RPM1, Proc. Natl. Acad. Sci. USA* **93**:15497–15502.

Leung, H., Borromeo, E.S., Bernardo, M.A., and Nottegghem, J.L., 1988, Genetic analysis of virulence in the rice blast fungus *Magnaporthe grisea, Phytopathology* **78**:1227–1233.

Mackill, D., and Bonman, J.M., 1986, New hosts of *Pyricularia grisea, Plant Dis.* **70**:125–127.

McDowell, J.M., and Dangl, J.L., 2000, Signal transduction in the plant immune response, *Trends Plant Sci.* **25**:79–82.

Meyers, B.C., Dickerman, A.W., Michelmore, R.W., Sivaramakrishnan, S., Sobral, B.W., and Young, N.D., 1999, Plant disease resistance genes encode members of an ancient and diverse protein family within the nucleotide-binding superfamily, *Plant J.* **20**:317–332.

Miki, D., Itoh, R., and Shimamoto, K., 2005, RNA silencing of single and multiple members in a gene family of rice, *Plant Physiol.* **138**:1903–1913.

Miki, D., and Shimamoto, K., 2004, Simple RNAi vectors for stable and transient suppression of gene function in rice, *Plant Cell Physiol.* **45**:490–495.

Nielsen, H., and Krogh, A., 1998, Prediction of signal peptides and signal anchors by a hidden Markov model, *Proc.Sixth Int. Conf. Int. Syst.Mol. Biol.* (ISMB 6), AAAS Press, Menlo Park, California, pp. 122–130.

Nielsen, H., Engelbrecht, J., Brunak, S., and von Heijne, G., 1997, Identification of prokaryotic and eukaryotic signal peptides and prediction of their cleavage sites, *Protein Eng.* **10**: 1–6.

Nimchuk, Z., Rohmer, L., Chang, J.H., and Dangl, J.L., 2001, Knowing the dancer from the dance: R-gene products and their interactions with other proteins from host and pathogen, *Curr. Opin. Plant Biol.* **4**:288–294.

Nitta, N., Farman, M., and Leong, S.A., 1997, Genome organization of *Magnaporthe grisea*: Integration of genetic maps, clustering of transposable elements, and identification of genome duplications and rearrangements, *Theor. Appl. Genet.* **95**:20–32.

Orbach, M.J., Farrall, L., Sweigard, J.A., Chumley, F.G., and Valent, B., 2000, A telomeric avirulence gene determines efficacy for the rice blast resistance gene Pi-ta, *Plant Cell* **12**: 2019–2032.

Peyyala, R., and Farman, M.L., 2006, *Magnaporthe oryzae* isolates causing grey leaf spot of perennial ryegrass possess a functional copy of the *AVRI-CO39* avirulence gene, *Mol. Plant Pathol.* **7**:157–165.

Rice Consortium Consortiums for Sequencing Rice Chromosomes 11 and 12, 2005, The sequence of rice chromosomes 11 and 12, rich in disease resistance genes and recent gene duplications. BMC Biology http://www.biomedcentral.com/1741-7007/3/20.

Rojo, E., Sharma, V.K., Kovaleva, V.K., Kovaleva, V., Raikhel, N.V., and Flectecher, J.C., 2002, *CLV3* is localized to the extracellular space, where it activates the Arabidopsis *CLAVATA* stem cell signaling pathway, *Plant Cell* **14**:969–977.

Rossi, M., Goggin, F.L., Milligan, S.B., Kaloshian, I., Ullman, D.E., and Williamson, V.M., 1998, The nematode resistance gene *Mi* of tomato confers resistance against the potato aphid, *Proc. Natl. Acad. Sci. USA* **95**:9750–9754.

Rouf Mian, M.A., Zwonitzer, J.C., Hopkins, A.A., Ding, X.S., and Nelson, R.S., 2005, Response of tall fescue genotypes to a new strain of Brome Mosaic Virus, *Plant Dis.* **89**:224–227.

Sakamoto, K., Tada, Y., Yokozeki, Y., Akagi, H., Hayashi, N., Fujimura, T., and Ichikawa, N., 1999, Chemical induction of disease resistance in rice is correlated with the expression of a gene encoding a nucleotide binding site and leucine-rich repeats, *Plant Mol. Biol.* **40**:847–855.

Silue, D., Tharreau, D., and Notteghem, J.L., 1992a, Evidence for a gene-for-gene relationship in the *Oryza sativa*-*Magnaporthe grisea* pathosystem, *Phytopathology* **82**:577–580.

Silue, D., Tharreau, D., and Notteghem, J.L., 1992b, Identification of *Magnaporthe grsiea* avirulence genes to seven rice cultivars, *Phytopathology* **82**:1462–1467.

Skinner, D.Z., Budde, A., Farman, M., Smith, R., Leung, H., and Leong, S.A., 1993, Genetic map, molecular karyotype and occurrence of repeated DNAs in the rice blast fungus *Magnaporthe grisea*, *Theor. Appl. Genet.* **87**:545–557.

Smith, J.R., and Leong, S.A., 1994, Mapping of a *Magnaporthe grisea* locus affecting rice (*Oryza sativa*) cultivar specificity, *Theor. Appl. Genet.* **88**:901–908.

Solomon, M., Belenghia, B., Delledonneb, M., Menachema, E., and Levine, A., 1999, The involvement of cysteine proteases and protease inhibitor genes in the regulation of programmed cell death in plants, *Plant Cell* **11**:431–444.

Song, W.-Y., Wang, G.-L., Chen, L.-L., Kim, H.-S., Pi, L.-Y., Holsten, T., Gardner, J., Wang, B., Zhai, W.-X., Zhu, L.-H., Fauquet, C., and Ronald, P., 1995, A receptor kinase-like protein encoded by the rice disease resistance gene, *Xa21*, *Science* **270**:1804–1806.

Sweigard, J.A., Carroll, A.M., Kang, S., Farrall, L., Chumley, F.G., and Valent, B., 1995, Identification, cloning and characterization of *PWL2*, a gene for host species specificity in the rice blast fungus, *Plant Cell* **7**:1221–1233.

Tai, T.H., Dahlbeck, D., Clark, E.T., Gajiwala, P., Pasion, R., Whalen, M.C., Stall, R.E., and Staskawicz, B.J., 1999, Expression of the *Bs2* pepper gene confers resistance to bacterial spot disease in tomato, *Proc. Natl. Acad. Sci. USA* **96**:14153–14158.

Tang, H., Zakaria, K., Lemaitre, B., and Hashimoto, C., 2006, Two proteases defining a melanization cascade in the immune system of Drosophila, *J. Biol. Chem.* **281**:28097–28104.

Tang, X., Frederick, R.D., Zhou, J., Halterman, D.A., Jia, Y., and Martin, G.B., 1996, Physical interaction of *AvrPto* and Pto Kinase, *Science* **274**:2060–2063.

Tosa, Y., Osue, J., Eto, Y., Tamba, H., Tanaka, K., Nakayashiki, H., Mayama, S., and Leong, S.A., 2005, Evolution of an avirulence gene *AVR1-CO39* concomitant with the evolution and differentiation of *Magnaporthe oryzae*, *Mol. Plant Microbe Interact.* **18**:1148–1160.

Valent, B., and Chumley, F.G., 1994, Avirulence genes and mechanisms of genetic instability in the rice blast fungus, in R. Zeigler, S. A. Leong, and P. Teng, eds., Rice Blast Disease, CABI, London, pp. 111–134.

Valent, B., Crawford, M.S., Weaver, C.G., and Chumley, F.G., 1986, Genetic studies of fertility and pathogenicity in *Magnaporthe grisea* (*Pyricularia grisea*), *Iowa State J. Res.* **60**:569–594.

Valent, B., Farrall, L., and Chumley, F., 1991, *Magnaporthe grisea* genes for pathogenicity and virulence identified through a series of backcrosses, *Genetics* **127**:87–101.

Van den Ackerveken, G.F., Van, J.M., Kan, J.A.L., and DeWit, P.G.M., 1992, Molecular analysis of the avirulence gene *avr9* of the fungal pathogen *Cladesporium fulvum* fully supports the gene-for-gene hypothesis, *Plant J.* **2**:359–366.

Van der Vossen, E.A.G., Rouppe van der Voort, J.N.A.M., Kanyuka, K., Bendahmane, A., Sandbrink, H., Baulcombe, D.C., Bakker, J., Striekema, W.J., and Klein-Lankhorst, R.M., 2000, Homologues of a single resistance gene cluster in potato confer resistance to distinct pathogens: a virus and a nematode, *Plant J.* **23**:567–576.

Wang, Z.X., Yano, M., Yamanouchi, U., Iwamoto, M., Monna, L., Hayasaka, H., Katayose, Y., and
    Sasaki, T., 1999, The *Pib* gene for rice blast resistance belongs to the nucleotide binding and
    leucine-rich repeat class of plant disease resistance genes, *Plant J.* **19**:55–64.

Whitham, S., McCormick, S., and Baker, B., 1996, The N gene of tobacco confers resistance to
    tobacco mosaic virus in transgenic tomato, *Proc. Natl. Acad. Sci. USA* **93**:8776–8781.

Xiao, S., Calis, O., Patrick, E., Zhang, G., Charoenwattana, P., Muskett, P., Parker, J.E., and
    Turner, J.G., 2005, The atypical resistance gene, *RPW8*, recruits components of basal defense
    for powdery mildew resistance in Arabidopsis, *Plant J.* **42**:95–110.

Yaegashi, H., 1978, Inheritance of pathogenicity in crosses of *Pyricularia* isolates from weeping
    lovegrass and finger millet, *Ann. Phytopathol. Soc. Jpn* **44**:626–632.

Yaegashi, H., and Asaga, K., 1981, Further studies on the inheritance of pathogenicity in crosses
    of *Pyricularia grisea* with *Pyricularia* sp. from finger millet, *Ann. Phytopathol. Soc. Jpn* **47**:
    677–679.

Yoshimura, S., Yamanouchi, U., Katayose, Y., Toki, S., Wang, Z.X., Kono, I., Kurata, N., Yano, M.,
    Iwata, N., and Sasaki, T., 1998, Expression of *Xa1*, a bacterial blight-resistance gene in rice, is
    induced by bacterial inoculation, *Proc. Natl. Acad. Sci. USA* **95**:1663–1668.

Index

Printed in The United States of America